15th European Workshop on Modern Developments and Applications in Microbeam Analysis (EMAS 2017 Workshop) & 7th Meeting of the International Union of Microbeam Analysis Societies (IUMAS-7)

IOP Conference Series: Materials Science and Engineering
Volume 304

Konstanz, Germany
7 – 11 May 2017

ISBN: 978-1-5108-5689-9
ISSN: 1757-8981

Printed from e-media with permission by:

Curran Associates, Inc.
57 Morehouse Lane
Red Hook, NY 12571

Printed by Curran Associates, Inc. (2018)

Institute of Physics
Dirac House, Temple Back
Bristol BS1 6BE UK

Phone: 44 1 17 929 7481
Fax: 44 1 17 920 0979

techtracking@iop.org

Table of contents

Volume 304

EMAS 2017 Workshop - 15th European Workshop on Modern Developments and Applications in Microbeam Analysis & IUMAS-7 Meeting - 7th Meeting of the International union of Microbeam Analysis Societies

7–11 May 2017, Konstanz, Germany

Accepted papers received: 12 January 2018
Published online: 25 January 2018

Preface

Papers

15th European Workshop on Modern Developments and Applications in Microbeam Analysis (EMAS 2017) / 7th Meeting of the International Union of Microbeam Analysis Societies (IUMAS-7) Konstanz, Germany, 7-11 May 2017

Michael B. Matthews[1], Edward P. Vicenzi[2], Silvia Richter[3], Xavier Llovet[4], Jan Dellith[5] and Miran Ceh[6]

[1] AWE Plc., Aldermaston, Reading RG7 4PR, Great Britain
[2] Smithsonian Institution, Museum Conservation Institute, 20746 Suitland, MD, U.S.A.
[3] RWTH Aachen, Gemeinschaftslabor für Elektronenmikroskopie (GFE), 552074 Aachen, Germany
[4] Universitat de Barcelona, Scientific and Technological Centres, Lluís Solé i Sabarís 1-3, 08028 Barcelona, Spain
[5] Leibniz Institute of Photonic Technologies, Dept. Instrumental Analytics, 07745 Jena, Germany
[6] Jožef Stefan Institute, Department for Nanostructured Materials, Centre for Electron Microscopy, Jamova cesta 39, 1000 Ljubljana, Slovenia

Preface

This volume of the IOP Conference Series: Materials Science and Engineering contains papers from the 15th Workshop of the European Microbeam Analysis Society (EMAS) on Modern Developments and Applications in Microbeam Analysis, organised in conjunction with the 7th Meeting of the International Union of Microbeam Analysis Societies (IUMAS), which took place from the 7th to the 11th of May 2017 at the Konzilgebäude, Konstanz, Germany.

The primary aim of this series of workshops is to assess the state-of-the-art and reliability of microbeam analysis techniques. The workshops also provide a forum where students and young scientists starting out on a career in microbeam analysis can meet and discuss with the established experts. The Konstanz EMAS-IUMAS meeting was organised within the usual EMAS format, comprising invited plenary lectures by internationally recognized experts, poster presentations by the participants and round table discussions on the key topics led by specialists in the field. The workshop was organised in collaboration with the Microprobe Division of the German Physical Society (DPG). IUMAS member societies co-funded 16 Early Career Scholars to attend the meeting in Konstanz. The technical programme included the following topics: pushing the limits - electron probe microanalysis, modelling, detector technologies, surface characterisation, cathodoluminescence, pushing the limits - general, and their applications.

As at previous workshops there was also a special oral session for young scientists. The best presentation by a young scientist was awarded with an invitation to attend the 2018 Microscopy and

IOP Conf. Series: Materials Science and Engineering **304** (2017) 011001 doi:10.1088/1757-899X/304/1/011001

Microanalysis meeting in Baltimore, Maryland. The prize went to Ery C. Hughes, University of Bristol (UK), for her presentation entitled: "*Low EPMA totals of hydrous basaltic glass: effect of sub-surface charging*" (by E.C. Hughes, B. Buse and S.L. Kearns).

Six half-day pre-meeting courses were held covering the following areas of microbeam analysis: introduction to EBSD, advanced EPMA including CalcZAF, electron microscopy maintenance, Monte Carlo simulation using PENEPMA, advanced spectral imaging, mapping and phase analysis, Monte Carlo simulation, and introduction to atom probe.

The continuing relevance of the EMAS-IUMAS meetings and the high regard in which they are held internationally can be seen from the fact that 93 posters from 20 countries were on display at the meeting and that the participants came from as far away as Australia, Brazil, Canada, China, Japan, and USA. A selection of participants with posters was invited to give a short oral presentation of their work in three dedicated sessions. The EMAS Young Scientist Poster Award was awarded with an invitation to participate in the 25th Australian Conference on Microscopy and Microanalysis (ACMM 25) in 2019 in Melbourne, Australia. The prize was awarded to Dr Joshua Einsle, University of Cambridge (UK), for his contribution entitled: "*The potential for mineralogical mapping through machine learning*" (by J.F. Einsle, R.J. Harrison, B. Martineau, I. Buisman, A.M. Piotrowski, S.M. Collins, Z. Saghi, D. Johnstone, A. Eggerman and P.A. Midgley). The EMAS Workshop Poster Prize was given to Nicole Wollschläger, Bundesanstalt für Materialforschung und –prüfung (BAM) (DE), for her contribution entitled: "*Characterisation of porous TiO_2 nanoparticle films using on-axis TKD in SEM - a new nano analytical tool for a large scale application*" (by N. Wollschläger, L. Palasse, I. Häusler, E. Ortel, K. Dirscherl and V.-D. Hodoroaba).

This proceedings volume contains the full texts of 8 of the invited plenary lectures and of 13 papers on related topics originating from the posters presented at the workshop. All the papers have been subjected to peer review by a least two referees.

November 2017

EMAS 2017 Workshop / IUMAS-7 Meeting, held at the Konzilgebäude, Konstanz, Germany, from 7 to 11 May 2017.

Acknowledgements

On behalf of the European Microbeam Analysis Society I would like to thank all the invited speakers, session chairs and members of the discussion panels for making the meeting such a great success. Special thanks go to Silvia Richter and Luc Van't dack who directed the organisation of the workshop giving freely of their time and talents. As was the case for previous workshops, the EMAS Board *in corpore* was responsible for the scientific programme. The Workshop also included a commercial exhibition where many leading instrument suppliers were represented. Several companies that exhibited provided financial support, either by sponsoring an event or by advertising.

Below, in alphabetical order, is a list of exhibiting companies and sponsors of the workshop.

- Ametek GmbH, Edax Business Unit
- Ametek GmbH, Cameca Business Unit
- Bruker Nano GmbH
- Deutsche Forschungsgemeinschaft (DFG)
- European Commission, JRC Directorate G – Nuclear Safety and Security (Germany)
- ibss Group, Inc.
- IfG - Institute for Scientific Instruments
- Jeol (Germany) GmbH
- Micro to Nano
- Oxford Instruments NanoAnalysis Ltd.
- Probe Software
- remX
- Target-Messtechnik
- TESCAN ORSAY HOLDING, a.s.
- Thermo Fisher Scientific

Michael B. Matthews
EMAS President

IOP Conf. Series: Materials Science and Engineering **304** (2017) 011002 doi:10.1088/1757-899X/304/1/011002

Peer review statement

All papers published in this volume of *IOP Conference Series: Materials Science and Engineering* have been peer reviewed through processes administered by the proceedings Editors. Reviews were conducted by expert referees to the professional and scientific standards expected of a proceedings journal published by IOP Publishing.

IOP Conf. Series: Materials Science and Engineering **304** (2017) 012001 doi:10.1088/1757-899X/304/1/012001

Trace element analysis by EPMA in geosciences: detection limit, precision and accuracy

V G Batanova[1,2], A V Sobolev[1,2,3] and V Magnin[1]

[1] University Grenoble Alpes, ISTerre, CNRS, IRD, IFSTTAR, 38000 Grenoble, France
[2] Vernadsky Institute of Geochemistry and Analytical Chemistry, ul. Kosygin 19, 119991 Moscow, Russia
[3] Institut Universitaire de France, 1 rue Descartes, 75231 Paris cedex 05, France

E-mail: valentina.batanova@univ-grenoble-alpes.fr

Abstract. Use of the electron probe microanalyser (EPMA) for trace element analysis has increased over the last decade, mainly because of improved stability of spectrometers and the electron column when operated at high probe current; development of new large-area crystal monochromators and ultra-high count rate spectrometers; full integration of energy-dispersive / wavelength-dispersive X-ray spectrometry (EDS/WDS) signals; and the development of powerful software packages. For phases that are stable under a dense electron beam, the detection limit and precision can be decreased to the ppm level by using high acceleration voltage and beam current combined with long counting time. Data on 10 elements (Na, Al, P, Ca, Ti, Cr, Mn, Co, Ni, Zn) in olivine obtained on a JEOL JXA-8230 microprobe with tungsten filament show that the detection limit decreases proportionally to the square root of counting time and probe current. For all elements equal or heavier than phosphorus ($Z = 15$), the detection limit decreases with increasing accelerating voltage. The analytical precision for minor and trace elements analysed in olivine at 25 kV accelerating voltage and 900 nA beam current is 4 - 18 ppm (2 standard deviations of repeated measurements of the olivine reference sample) and is similar to the detection limit of corresponding elements. To analyse trace elements accurately requires careful estimation of background, and consideration of sample damage under the beam and secondary fluorescence from phase boundaries. The development and use of matrix reference samples with well-characterised trace elements of interest is important for monitoring and improving of the accuracy. An evaluation of the accuracy of trace element analyses in olivine has been made by comparing EPMA data for new reference samples with data obtained by different in-situ and bulk analytical methods in six different laboratories worldwide. For all elements, the measured concentrations in the olivine reference sample were found to be identical (within internal precision) to reference values, suggesting that achieved precision and accuracy are similar. The spatial resolution of EPMA in a silicate matrix, even at very extreme conditions (accelerating voltage 25 kV), does not exceed 7 - 8 µm and thus is still better than laser ablation inductively coupled plasma mass spectrometry (LA-ICP-MS) or secondary ion mass spectrometry (SIMS) of similar precision. These make the electron microprobe an indispensable method with applications in experimental petrology, geochemistry and cosmochemistry.

IOP Conf. Series: Materials Science and Engineering **304** (2017) 012001 doi:10.1088/1757-899X/304/1/012001

1. Introduction

There is no rigorous definition of a trace element, and typically accepted definitions depend on the field of science. For example, in analytical chemistry, a trace element is one whose concentration is less than 100 parts per million (ppm) or less than 100 micrograms per gram; while in geochemistry, a trace element has a concentration less than 1000 ppm or 0.1 wt% of a rock's mass and substitutes for major elements in the rock-forming minerals. Reed [1] defines a trace element in wavelength dispersive X-ray spectrometry (WDS) microprobe analysis as an element with a concentration similar to the detection limit. However this definition is too complicated because the detection limit depends on analytical conditions, atomic number and measured X-ray line. Here we propose to consider an element with concentration below 1,000 ppm as a trace element and an element with concentration between 1,000 ppm and 1 wt% as a minor element.

Wavelength dispersive X-ray spectrometry electron probe microanalysis (EPMA) of minor and trace elements has several important advantages compared with other in-situ methods such as secondary ion mass spectrometry (SIMS), laser ablation inductively coupled plasma mass spectrometry (LA-ICP-MS) and micro particle-induced X-ray emission (μ-PIXE). These advantages are: the highest spatial resolution (from 1 to several micrometres); non-destructive analysis; well-developed matrix correction procedures and relatively low cost. A significant disadvantage is the sensitivity of EPMA, which is not enough to analyse elements with concentration below 10 ppm level with sufficient precision.

Traditionally EPMA has been considered as an ideal method for in-situ quantitative analysis of major and minor elements. Before the start the third millennium, only a few publications had considered the use of the microprobe for trace element analysis (e.g., [2-7]). However, significant progress in instrumental development during the last decade has allowed the EPMA to be successfully used for trace-element analyses (down to 10 ppm). These improvements include an improved stability of spectrometers and the electron column when operated at high probe current and the development of new crystal monochromators with larger areas, spectrometers with ultra-high count rates and powerful software packages. Moreover, full integration of energy-dispersive X-ray spectrometry (EDS) / WDS signals allows quantitative data for major, minor and trace elements to be collected simultaneously in a single spot. As a result EPMA is successfully used to measure trace elements in monazite (e.g., [8-10]), quartz [11-15], olivine [16-22], sulphides [23], basaltic glasses [17, 24, 25], garnet [7, 26, 27], gold [28], Fe- and Mn-oxides in soil [29], rutile [30], diamonds [31], etc.).

Theoretical aspect of trace-element analysis by EPMA have been considered in comprehensive reviews by Reed [1] and Jercinovic *et al.* [10]. In this study we concentrate on practical aspects of trace element analysis in the geosciences. All analyses presented here were produced on the JEOL JXA-8230 microprobe with tungsten filament, 5 WD and SDD ED spectrometers at ISTerre, University Grenoble Alpes, Grenoble, France.

2. Detection limit and counting statistics

The current challenge for the trace-element measurements is to decrease the detection limit (minimum detectable concentration) and to improve analytical precision (reproducibility) and accuracy.

Liebhafsky *et al.* [32] suggested that an element can be considered to be present if the value of peak intensity (N_P) exceeds the intensity of the background (N_B) by $3(N_B)^{0.5}$. The concentration at the limit of detection (C_{DL}) is expressed by different equations, among which the most frequently used are equations of [33-37]. Analytical equations show that the detection limit (DL) varies in reverse proportion to the square root of the peak-to-background ratio (P/B), the net counting rate on the peak position (P/t), and the counting time (t). Increasing of any of these parameters by a factor of 4 will decrease DL by a factor of 2. The most easily controlled parameter is counting time. Other well-known approaches to increase the characteristic X-ray signals (P/B and P/t) are to increase the beam current and the accelerating voltage (e.g., [1, 8-10, 13, 38]). Figure 1 shows the detection limit for Zn in the San Carlos olivine reference sample as function of beam current and peak counting time at a beam acceleration voltage of 25 kV. The detection limit was determined with JEOL software using background statistics and the 3-sigma criterion [35-36]:

$$C_{DL} = 3 \text{ x } [(2N_b/T_bA_{samp})^{0.5}/((N_{std}/C_{std}) \text{ x } (i_{samp}/i_{std}))] \text{ x } 10{,}000 \quad (1)$$

where N_b is average intensity of background (cps/μA) measured on the sample, T_b is total counting time of background (s) for the sample, A_{samp} is the number of accumulation on the sample, N_{std} is net intensity on the standard (cps/μA), C_{std} is concentration in the standard (wt%), i_{samp} and i_{std} are the probe currents measured on the standard and sample (μA). We use the JEOL equation defining the detection limit, because it is determined for each analytical point by JEOL software. Obtained values are similar to those given by all equations, which use 3-sigma criterion and do not use matrix correction [32, 34-37]. To compare with detection limit equations that take into account the matrix correction [e.g., 33], one has to multiply the detection limit values obtained by eq. (1) by matrix correction coefficients. For most elements in olivine, such corrections do not exceed 10 % relative and are thus well within statistical error of the background intensity. The two exceptions are Al and P, which have ZAF correction coefficients of 1.9 and 1.6 respectively. Therefore, the detection limits for these elements are about twice and 1.5 times larger, respectively, than those obtained using the eq. (1). All relations between detection limits and beam current, counting time and accelerating voltage that are considering here are equal for all models.

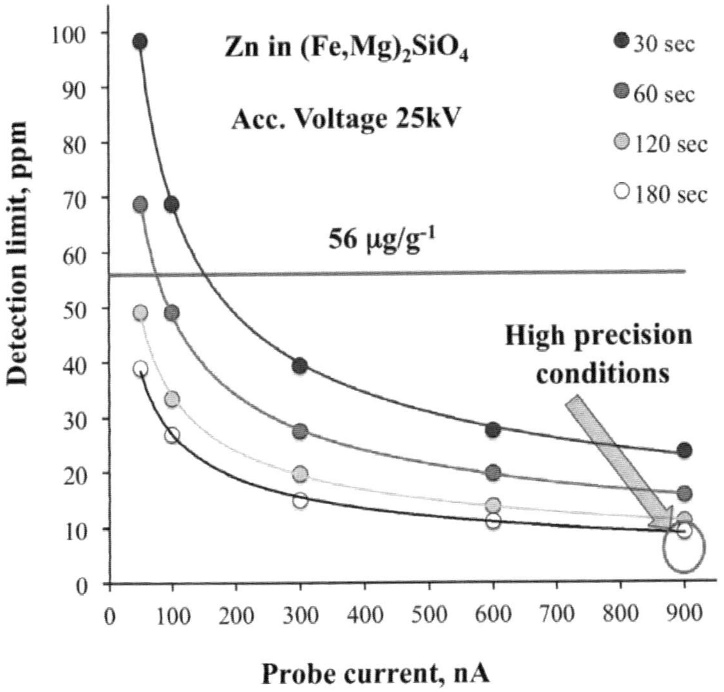

Figure 1. Detection limit for Zn in San Carlos olivine reference sample (USNM111312-44) as a function of peak counting time and beam current.

The detection limit of Zn in olivine (figure 1) behaves in accordance with counting-statistics theory: it decreases as the square root of counting time and beam current. The accepted concentration of Zn in the sample is 56 ppm (green line on figure 1, [19, 39]). Precise analysis at such low concentrations requires the combination of high probe current with long counting time (180 s and higher). Some authors have recommended to divide the long time into relatively short peak and background measurements to minimise the effect of instrumental drift or phase instability (e.g., [1, 20, 24, 40]). The maximum precision in the determination of the background-corrected peak intensity is obtained with equal counting times for peak and background [1].

For heavy elements with atomic number (Z) > 10, the peak intensity and P/B ratio increase with increasing accelerating voltage, which therefore decreases the detection limit (e.g., [1, 10]. However, in cases of high absorption, the peak intensity passes through a maximum and then decreases. The experimental study of 10 elements (Na, Al, P, Ca, Ti, Cr, Mn, Co, Ni, Zn) in olivine shows that, for all elements starting from phosphorus (Z = 15), the detection limit decreases with increasing accelerating voltage. The behaviour of these elements is in accordance with counting statistics: DL decreases proportionally to the parameters (counting time, probe current and accelerating voltage), which leads to an increase in the peak intensity and P/B ratio (figures 2a and 2c).

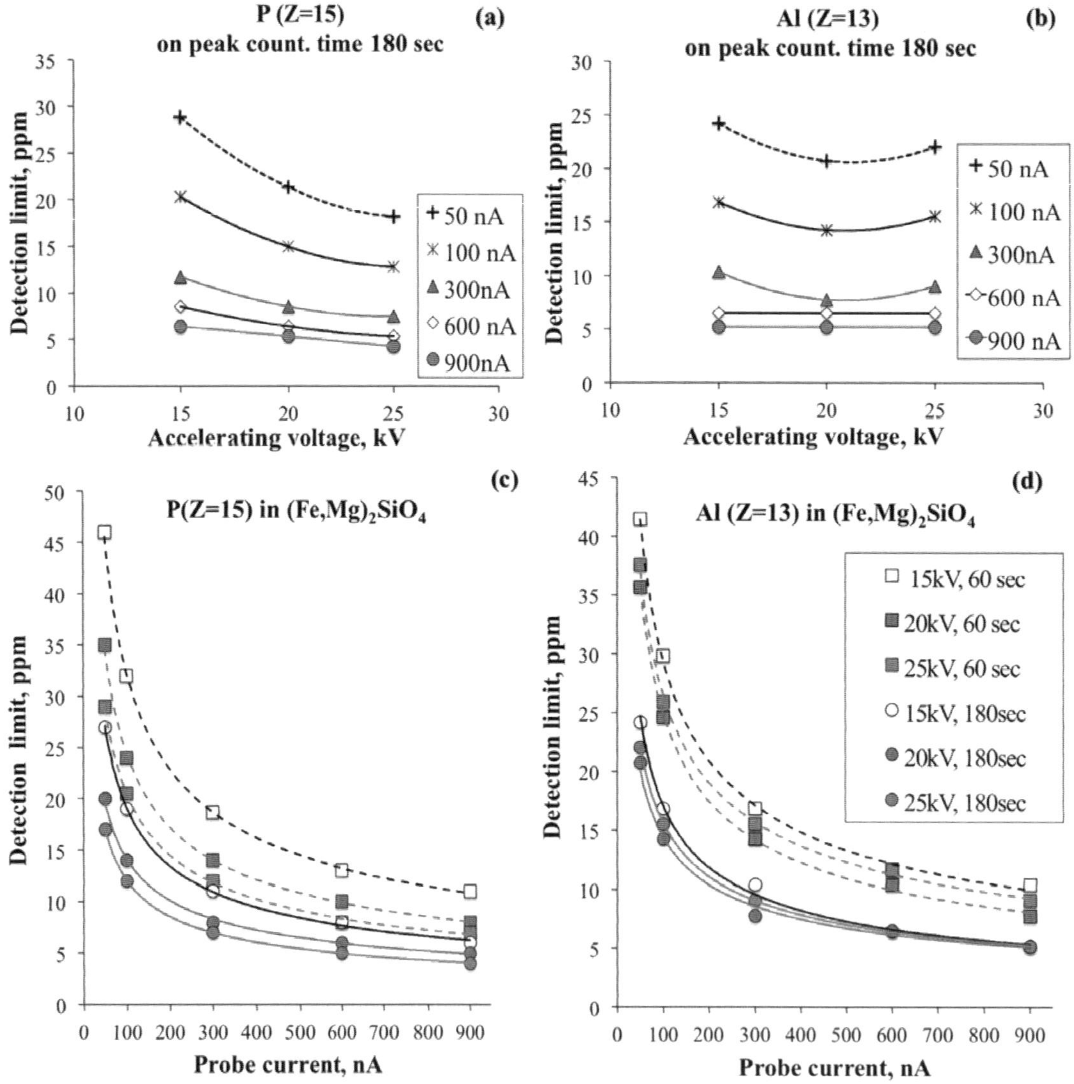

Figure 2 a-d. Detection limits for Al and P in olivine $(Fe,Mg)_2SiO_4$ matrix as a function of accelerating voltage , beam current and counting time (after [19]).

IOP Conf. Series: Materials Science and Engineering **304** (2017) 012001 doi:10.1088/1757-899X/304/1/012001

The relationship between accelerating voltage, beam current and detection limit is more complex for elements with low atomic number and, therefore, low excitation energy and high absorption coefficients (Na, Al). The detection limit at beam current up to 300 nA and long counting time (180 s on peak) decreases as accelerating voltage changes from 15 to 20 kV but then drastically increases at 25 kV (figure 2b). However, when a beam current higher than 600 nA is combined with a long counting time, the detection limit remains constant as the accelerating voltage is increased from 15 to 25 kV. This effect is not observed for short (60 s) counting times: under these conditions the detection limit increases from 20 kV to 25 kV for all probe currents (figure 2d).

Detection limits approaching the 1 ppm level can theoretically be obtained using extreme conditions (high accelerating voltage and beam current, and long counting time) [1]. Experimentally this was shown for S in steel with accelerating voltage = 20 kV and probe current = 2 μA [41].

3. Spatial resolution

It is well known that the spatial resolution decreases as accelerating voltage increases (e.g., [1, 10]). Figure 3 shows an electron flight Monte Carlo simulation of the electron interaction and X-ray generation as well as emission volumes for Al-Kα and Ni-Kα in $(Fe,Mg)_2SiO_4$ at 15, 20 and 25 kV [19]. For Ni-Kα, the X-ray generation and emission volumes have roughly equal diameters ≈ 2 μm, 4 μm and 6 μm at 15 kV, 20 kV and 25 kV, respectively. For Al-Kα, the X-ray emitted area is restricted to a depth of 2 μm for 15 kV, 20 kV, and 25 kV because of high absorption. According to Jercinovic *et al* [10] the spatial resolution of analysis (analytical resolution, AR) can be expressed as:

$$D_{AR} = (D^2_{beam} + D^2_{emission})^{0.5} \quad (2)$$

where D_{beam} is the beam diameter and $D_{emission}$ is the emission volume.

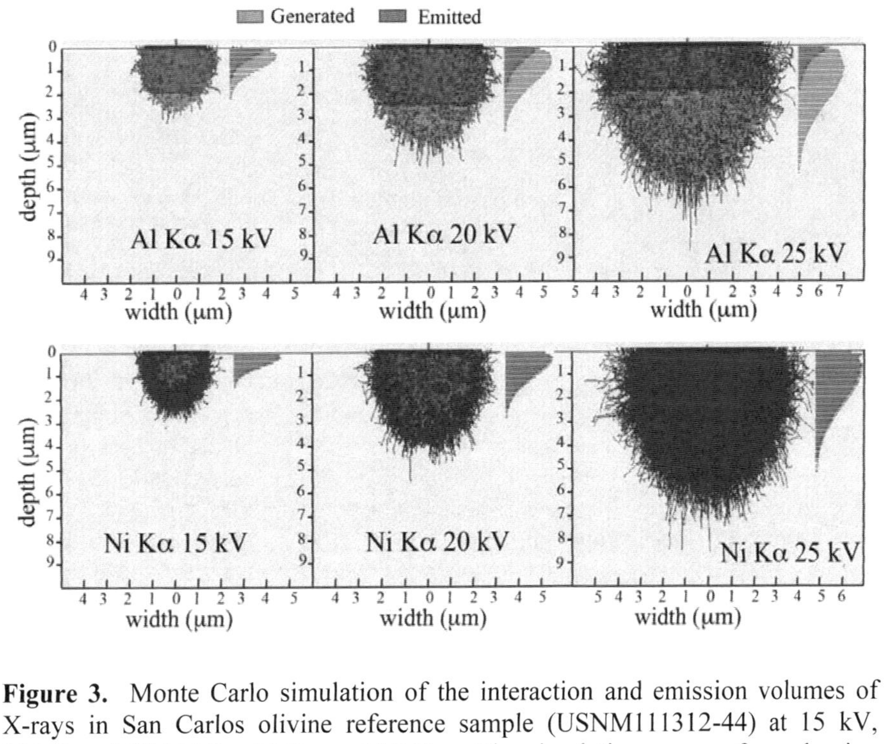

Figure 3. Monte Carlo simulation of the interaction and emission volumes of X-rays in San Carlos olivine reference sample (USNM111312-44) at 15 kV, 20 kV and 25 kV for Al-Kα and Ni-Kα. The simulation was performed using electron flight simulator (Small World, Inc.) and consists of 32,000 trials (after [19]). Printed with the permission of Chemical Geology.

Using this equation, the emission volumes listed above, and a beam diameter of 2 μm [19], we obtained a maximum diameter of the analysed area (D_{AR}) of 7.3 μm of Al-Kα for an accelerating voltage of 25 kV and a beam current of 900 nA. For accelerating voltages of 15 kV and 20 kV and a beam current of 900 nA, the diameter of the analysed area is correspondingly smaller, at 3.2 and 5.1 μm (beam diameter 1 μm [19]).

Nonetheless, the spatial resolution of EPMA of silicate minerals, even at a high accelerating voltage 25 kV, remains significantly better than with the LA-ICP-MS or SIMS. However, in the case of analysis of very small phases, the analyst has to keep accelerating voltage as low as possible (usually around 2 times of excitation energy of analysing line) and has to choose appropriate analysis times and probe currents in order to attain the required detection limit.

4. Stability, coating and optimal conditions for measurements

As discussed above, the WDS-EPMA technique requires measuring trace elements with high probe currents (hundreds of nA) and high accelerating voltages coupled with long counting times (more than few minutes) (e.g., [1, 8-10]). Thus instrumental and sample stability during long acquisition is obligatory for accurate trace element analysis. The new generation of EPMA machines provides a stable beam at high probe current for long periods. However high current beam produces significant heating of the sample surface, especially if the beam is focussed. This may damage the sample. Conductive (for both electrons and heat) materials (metals, alloys, etc.) are stable under beams with high current densities. However, the mineral phases we study in geosciences are insulators and require a conductive coating. Although various different metals (Au, Al, Ag, Cu, Be, etc.) can be used in special cases (e.g., [10] and references therein), carbon is still the most appropriate coating for EPMA. When covered with a carbon coating with thickness of about 20 - 25 nm, minerals such as olivine, orthopyroxene, some garnets, zircon, Fe and Fe-Ni sulphides remain stable under high beam currents (900 nA) during long counting time (figure 4).

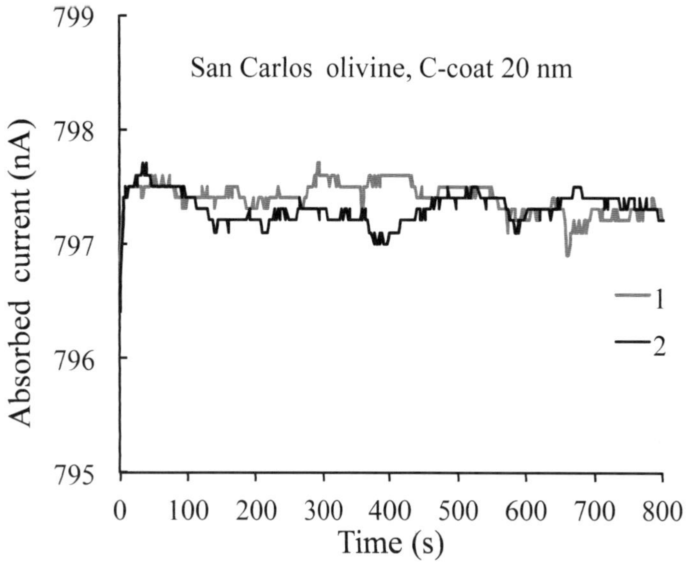

Figure 4. Absorbed current in San Carlos olivine reference sample (USNM111312-44). Beam diameter 2 μm, accelerating voltage 25 kV, beam current 900 nA, total time of beam exposure 12 min Different colour lines (1, 2) correspond to repeated measurements in different stage position within the same olivine fragment.

Figures 4 and 5 show absorbed current and signals of the trace elements during 12-minute analyses of olivine at 25 kV and 900 nA and a beam diameter of 2 μm. After 12 min of beam exposure, minor degradation of the carbon coat and slight surface damage was noted [19]. However, these effects do not have any serious impact on the stability of the absorbed current and the signal of the analysed elements. The measured absorbed current remains constant (better than 0.05 % relative) with a 900 nA beam on olivine for an exposure time of over 720 s, the duration of a single analysis (figure 5a). The X-ray count rates of all measured trace elements show no systematic variation for exposure times exceeding 720 s (figures 5a and 5b) – 2 standard relative errors of averages are 0.1 - 0.2 % for Ni, Ti, Ca, Zn, Al, Mn, 0.3 % for Co, 0.4 % for Na and 0.7 % for P. No significant changes in X-ray counting rates occur during the first minute of analysis, even for Na (figure 5b). In addition, to estimate a possible variation and dispersion with reference to the value of the peak intensity, we normalised all intensities registered after 20 second to the mean intensity during first 20 seconds. The averages and standard deviations of these values for 700 seconds of acquisition are: Al 1.012 ± 0.031; Zn 1.001 ± 0.033; Ca 1.006 ± 0.015; Ti 1.000 ± 0.016; Ni 1.001 ± 0.014; Na 0.981 ± 0.061; Co 1.003 ± 0.042; P 0.989 ± 0.100; Mn 1.007 ± 0.024. These data indicate that deviations of normalised averages from unity are much less than their standard deviations for all measured trace elements. Also, the largest deviations from unity found for Na (2 % relative) and for Al and P (1 % relative) are equal to concentrations of 1 ppm, 2 ppm and 0.2 ppm correspondingly. We thus conclude that there are no statistically significant changes in trace element concentrations during 720 s acquisition on magnesian olivine at 900 nA beam current, 25 kV acceleration voltage and beam size of 2 μm.

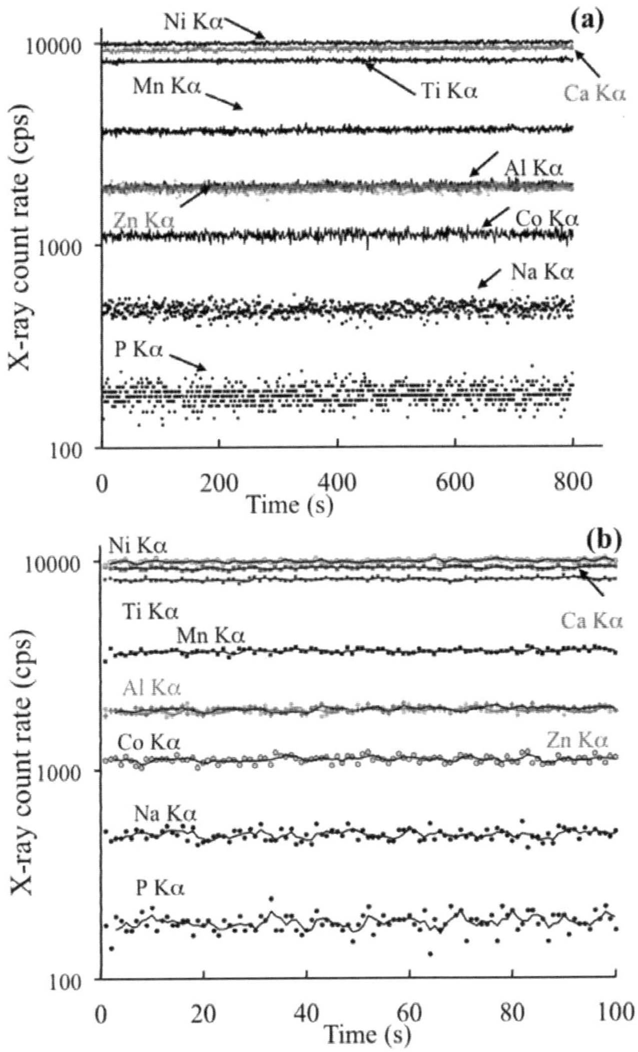

Figure 5. a) Na-Kα, P-Kα, Al-Kα, Co-Kα, Zn-Kα, Ca-Kα, Ti-Kα, Ni-Kα, and Mn-Kα line intensities as a function of time. Beam diameter 2 μm, accelerating voltage 25 kV, beam current 900 nA, total time of beam exposure 12 min (modified after [19]), and b) first 100 s of (a); lines correspond to moving 3-point averages.

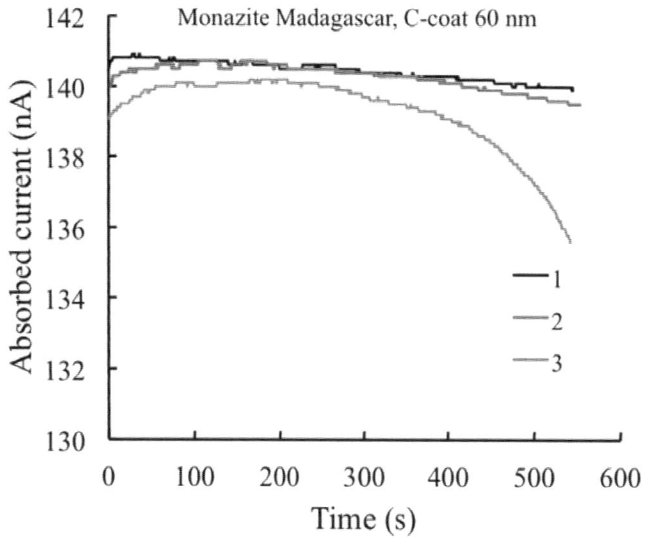

Figure 6. Absorbed current in Madagascar monazite with C-coat 60 nm as a function of time. Beam diameter 1 μm, accelerating voltage 15 kV, beam current 200 nA, total time of beam exposure 8 min. The different colour lines (1 - 3) correspond to repeated measurements in different stage position within the same monazite fragment.

We also checked the stability of major elements (measured by EDS) during typical 12-minute analyses through a series of four 3-minute analyses on the same point on San Carlos olivine. The data show no statistical differences in the measured concentrations [19] indicating that there are no significant time-dependent effects during the course of an analysis. In addition, increasing the beam diameter decreases the current density and thus improves phase stability.

Jercinovic *et al.* [8-10] considered in detail the stability of monazite with different types of coating. For the analysis of small (5 - 10 μm) monazite grains, for which the beam diameter should not exceed 1 μm, a moderate accelerating voltage (15 kV) and thick carbon coating was recommended [42]. Our experiments with the Madagascar Manangoutry monazite reference sample [43, 44] showed that the stability of this monazite improves with a carbon coating of 60 nm (figure 6). The average of normalised peak intensities (figure 7, *I*/*I*mean20, see explanation above): Dy-Lβ 1.010 ± 0.061; U-Mβ 1.006 ± 0.056, Pb-Mβ 1.009 ± 0.067 and Y-Lα 1.042 ± 0.144.

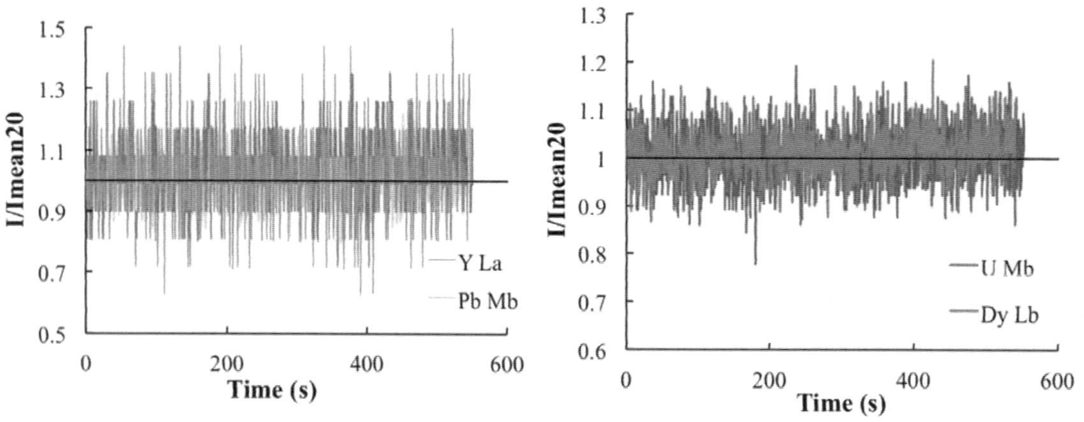

Figure 7. Normalised intensities of Dy-Lβ, U-Mβ, and Pb-Mβ, Y-Lα lines in Madagascar monazite with C-coat 60 nm as a function of time.

One should keep in mind that the standards have to be coated in the same way as samples and have a similar thickness of coating material.

Briefly, the optimal conditions for trace element analysis (accelerating voltage, beam current, counting time) should be selected taking into account: 1) the size, composition and stability of the analysing phase under the electron beam; 2) the number of trace and minor elements to be analysed and the required detection limits and precision; 3) the opportunity to use EDS for simultaneous measurement of major elements on the same point.

For phases that are unstable during long exposure under high beam current, the following measures are used in many laboratories: 1) integration (combination) of intensities using multiple spectrometers; 2) increasing the beam diameter; 3) fractional counting times at several sample positions; 4) use of a thick carbon coating or thin Au coating (e.g., [1, 8-10, 13, 42, 45]); 5) use of an iterative scheme of sample exposure in which the electron beam is cut for 10 seconds after each 10 second counting interval, allowing the sample to cool [24].

5. Accuracy and precision of trace element analysis

Given that the analytical precision in EPMA is a function of X-ray counting statistics for both peak and background, all factors that were considered above (Section 2) for the improving detection limit during trace element analysis are applicable to the precision. Mechanical reproducibility of the spectrometers is usually considered to be the limiting factor in precise measurements. Repeated measurements of trace and minor elements in olivine reference samples show that the detection limit of an element is equal to the analytical precision, calculated as 2.65 ± 0.03 standard deviations (figure 8).

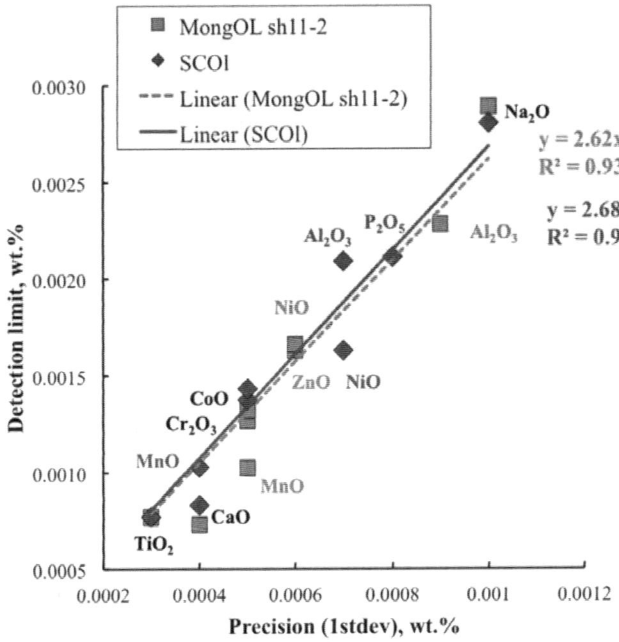

Figure 8. Relationship between detection limit (in oxides of Al, Ti, Ca, P, Na, Cr, Ni, Mn, Zn, Co) calculated by equation (1) of JEOL property software for 3 standard errors and multiplied by the ZAF correction coefficients, and reproducibility (precision) calculated as the standard deviation of 44 analyses of San Carlos olivine and 24 analyses of MongOl sh11-2 olivine reference samples measured as unknowns after each 30 - 40 analyses over a period of several days.

The accuracy of measurement by WDS for trace as well as for major elements depends on many factors including sample preparation, the type and thickness of the coating, stability under the electron beam, accurate knowledge of the composition of the primary standards, and the quality of algorithms used for matrix (i.e., the ZAF, $\varphi(\rho z)$) and interference corrections (e.g., [1, 10, 45-47]).

Jercinovic *et al.* [10] showed that the accuracy of EPMA decreases at low concentrations because most random and systematic errors encountered during major element analysis are magnified at trace level and specific errors appear. Among the special sources of errors during trace element analysis, the most important are spectral background subtraction, secondary fluorescence from phase boundaries (e.g., [48-50]), surface contamination from coating or sample preparation (e.g., [19]) and beam damage caused by high beam current and long counting times [1, 10].

5.1. Accuracy: background estimation and blank correction

One important source of systematic error during trace element analysis is poor estimation of backgrounds, because for trace elements, the peak-to-background ratio is significantly lower than for major elements [1, 10, 13, 45]. The general recommendation for accurate background estimation is a detailed WD intensity scan on the both sides of the peak of interest for the analysed sample as well as for a blank sample [1, 10, 13, 45]. The detailed WD intensity scan allows the analyst to select positions for background measurement and the appropriate method of background interpolation (classic linear, exponential or multipoint [13]. It also allows the analyst to avoid interferences on background, background discontinuities and other background artefacts such as "negative peaks" or "holes" described, for example, for LIF crystal (e.g., [1, 10, 12, 13]. Incorrect evaluation of the curvature of the continuum or different sorts of discontinuities of the continuum in the region close to the peak can bias apparent peak intensities by tens of ppm.

Figures 9 and 10 show WD intensity scans on both sides of Al-Kα on TAPJ four crystal spectrometer (Rowland circle 140 mm, P_{10} gas flow counter), and Ti-Kα for H-type (Rowland circle 100 mm) and L-type (Rowland circle 140 mm) WDS within olivine.

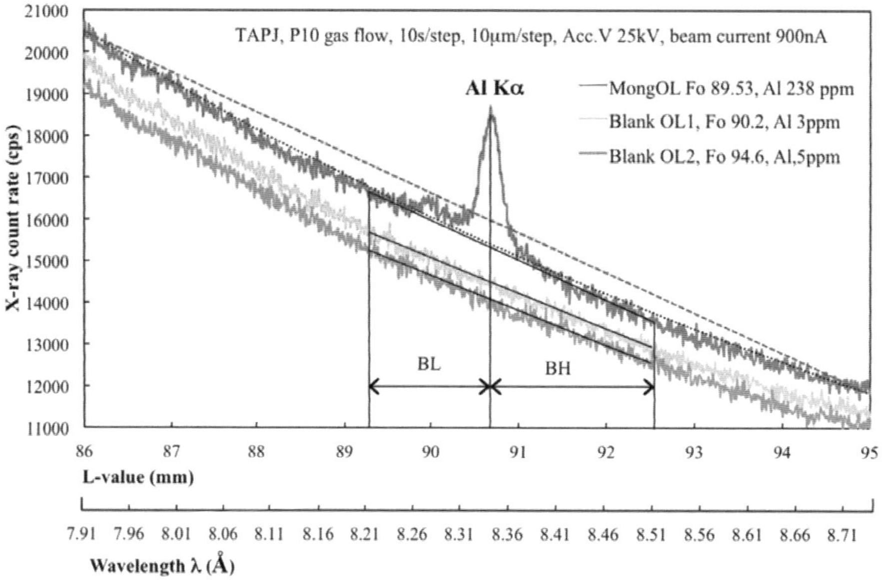

Figure 9. WD intensity scan on both sides of the Al-Kα line (TAP, P10 gas flow) in MongOl sh11-2 olivine and "blank" (Al-free) olivines with different fosterite (Fo) contents. In addition to the conditions shown on figure, counter differential mode was used (base level 0.7 V, window 9.3 V). The dashed straight line highlights the curvature of the continuum and the possibility of overestimating background intensities and underestimating Al-Kα peak intensities. The dotted line shows polynomial interpolation (e.g., [13]). BH and BL indicate the possible positions of background measurements for two points of linear interpolation model.

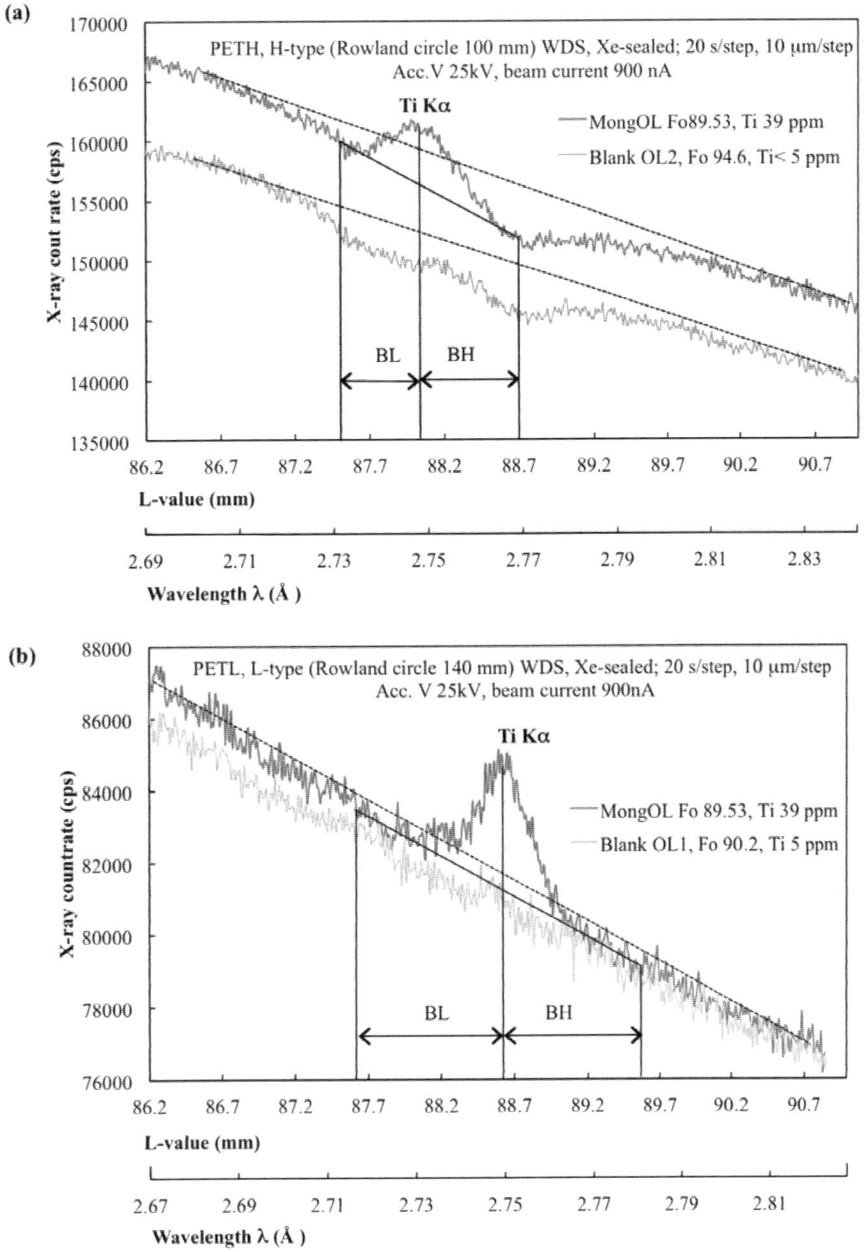

Figure 10. WD intensity scan on both sides of the Ti-Kα line for: a) H-type and b) for L-type WDS for MongOl sh11-2 olivine and "blank" olivines. In addition to the conditions shown on figures 10a and 10b, counter differential mode was used (base level 2 V, window 8 V). The dashed straight line illustrates the curvature of the continuum and the possibility of overestimating background intensities and underestimating Ti-Kα peak intensities. BH and BL indicate possible positions of background measurements for two points of linear interpolation.

The scans for Al-Kα in olivine do not reveal crystal diffraction artefacts but show the significant concave curvature of the X-ray continuum (figure 9). To avoid the negative net counts at the Al-Kα peak position, which can arise when using the classic linear background model, one should choose background positions close to the peak (BL, BH on figure 9) or select a polynomial model for background interpolation (e.g., [13]).

The position Ti-Kα is close to the spectrometer drive limit on H-type WDS. This probably results in nonlinearity of the continuum in the peak region, which is more visible on the scan of "blank" olivine (figure 10a). Use of the blank sample is important to quantify this effect (e.g., [1]). Quantitative analysis of blank olivine should produce a zero concentration when background positions are properly selected (BL, BH figure 10b). The negative concentration values obtained on the blank sample indicate problems that must be resolved before proceeding to unknowns. Minor inaccuracies from unknown sources can be corrected using a blank correction scheme. Another important task that the detailed WDS intensity scans helps to resolve is the choice of crystal. For analysis of trace concentrations of Ti (figures 10a and 10b), the L-type spectrometer is preferable because it avoids problems with linearity of continuum in the Ti-Kα region.

Classical linear interpolation of the background may be a cause of errors for complex systems such as the measurement of Pb-Mα in monazite (e.g., [1, 8-9, 13]. Possible approaches in such situations are the use of exponential or multi-point background fits, as have been developed and incorporated in Probe Software [51], as well as the use of the mean atomic number (MAN) background correction procedure based on the modelling of the continuum using Kramer's equation (e.g., [52-54]).

When the presence of a neighbouring peak rules out background measurement on one side, and only the background on the other side can be used for correction, special correction procedures are required (e.g., [1]). Sometimes an empirical solution provides very good results, as in the case of fluorine (F) analysis in Fe-bearing minerals and glasses, where the peak of F-Kα position on the sensitive LDE1 (JEOL) PC1 (CAMECA) crystals is on the "shoulder" of the Fe-Lα_1 line [24, 25].

5.2. *Accuracy: analysis of reference sample*

Conventional methods for trace element analysis of geological samples include running reference samples and blank samples as unknowns together with analysed samples. This allows the operator to monitor short-term systematic errors and correct for instrumental drift during long analytical sessions.

Our routine EPMA procedure includes evaluation of accuracy and precision and the monitoring of instrumental drift (e.g., [17, 19]). When analysing olivine using the Trace protocol, we usually run the olivine reference sample (San Carlos olivine USNM111312-44 [19, 39, 55]) as an unknown, three times for every 30 - 40 measurement points. We only run a blank sample as an unknown three times at the beginning and at the end of each session. All measurements of major and trace elements are corrected for deviation from San Carlos olivine reference values if the deviation is higher than 2 sigma errors. The analytical precision (reproducibility) of olivine analyses, established by repeated measurement of olivine standards, is 200 - 300 ppm (2 standard deviations) for the major element equivalent of Fo and around 4 to 18 ppm for trace elements [19]. These values are proportional to the detection limits (figure 8). We note that all three fragments of San Carlos olivine USNM111312-44 reference sample available in our laboratory are similar in composition within the errors reported above, which justifies their use as precision monitors during our Trace protocol. For concentrations of Si, Fe and Mg in the San Carlos olivine USNM111312-44 reference sample, we use data of Jarosewich *et al.* [55], while for trace and minor elements we use values from Batanova *et al.* [19]. The reliability of these values is proven by the good agreement between our data on the composition of olivine reference sample MongOl sh11-2 with data obtained by other in-situ and bulk chemical methods (see below).

More than 240 fragments of natural olivine sample (MongOl sh11-2) have been analysed using our EPMA-Trace method [19], and 120 of them were also analysed by LA-ICP-MS, SIMS and EPMA in six different laboratories world-wide [56]. The homogeneity of this olivine reference sample, within quoted standard errors (except for Al and P), is established by EPMA and LA-ICP MS data of individual fragments. In addition, this olivine has been analysed with bulk analytical techniques (solution ICP-MS, isotope dilution and XRF). As a result of this work, preliminary reference values for major and trace elements have been estimated. Figure 11 shows comparison of data for Al and Ti obtained by our EPMA-Trace method with data obtained by other in-situ and bulk chemical methods.

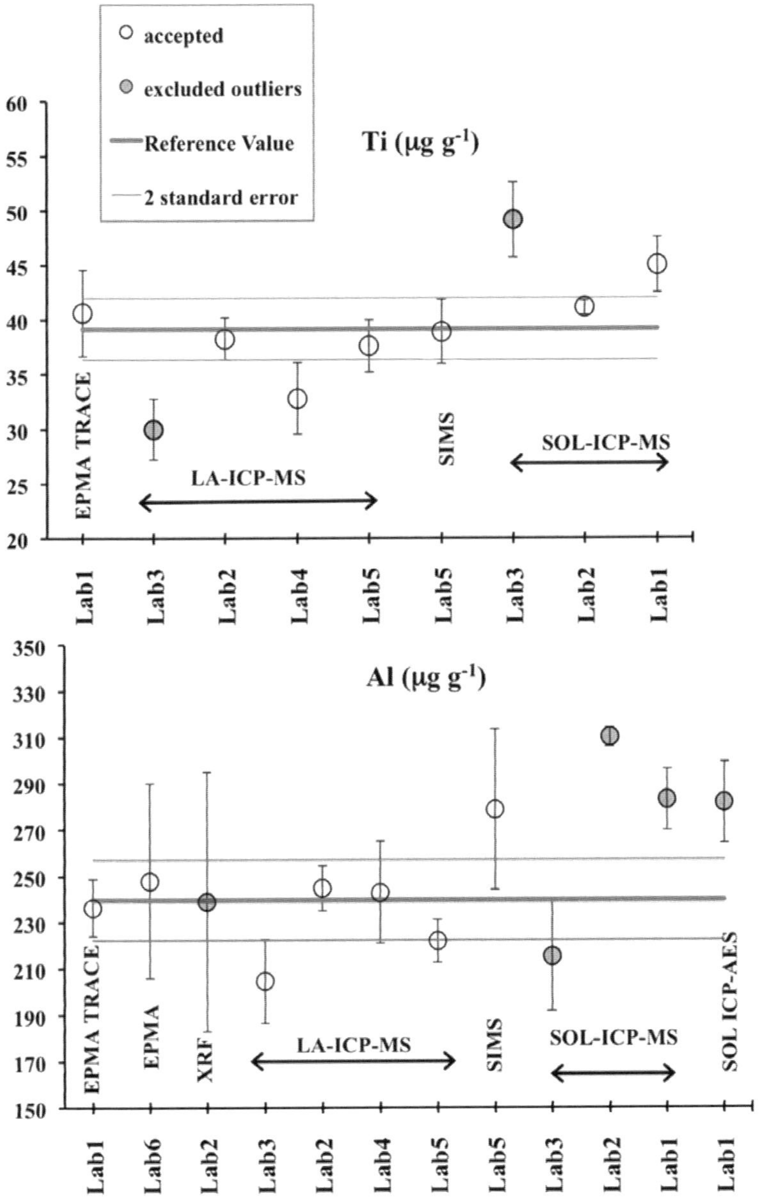

Figure 11. Trace element (Al and Ti) abundances of the new olivine reference sample MongOl sh11-2 obtained by different laboratories and analytical techniques (*in-situ* and bulk) [56]. Each value represent the mean of 120 analysis for EPMA, LA-ICP-MS and SIMS, and 1 - 3 multiply analysed aliquots for XRF, and ICP-MS. Error bars represent reported 2 standard deviations of mean.

The reference values were determined as an average of accepted values. We excluded from the averages values with 2 standard deviations outside ± 2 standard errors interval of whole population. In addition, we excluded bulk analyses for Al (figure 11) due to presence of micro inclusions of Al-rich spinel.

There is a good agreement between EPMA and other in-situ and bulk methods for 10 trace and minor elements (Na, Al, P, Ca, Ti, Cr, Mn, Co, Zn and Ni). For each element, the measured concentrations plot on the one-to-one line within the 2 standard deviations internal precision for EPMA and reference values, as shown by the error bars (figure 12). This suggests that our EPMA-Trace method for olivine is free of systematic errors and that its precision and accuracy are similar.

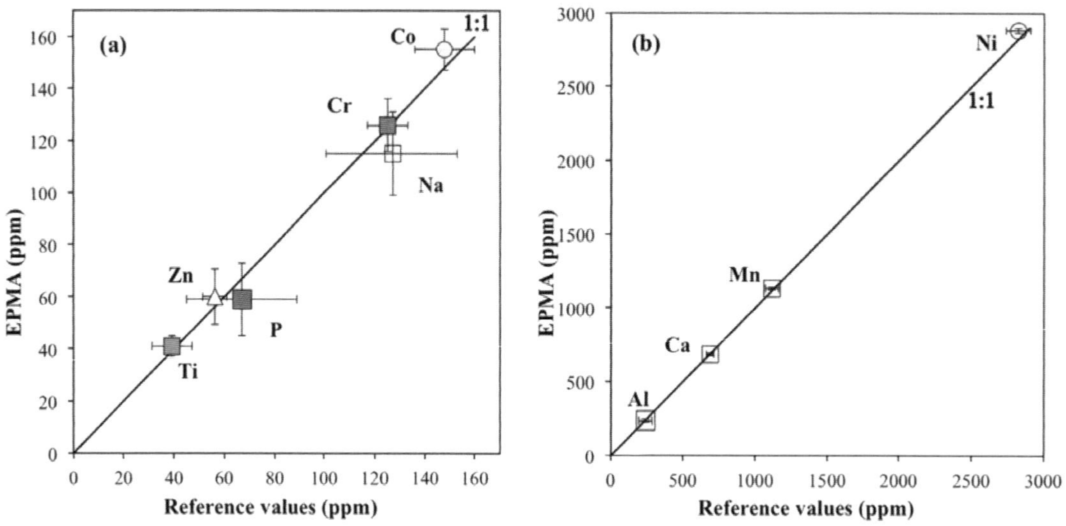

Figure 12. Comparison of trace element analyses of the new reference olivine sample MongOl sh11-2 obtained by EPMA-Trace and reference values estimated on the basis of inter laboratory *in-situ* and bulk analytical techniques [56]. Error bars for each element indicate 2 standard deviations.

6. Concluding remarks

In this study we concentrate on practical aspects of trace (below 1,000 ppm) and minor (between 1,000 ppm and 1 wt%) element analysis of geological samples. All our data and experiments were produced on the JEOL JXA-8230 microprobe with a tungsten filament, 5 WDS and SDD EDS spectrometers at ISTerre, University Grenoble Alpes, Grenoble, France. We found that:

1. The analysis of 10 elements (Na, Al, P, Ca, Ti, Cr, Mn, Co, Ni, Zn) in olivine shows that detection limits decrease proportionally to the square root of counting time and probe current. For all elements with atomic mass similar or greater than phosphorus ($Z = 15$), the detection limits decrease with increasing accelerating voltage. The lowest detection limit achieved in this work is 4 ppm for Ti in olivine at 25 kV acceleration voltage, 900 nA probe current and 180 s counting time for both peak and background. Repeated measurements of trace and minor elements in olivine reference samples show that the detection limit of element is equal to the analytical precision calculated as 2.65 ± 0.03 standard deviations.
2. The spatial resolution of high-precision quantitative electron microprobe analysis of silicate minerals, even at very extreme conditions (accelerating voltage = 25 kV), does not exceed 7 - 8 μm and thus is significantly better than LA-ICP-MS or SIMS of similar precision.
3. Some mineral samples with carbon-coated surfaces (thickness about 20 - 25 nm) remain stable under high beam currents (900 nA) during long counting times. Among these minerals are olivine, orthopyroxene, Fe and Fe-Ni sulphides, some garnets and zircon.

4. The accuracy of measurement by WDS for trace, minor and major elements can be evaluated by measuring reference samples with well-defined compositions similar to those of the samples. An important source of systematic error for trace elements is poorly estimated backgrounds. To obtain accurate background estimations requires a detailed WD intensity scan on the both sides of the peak of interest for both the analysed sample and a blank sample.
5. Our preferred method for trace element analysis of geological materials includes running reference samples and blank samples as unknowns together with analysed samples. This allows us to monitor short terms systematic errors and to correct for instrumental drift during long analytical sessions. Full integration of EDS/WDS signals allows us to obtain quantitative data for major, minor and trace elements simultaneously in a single spot.
6. Our routine EPMA protocol for olivine includes evaluation of accuracy and precision as well as monitoring of instrumental drift by running one or more olivine reference samples as unknowns. The analytical precision (reproducibility) of olivine analyses is 200 - 300 ppm (2 standard deviations) for the major element equivalent of Fo and between 6 and 20 ppm for trace elements. The precision is equal to the detection limits of corresponding elements. For each of 10 trace and minor elements (Na, Al, P, Ca, Ti, Cr, Mn, Co, Zn and Ni), the measured concentrations in the olivine reference was found to be same, within the 2 standard deviation internal precision, to concentrations in reference values, suggesting that our method for olivine is free of systematic errors and that accuracy is similar to precision.
7. The advantages of EPMA microanalysis of minor and trace elements over other in-situ methods such as SIMS, LA-ICP-MS and μ-PIXE are: 1) a high spatial resolution (1 - 10 μm); 2) non-destructive analysis; 3) a well-developed matrix correction procedure, and 4) relatively low cost. This makes EPMA an indispensable method with many applications in experimental petrology, geochemistry and cosmochemistry.
8. The major disadvantage of EPMA is current sensitivity, which is not sufficient to analyse elements below 10 ppm concentration levels with sufficient precision. Further advance of the EPMA technique for trace-element analysis requires the development of new ultra-high sensitive spectrometers and detectors, new background acquisition software, and development of reference materials with matrix compositions similar to analysed samples and with well-characterised trace element compositions.

Acknowledgments

VB thanks the European Microbeam Analysis Society for the invitation to the EMAS 2017 Workshop in Konstanz, Germany. We thank N. Arndt for help in improving the English and the clarity of the manuscript. The constructive critical comments of two anonymous reviewers and discussion with V.N. Korolyuk significantly improved the paper. The study was supported by a grant from Labex OSUG@2020 (Investissements d'avenir – ANR10 LABX56) and Russian Science Foundation grant 14-17-00491 to AVS.

References

[1] Reed S J B 2000 *Mikrochim. Acta* **132** 145-151
[2] Merlet C and Bodinier J L 1988 *Chem. Geol.* **70** 172-172
[3] Merlet C and Bodinier J L 1990 *Chem. Geol.* **83** 55-69
[4] Fialin M, Remy H, Richard C and Wagner C 1999 *Am. Mineral.* **84** 70-77
[5] Williams M L, Jercinovic M J and Terry M P 1999 *Geology* **27** 1023-1026
[6] Laputina I P, Batyrev V A and Yakushev A I 1999 *J. Anal. At. Spectrom.* **14** 465-469
[7] Pyle J M and Spear F S 2000 *Contrib. Mineral. Petrol.* **138** 51-58
[8] Jercinovic M J and Williams M L 2005 *Am. Mineral.* **90** 526-546
[9] Jercinovic M J, Williams M L and Lane E D 2008 *Chem. Geol.* **254** 197-215
[10] Jercinovic M J, Williams M L, Allaz J and Donovan J J 2012 *IOP Conf. Ser.:Mat. Sci. Engng.* **32** 012012

[11] Rusk B G, Reed M H, Dilles J H and Kent A J R 2006 *Am. Mineral.* **91** 1300-1312
[12] Wark D A and Watson E B 2006 *Contrib. Mineral. Petrol.* **152** 743-754
[13] Donovan J J, Lowers H A and Rusk B G 2011 *Am. Mineral.* **96** 274-282
[14] Kronz A, Kerkhof A M and Muller A 2012 *Quartz: deposits, mineralogy and analytics;* (Goetze J and Moeckel R; eds.) (Springer: Berlin, Heidelberg) 191-217
[15] Audetat A, Garbe-Schonberg D, Kronz A, Pettke T, Rusk B, Donovan J J and Lowers H A 2015 *Geostand. Geoanal. Res.* **39** 171-184
[16] Sobolev A V, Hofmann A W, Sobolev S V and Nikogosian I K 2005 *Nature* **434** 590-597
[17] Sobolev A V, Hofmann A W, Kuzmin D V, Yaxley G M, Arndt N T, Chung S L, Danyushevsky L V, Elliott T, Frey F A, Garcia M O, Gurenko A A, Kamenetsky V S, Kerr A C, Krivolutskaya N A, Matvienkov V V, Nikogosian I K, Rocholl A, Sigurdsson I A, Sushchevskaya N M and Teklay M 2007 *Science* **316** 412-417
[18] Sobolev N V, Logvinova A M, Zedgenizov D A, Pokhilenko N P, Malygina E V, Kuzmin D V and Sobolev A V 2009 *Lithos* **112** 701-713
[19] Batanova V G, Sobolev A V and Kuzmin D V 2015 *Chem. Geol.* **419** 149-157
[20] Korolyuk V N and Pokhilenko L N 2016 *Rus. Geol. Geophys.* **57** 1750-1758
[21] Sobolev A V, Asafov E V, Gurenko A A, Arndt N T, Batanova V G, Portnyagin M V, Garbe-Schonberg D and Krasheninnikov S P 2016 *Nature* **531** 628-632
[22] Zamboni D, Trela J, Gazel E, Sobolev A V, Cannatelli C, Lucchi F, Batanova V G and De Vivo B 2017 *Lithos* **272** 185-191
[23] Gervilla F, Cabri L J, Kojonen K, Oberthur T, Weiser T W, Johanson B, Sie S H, Campbell J L, Teesdale W J and Laflamme J H G 2004 *Microchim. Acta* **147** 167-173
[24] Witter J B and Kuehner S M 2004 *Am. Mineral.* **89** 57-63
[25] Zhang C, Koepke J, Wang L X, Wolff P E, Wilke S, Stechern A, Almeev R and Holtz F 2016 *Geostand. Geoanal. Res.* **40** 351-363
[26] Lavrent'ev Y G, Korolyuk V N, Usova L V and Logvinova A M 2006 *Rus. Geol. Geophys.* **47** 1075-1078
[27] Borghi A, Cossio R, Olmi F, Ruffini R and Vaggelli G 2002 *Mikrochim. Acta* **139** 17-25
[28] Gauert C, Schannor M, Hecht L, Radtke M and Reinholz U 2016 *Geostand. Geoanal. Res.* **40** 267-289
[29] Neaman A, Martinez C E, Trolard F and Bourrie G 2008 *Appl. Geochem.* **23** 778-782
[30] Spear F S, Wark D A, Cheney J T, Schumacher J C and Watson E B 2006 *Contrib. Mineral. Petrol.* **152** 375-385
[31] Weiss Y, Griffin W L, Elhlou S and Navon O 2008 *Chem. Geol.* **252** 158-168
[32] Liebhafsky H A, Pfeiffer H G and Zemany P D 1960 in: *2nd International Symposium: X-ray Microscopy and X-ray Microanalysis.* (Engstrom A, Cosslet V and Pattee H; eds.) (Elsevier, Amsterdam) 321
[33] Ancey M, Bastenaire F and Tixier R 1978 in: Proc. *Summer School St. Martin-d'Herres.* (Maurice F; ed.) (Orsay, France: Les Editions de Physique) 319-343
[34] Ziebold T O 1967 *Anal. Chem.* **39** 858-861
[35] Sato A, Mori N, Takakura M and Notoya S 2007 *JEOL News* **42** 46-52
[36] Goldstein J, Newbury D, Joy D, Lyman C, Echlin P, E. L, Sawyer L and Michael J 2003 *Scanning electron microscopy and X-ray microanalysis.* (New York, NY: Springer) 690
[37] Toya T and Kato A 1983 *Practical techniques for microprobe analysis.* (Tokyo, Japan: JEOL Training Center)
[38] Lavrent'ev Y G 2010 *X-ray Spectrom.* **39** 37-40
[39] De Hoog J C M 2008 *Geochim. Cosmochim. Acta* **72** A208-A208
[40] Korolyuk V N and Pokhilenko L N 2014 *X-ray Spectrom.* **43** 353-358
[41] Mori N, Kamiyama R, Tanaka T, Kimura T, McSwiggen P, Onodere H and Nielsen C 2013 *Microsc. Microanal.* **19** (Suppl. 2) 1432

[42] Montel J M, Razafimahatratra D, Ralison B, De Parseval P, Thibault M and Randranja R 2011 *Europ. J. Mineral.* **23** 745-757 [43] Montel J M, Foret S, Veschambre M, Nicollet C and Provost A 1996 *Chem. Geol.* **131** 37-53
[44] Seydoux-Guillaume A M, Wirth R, Deutsch A and Scharer U 2004 *Geochim. Cosmochim. Acta* **68** 2517-2527
[45] Rinaldi R and Llovet X 2015 *Microsc. Microanal.* **21** 1053-1069
[46] Lifshin E and Gauvin R 2001 *Microsc. Microanal.* **7** 168-177
[47] Gunn J S, Harrowfield I R, Proctor C H and Thresher R E 1992 *J. Exp. Mar. Biol. Ecol.* **158** 1-36
[48] Dalton J A and Lane S J 1996 *Am. Mineral.* **81** 194-201
[49] Llovet X, Galan G 2003 *Am. Mineral.* **88** 121-130
[50] Llovet X, Pinard P T, Donovan J J and Salvat F 2012 *J. Phys. D:Appl. Phys.* **45**
[51] Probe software Inc., 885 Crest Drive, Eugene, OR 97405, USA
[52] Donovan J J and Tingle T N 1996 *J. Microsc. Microanal.* **2** 1-7
[53] Laputina I P and Batyrev V A 1998 *Mikrochim. Acta* **Suppl. 15** 247-252
[54] Donovan J J, Singer J W and Armstrong J T 2016 *Am. Mineral.* **101** 1839-1853
[55] Jarosewich E J, Nelen J A and Norberg J A 1980 *Geostand. Newslett.* **4** 43-47
[56] Batanova V, Sobolev A, Thompson J M, Danyushevsky L V, Goemann K, Portnyagin M V, Garbe-Schonberg D, Hauri E H, Kimura J-I, Chang Q, Senda R, Chauvel C, Campillo S and Ionov D 2017 *Goldschmidt Conference.* (Paris, France)

Measurements of the quantitative lateral analytical resolution at evaporated aluminium and silver layers with the JEOL JXA-8530F FEG-EPMA

D Berger and J Nissen

Technische Universität Berlin, Zentraleinrichtung Elektronenmikroskopie ZELMI, Straße des 17. Juni 135, 10623 Berlin, Germany

E-mail: dirk.berger@tu-berlin.de; joerg.nissen@tu-berlin.de

Abstract. The studies in this paper are part of systematic investigations of the lateral analytical resolution of the field emission electron microprobe JEOL JXA-8530F. Hereby, the quantitative lateral resolution, which is achieved in practise, is in the focus of interest. The approach is to determine the minimum thickness of a metallic layer for which an accurate quantitative element analysis in cross-section is still possible. Previous measurements were accomplished at sputtered gold (Z = 79) layers, where a lateral resolution in the range of 140 to 170 nm was achieved at suitable parameters of the microprobe. To study the Z-dependence of the lateral resolution, now aluminium (Z = 13) resp. silver (Z = 47) layers with different thicknesses were generated by evaporation and prepared in cross-section subsequently by use of a focussed Ga-ion beam (FIB). Each layer was analysed quantitatively with different electron energies. The thinnest layer which can be resolved specifies the best lateral resolution. These measured values were compared on the one hand with Monte Carlo simulations and on the other hand with predictions from formulas from the literature. The measurements fit well to the simulated and calculated values, except the ones at the lowest primary electron energies with an overvoltage below ~ 2. The reason for this discrepancy is not clear yet and has to be clarified by further investigations. The results apply for any microanalyser – even with energy-dispersive X-ray spectrometry (EDS) detection – if the probe diameters, which might deviate from those of the JEOL JXA-8530F, at suitable analysing parameters are considered.

1. Introduction

In contrast to the qualitative lateral analytical resolution, where the analytical issue is to recognise the smallest features by element distributions, the quantitative lateral analytical resolution is of great significance for the determination of the accurate element composition of small features. The first attempt to define the quantitative resolution are measurements of the rise of the X-ray intensity of an element as edge profile at the interface of two different materials from 1 to 99 % X-ray intensity. Measurements at an Au-Si interface resulted in a quantitative lateral resolution of 90 nm [1]. Since resolution measurements by edge profiles depends on the adjacent elements, it is more useful to determine the quantitative resolution from the smallest feature for which an accurate quantitative element analysis is still feasible. The limiting parameters hereby are mainly the mean atomic number Z of the feature elements, the primary electron energy E_0 and the beam diameter which might be rather large for low E_0 and suitable beam currents (figure 1). To study the Z- and E_0-dependence of the analytical lateral resolution, features were generated by vapour deposition of metallic layers with different atomic number and then analysed quantitatively by varying the primary electron energy.

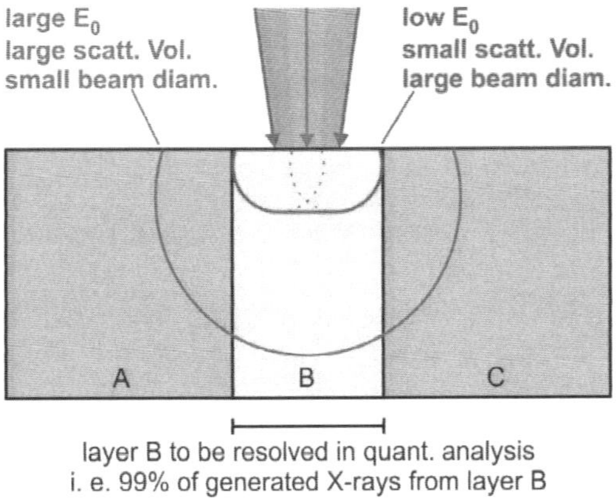

Figure 1. Influence of primary electron energy E_0 and beam diameter on X-ray emitting volume and quantitative lateral analytical resolution.

2. Experimental and results

2.1. Deposition and preparation of the layers

For the experimental determination of the quantitative analytical lateral resolution under different measurement conditions, aluminium layers with different thicknesses (3850, 940, 445, 345, 280 and 260 nm) were generated by evaporation on silicon wafer substrates. The cross-sectional preparation of the specimens was done first by cleaving and then by cutting an about 50 - 100 µm wide cross-section with a rough (30 kV, 21 nA) focussed Ga-ion beam (FEI Helios Nano Lab600 FIB). Second, free spaces are milled adjacent to the cross-section, if necessary, to avoid fluorescence effects, see figure 2.

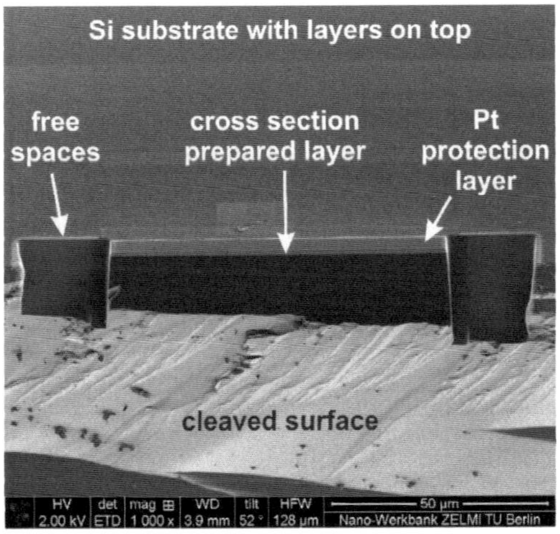

Figure 2. SEM image of a cleaved Si substrate with ion-beam prepared cross-section surface and the free spaces. The evaporated Al layer on top of Si is too thin to be resolved in this image.

In the last preparation step a gentle ion beam (30 kV, 90 pA) is used to polish the cross-section surface with low implantation of Ga into the Al layer. The cross-section polishing process requires a mechanical protection layer consisting of Ga, Pt and C which is apparent on top of the aluminium layer. Figure 3 shows high resolution SE-images of the cross-section of the 6 layers.

IOP Conf. Series: Materials Science and Engineering **304** (2017) 012002 doi:10.1088/1757-899X/304/1/012002

Figure 3. Cross-section images of the test specimens with different aluminium layers from 3850 to 260 nm thickness.

The preparation of the silver layers was accomplished in the same way as the aluminium layers with one difference. To avoid fluorescence effects in the Si-substrate and the FIB protection layer, two additional palladium layers were deposited, adjacent to the silver layer. Ag-Lα X-ray with energy of 2.984 keV cannot excite Pd-Lα X-ray by fluorescence due to the higher critical ionisation energy of Pd-Lα X-ray at 3.172 keV. Figure 4 shows the layer structure with the 3 thinnest silver layers.

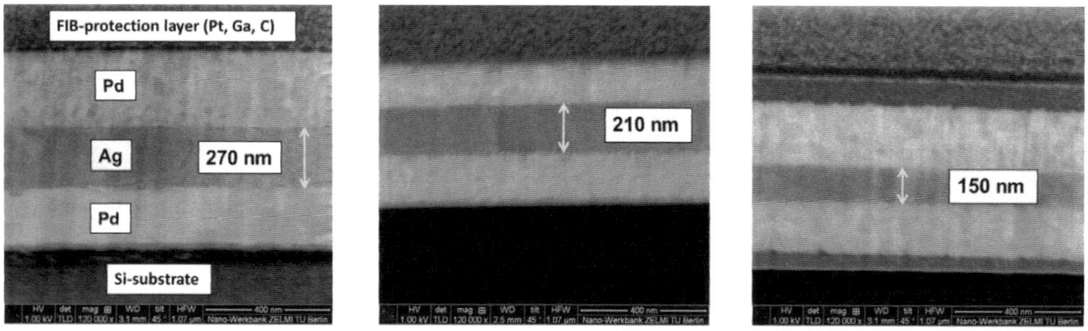

Figure 4. Cross-section images of the test specimens with different silver layers from 270 to 150 nm thickness.

The density of both the Al and the Ag-layers were determined by the weighing method [2] to (2.68 +/- 0.02) g/cm^3 and (10.4 +/- 0.1) g/cm^3 respectively. These density values are close to the bulk material densities indicating that the evaporated layers are not porous. Finally, the density values are verified by analysing quantitative measurements with the thin film software STRATAGEM.

2.2. Measured quantitative lateral resolution

Each test specimen is analysed quantitatively for different electron energies using the 3850 nm Al layer and the 2000 nm Ag layer as standard respectively. By applying an appropriate probe current, the beam diameter is chosen to be smaller than about 50 nm, see figure 1. Beam diameters are determined in correspondence with ISO/TS 24597 by the analysis of the contrast profile in images of gold insulars on carbon [1]. Wavelength-dispersive X-ray spectrometry (WDS) measurements are done with the Al Kα-line on the TAP crystal and the Ag Lα-line on the PET crystal. The acquisition time (20 s for the peak and 10 s for the background) and the beam currents are chosen to minimise drift effects at a reasonable counting statistics.

Table 1 summarizes the results of the element quantification for the aluminium layers at the electron energies 15, 8, 7, 5, 4 and 3 keV. Even lower energies are not practical due to the decreasing image quality, which makes the beam adjustment more difficult, and due to the too low overvoltage. To get sufficient X-ray intensity, the appropriate overvoltage to excite Al-Kα X-ray (critical ionisation energy: 1.559 keV) should not be below ~ 2 [3].

Table 1. Results of the quantitative measurements of the aluminium layers in wt% Al.

Al layer thickness [nm]	15 keV	8 keV	7 keV	5 keV	4 keV	3 keV
3850 (Reference)	100.00	100.00	100.00	100.00	100.00	100.00
940	/	97.82	100.08	99.96	100.16	99.92
445	/	/	89.76	99.68	99.75	99.95
345	/	/	/	95.18	99.89	99.87
280	/	/	/	/	97.93	99.61
260	/	/	/	/	/	98.95
Beam diameter d_0 [nm]	**30**	**43**	**33**	**38**	**45**	**52**
Probe current [nA]	**5**	**5**	**1**	**1**	**1**	**1**

The values in the table are the averages of 10 measurements in weight percent aluminium (wt% Al). A quantification result is expected to be accurate if it is in the range of (100.0 +/- 0.5) %, indicating that the source volume of the emitted X-rays is completely inside the Al layer within statistical fluctuations. The 280-nm layer is the thinnest which can be resolved (99.61 wt% Al). Thus, for aluminium a lateral quantitative resolution in the range of 260 to 280 nm is achieved at 3 keV electron energy.

The quantification results for silver are shown in table 2. Here, the lowest practical electron energy to excite Ag-Lα X-ray (critical ionisation energy: 3.352 keV) is 5 keV (overvoltage: 5/3.352 = 1.49). The thinnest layer, for which an accurate analysis is still feasible here, is 210 nm (100.29 wt% Ag). The best resulting lateral resolution for silver is the range of 150 to 210 nm at 5 keV.

Table 2. Results of the quantitative measurements of the silver layers in wt% Ag.

Ag layer thickness [nm]	15 keV	8 keV	7 keV	6 keV	5 keV
2000 (Reference)	100.00	100.00	100.00	100.00	100.00
870	98.77	99.82	100.08	99.96	100.16
270	/	96.21	99.82	99.65	100.28
210	/	/	96.78	99.71	100.29
150	/	/	/	/	95.79
Beam diameter d_0 [nm]	**30**	**43**	**33**	**35**	**38**
Probe current [nA]	**5**	**5**	**1**	**1**	**1**

2.3. Monte Carlo simulation of the quantitative lateral resolution

The electron scattering in aluminium and silver for electron energies between the primary energy (15, 8, 7, 6, 5, 4 and 3 keV) and the critical ionisation energy (1.559 keV for Al-K and 3.352 keV for Ag-L) is simulated with the programme CASINO (version 2.48) [4]. Figures 5 and 6 show the resulting plots of the electron paths of primary electrons (blue) and backscattered electrons (red). For the determination of the quantitative analytical resolution 99 % of the 10,000 simulated trajectories were taken into account by counting by hand. The width of the interaction volume of these 99 % of electron paths is defined as the simulated lateral resolution, see indications in figures 5 and 6. The self-absorption of Al-Kα in Al and Ag-Lα in Ag is proofed to be very small so that the range of the electron paths is a very well measure for the range of emitted X-rays.

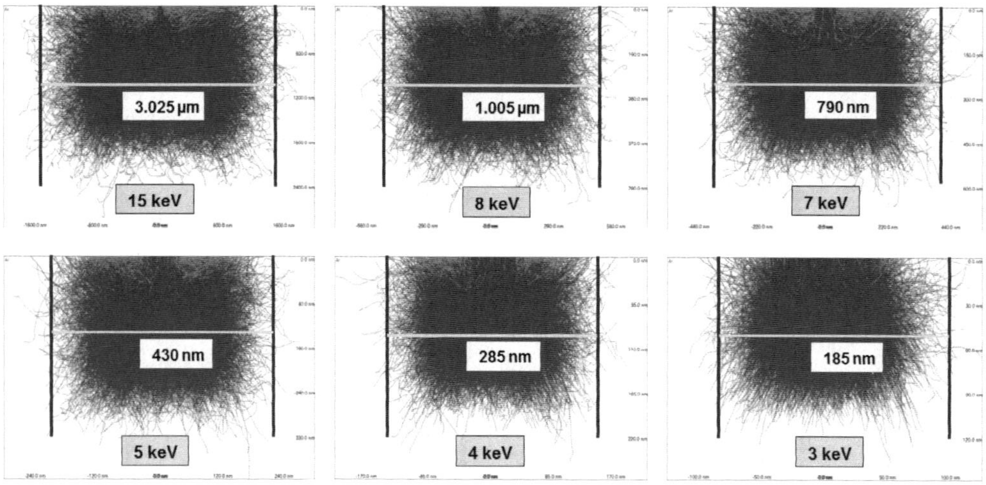

Figure 5. Monte Carlo simulations of scattered electrons in Al with different electron energies.

IOP Conf. Series: Materials Science and Engineering **304** (2017) 012002 doi:10.1088/1757-899X/304/1/012002

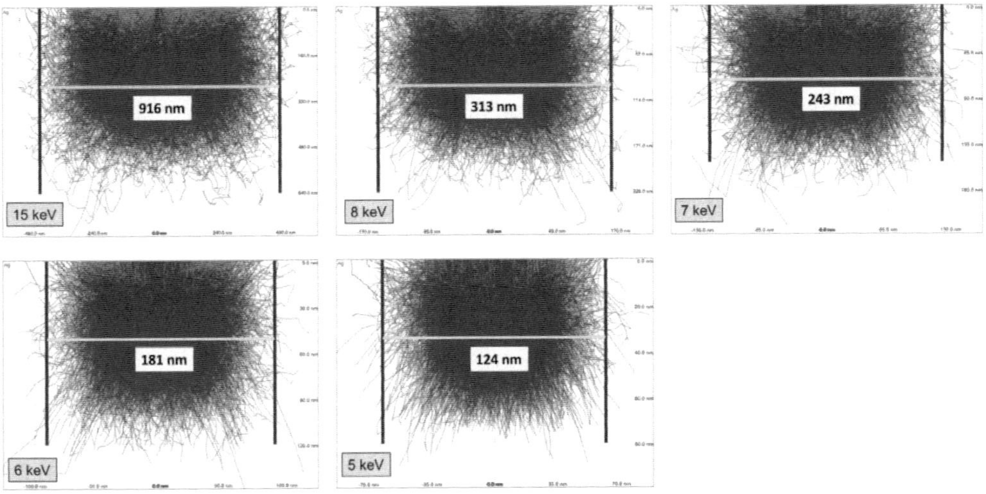

Figure 6. Monte Carlo simulations of scattered electrons in Ag with different electron energies.

2.4. Calculation of the quantitative lateral resolution

In the last step the quantitative lateral resolution R_s was calculated firstly by the eq. (1) from Merlet and Llovet [5] and secondly by the eq. (2) from Willich and Bethke [6]:

$$R_S = \sqrt{4 \cdot (R_x - Z_m)^2 + d_0^2} \tag{1}$$

$$R_S = R_x + d_0 \tag{2}$$

with R_S is the lateral analytical resolution in nm, R_x is the ionisation range, d_0 the beam diameter and Z_m the depth of the maximum of the X-ray depth distribution [7].

Figure 7 illustrates that equation (1) is based on the convolution of the interaction volume (its radius is estimated by R_x - Z_m) with the beam diameter which is obtained from measurements. In contrast, eq. (2) regards a linear sum between the ionisation range and the beam diameter.

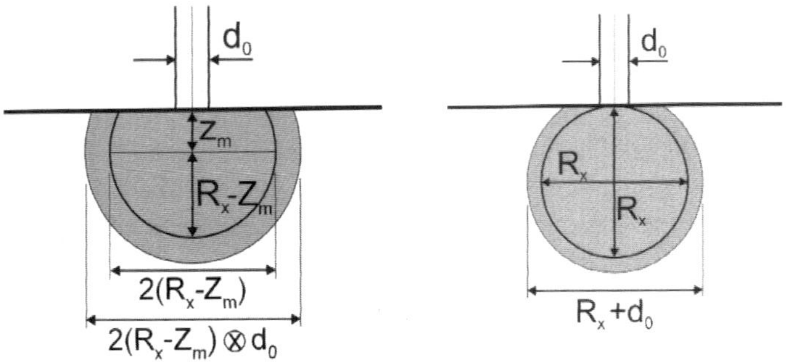

Figure 7. Schematic illustration of parameters specifying the lateral resolution in eqs. (1) and (2).

For aluminium table 3 and figure 8 represent the comparison of the simulated, calculated and measured lateral resolutions as a function of the excitation energy E_0. Since the measurements are based on 6 specimens with different layer thickness, the measured lateral resolutions are specified as a range.

Table 3. Comparison of simulated, calculated and measured quantitative lateral resolutions for Al and different primary electron energies.

Electron energy	Lateral resolution from MC simulation [nm]	R_x [nm]	Z_m [nm]	Calculated lateral resolution R_s with formula (1) [nm]	Calculated lateral resolution R_s with formula (2) [nm]	Measured lateral resolution [nm]
15 keV	3025	2499	548	3902	2529	940-3850
8 keV	1005	823	202	1242	866	940-3850
7 keV	790	645	163	965	678	445-940
5 keV	430	340	92	497	378	345-445
4 keV	285	216	61	313	261	280-345
3 keV	185	111	164	164	163	260-280

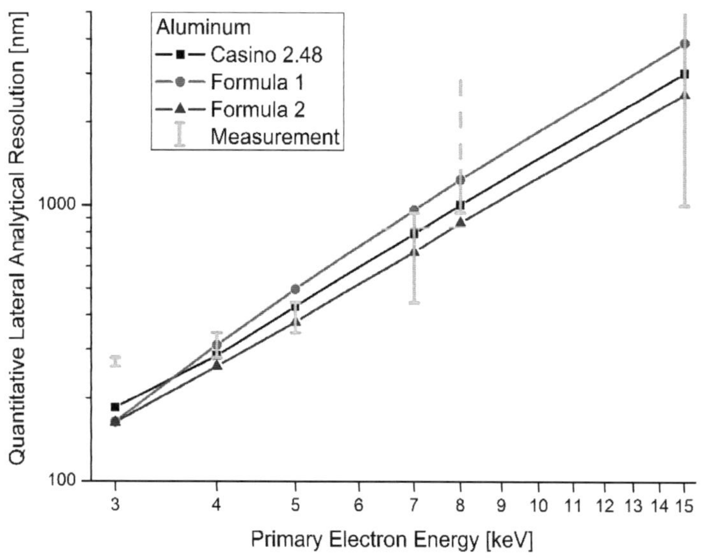

Figure 8. Comparison of the simulated, calculated and measured quantitative lateral resolutions for Al and different primary electron energies.

The graph elucidates the improvement of the resolution with decreasing excitation energy due to the decreasing source volume of emitted X-rays. Another remarkable result is the good correlation between measurement and Monte Carlo simulation, except at the primary electron energy of 3 keV for which the overvoltage is below 2. The reasons for this discrepancy are not clear.

The calculated values deviated from the measurements in the same manner at 3 keV. For higher electron energies there is the tendency that eq. (1) predicts a little bit too large values and eq. (2) a little bit too low values.

In the case of silver (table 4 and figure 9) the results from measurement and Monte Carlo simulation coincide as well as for aluminium. In addition, the calculated value from eq. (1) are in good agreement with both whereas eq. (2) predicts rather to small values again.

Table 4. Comparison of simulated, calculated and measured quantitative lateral resolutions for Ag and different primary electron energies.

Electron energy	Lateral resolution from MC simulation [nm]	R_x [nm]	Z_m [nm]	Calculated lateral resolution R_s with formula (1) [nm]	Calculated lateral resolution R_s with formula (2) [nm]	Measured lateral resolution [nm]
15 keV	916	670	70	1201	700	870-2000
8 keV	313	193	24	340	236	270-870
7 keV	243	142	18	250	175	210-270
5 keV	181	96	13	170	131	150-210
3 keV	124	55	7	104	93	150-210

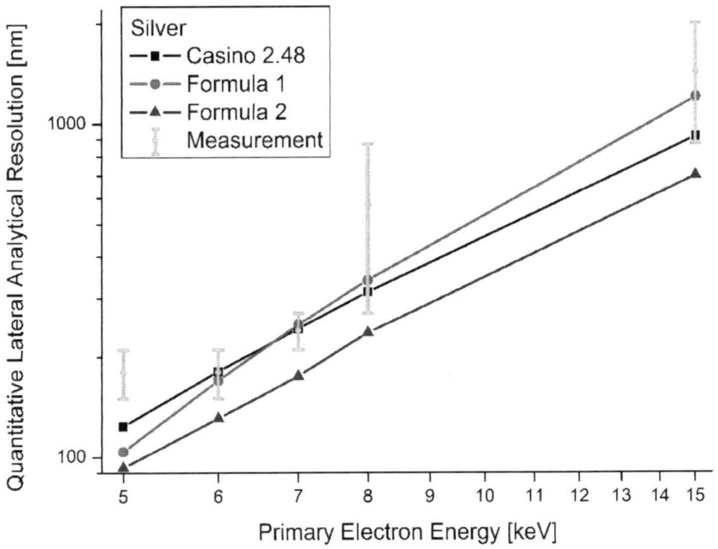

Figure 9. Comparison of the simulated, calculated and measured quantitative lateral resolutions for Ag and different primary electron energies.

Nevertheless, the discrepancy between simulation and calculation to measurement are once more obvious for the lowest primary electron energy. Since the electron optical properties and the stability of the used microprobe are perfect for the electron energy of 5 keV, as can be seen from the consistent experimental data for 5 keV in the aluminium measurements, there is no reason to doubt the accurateness of the measurements. Therefore, the most probable reason for the discrepancy is the wrong estimation of the scattering volume due to inaccurate values of R_x and Z_m in the calculation and the scattering cross-sections in the Monte Carlo simulations which show up more clear if the primary electron energy is close to the ionisation energy with an overvoltage well below or close to 2.

2.5. Z-dependence of the quantitative lateral resolution

From the results of this and the previous paper [8] the atomic number (Z)-dependence of the quantitative lateral resolution can be derived for the first time. Figure 10 shows this Z-dependence for different primary electron energies.

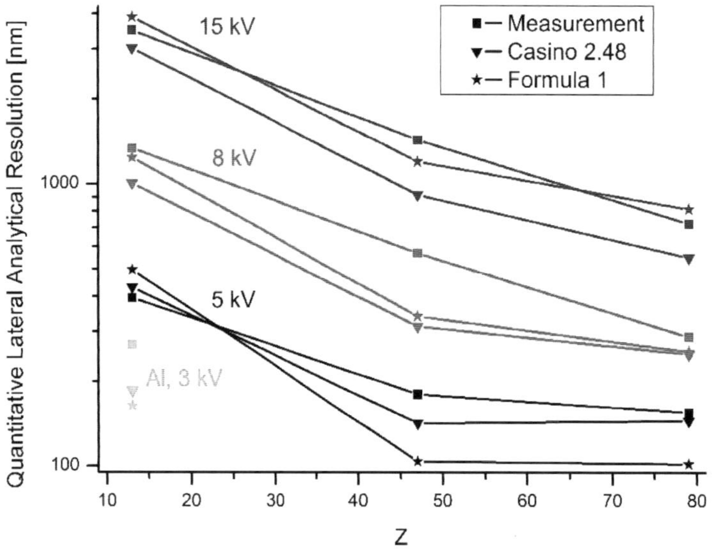

Figure 10. Quantitative lateral analytical resolution as a function of atomic number at 15, 8 and 5 keV primary electron energy at suitable measurement conditions.

On one hand the resolution improves with decreasing energy, due to the reduction of the electron interaction volume inside the material. Obviously, the reduction of the interaction volume is not compensated completely by the increase of the beam diameter at low primary electron energies. On the other hand, the resolution becomes worse with lower atomic number owing to the lower density of the material and hence the increasing spread of electrons. However, at 5 keV there is no significant difference between Au (79) and Ag (47). For the Ag measurement, the overvoltage is low in relation to the critical ionisation energy of the Ag-Lα X-ray line at 3.352 keV. Hence, the range of excited Ag-Lα X-rays is small. For the Au measurement, the Au-Mα X-ray line was used (ionisation energy of 2.31 keV) for which the overvoltage is larger. Therefore, the decrease of the X-ray emitting volume due to higher Z is compensated by its increase due to the higher overvoltage. It should be noted again that the further decrease of the primary electron energy would on one hand increase the resolution but on the other hand worsens the counting statistics to unusable values. In addition, it should be kept in mind that the shown *Z*-dependence relies on different X-ray lines with different absorption properties. The aim of this investigation is rather to determine the thickness of layers for which an accurate element quantification is possible than to compare the lateral resolution for different X-ray lines.

3. Conclusion

In this paper measurements of the quantitative lateral analytical resolutions at thin test specimens with usable beam currents and acquisition times are presented. The experiments are accomplished at metallic layers with different thicknesses and different atomic numbers. For silver the lowest quantitative lateral resolution in the range of 150 to 210 nm is obtained for 5 keV and for aluminium 260 to 280 nm for 3 keV. These results are in good agreement with those previously obtained at gold (140 to 170 nm at 5 keV).

The experimental results are in accordance with Monte Carlo simulations and analytical calculations for most primary electron energies, but differ significantly for low energies for which the overvoltage is below or close to two. The cause for this discrepancy is not clear yet and has to be clarified by further investigations, in particular because the best quantitative analytical resolution is obtained at low energies for which the discrepancy is the largest.

IOP Conf. Series: Materials Science and Engineering **304** (2017) 012002 doi:10.1088/1757-899X/304/1/012002

Acknowledgements
The financial support by the Europäische Fonds für Regionale Entwicklung (EFRE) within the projects "Nano-Analytiklabor" and "Nano-Werkbank" is gratefully acknowledged. We thank Xavier Llovet for fruitful discussion and the support in applying eq. (1).

References
[1] Berger D and Nissen J 2014 *IOP Conf. Ser.: Mater. Sci. Engng.* **55** 012002
[2] Berger D and Galbert F 2011 *MC2011 Conf. Proc.* (Kiel, Germany) ISBN 978-3-00-033910-3 IM7.P208
[3] Carpenter P K and Jolliff B L 2015 *Microsc. Microanal.* **21** (Suppl. 3) 1443-1444
[4] Drouin D, et al. 2007 *Scanning* **29** 92-101
[5] Merlet C and Llovet X 2012 *IOP Conf. Ser.: Mater. Sci. Engng.* **32** 012016
[6] Willich P and Bethke R 1996 *Mikrochim Acta* **Suppl. 13** 631
[7] Merlet C 1994 *Mikrochim. Acta* **114/115** 363-376
[8] Berger D and Nissen J 2016 *IOP Conf. Ser.: Mater. Sci. Engng.* **109** 012001

IOP Conf. Series: Materials Science and Engineering **304** (2017) 012003 doi:10.1088/1757-899X/304/1/012003

Understanding deformation with high angular resolution electron backscatter diffraction (HR-EBSD)

T B Britton and J L R Hickey

Imperial College London, Department of Materials, Exhibition Road, London SW7 2AZ, Great Britain

E-mail: b.britton@imperial.ac.uk

Abstract. High angular resolution electron backscatter diffraction (HR-EBSD) affords an increase in angular resolution, as compared to 'conventional' Hough transform based EBSD, of two orders of magnitude, enabling measurements of relative misorientations of 1×10^{-4} rads (~ 0.006°) and changes in (deviatoric) lattice strain with a precision of 1×10^{-4}. This is achieved through direct comparison of two or more diffraction patterns using sophisticated cross-correlation based image analysis routines. Image shifts between zone axes in the two-correlated diffraction pattern are measured with sub-pixel precision and this realises the ability to measure changes in interplanar angles and lattice orientation with a high degree of sensitivity. These shifts are linked to strains and lattice rotations through simple geometry. In this manuscript, we outline the basis of the technique and two case studies that highlight its potential to tackle real materials science challenges, such as deformation patterning in polycrystalline alloys.

1. Introduction

Electron diffraction offers an opportunity to study the internal structure of materials. In the scanning electron microscope (SEM), we can perform electron diffraction using backscattering electrons.

In electron backscatter diffraction (EBSD), these electrons enter the sample and scatter through electron-matter interactions [1]. If the sample is tilted to a high angle, a high fraction of these scattering electrons escape, and if their last scattering angle is favourably oriented with respect to the crystal lattice they will diffract according to Bragg's law [2]. In the kinematic approximation of diffraction, edges to Kikuchi bands are formed at an angle $\pm \theta$ with respect to the perpendicular of the plane normal of the diffracting plane [3]. The EBSD pattern contains many bands, which are located on the phosphor screen as a function of the microscope settings (i.e., voltage), crystal orientation, lattice parameters, and the location of the imaging screen (i.e., the pattern centre) [4]. The geometry of the electron scattering process results in the formation of a direct projection of the diffracting lattice planes.

A change in band width (e.g., due to a change in a hydrostatic strain state) can occur due to the change in interplanar spacing affecting Bragg's law but the conventional EBSD geometry is relatively insensitive to this [5]. A change in interplanar angles will cause the bands to move across the diffracting pattern and this is more obvious and, therefore, measurable. For a deformation experiment, these changes can be induced by the presence of a (relative) deviatoric strain, and furthermore the bands may move across the screen due to the crystal rotating due to the presence and movement of dislocations during crystal slip. The presence of stored dislocations which result in local orientation gradients, can be linked through Nye's analysis [6] and, therefore, high precision measurements of spatial rotation gradients can be used to estimate the lower bound of the stored dislocation content [7, 8].

High resolution electron backscatter diffraction increases the relative angular resolution of the technique by two orders of magnitude, through comparison of two diffraction patterns to extract subtle changes. The relative merits of EBSD and HR-EBSD are summarised in table 1.

Table 1. A relative comparison of conventional (i.e., Hough-based) and high resolution (i.e., cross-correlation based) electron backscatter diffraction.

	EBSD	HR-EBSD
Absolute orientation	~ 2° [9]	No
Misorientation	~ 0.1 to 0.5° [10, 11]	~ 0.006° ($1x10^{-4}$ rads) [7]
GNDs @ 1 μm step	> $3x10^{13}$	> $3x10^{11}$
GNDs @ 100 nm step **in lines / m²** (b = 0.3 nm) [12, 13]	> $3x10^{14}$	> $3x10^{12}$
Relative elastic strain	No	Deviatoric strain ± $1x10^{-4}$
Relative residual stress (Type III – within grain)	No	Using anisotropic Hooke's law ± ~20 MPa (when E = 200 GPa)
Example tasks:	Microstructure, Texture, Grain size	Deformation i.e., elastic strain, misorientation and residual dislocation content

Conventional, Hough-based, EBSD uses band identification, through peak identification in Hough space, and subsequent indexing to identify the crystal orientation [4]. Work by Humphrey's indicates that the effective precision of this approach for misorientation analysis is ~ 0.5° (i.e., 0.0087 rads) [14]. The high resolution EBSD analysis, introduced principally by Wilkinson, Meaden and Dingley [7, 8] in 2006 directly compares the two diffraction patterns using cross-correlation based image analysis to precisely measure the change in interplanar angles and results in an increase in measurement precision of two orders of magnitude, i.e., 0.0001 rads [7, 8].

These subtle shifts can be ascribed to lattice deformations resulting from a change in the lattice parameters due to a deviatoric change in lattice strain and therefore enable, using Hooke's law and a plane stress assumption, the residual stress variations between points within a single grain to be evaluated with a precision of ~20 MPa (for a material with a Young's modulus of 200 GPa).

The HR-EBSD technique has been widely applied across a range of semiconductor [7, 15], superconductor [16, 17], metals (e.g., [18-20]) and geological systems [21], often within the scope of understanding local changes in lattice parameter and especially within the field of deformation analysis.

In this manuscript, we will briefly introduce the key methodology of the technique. This will be followed by an example showing its efficacy at resolving high resolution orientation gradients in deformed copper and focus on studying the storage of geometrically necessary dislocations around an indent made in interstitial free (IF) steel.

2. Methodology

HR-EBSD involves analysis of diffraction patterns that are captured using the conventional EBSD set-up, with the camera inserted as for a texture or grain mapping experiment. The only differences are that the quality of the sample preparation must be higher (as surface scratches or deformation is more evident when the precision of the experiment is higher) and that the best quality patterns are captured. The quality of the patterns is important as signal to noise ratio and angle subtended per pixel (i.e., the pattern resolution) affects the resolution of the result [13, 22, 23]. HR-EBSD can be conducted both with so called 'fast' cameras or high binning modes of more 'sensitive' cameras [23], yet for the ultimate resolution often the images are not binned and often long (up to 1 s) exposures with moderate probe currents (~10 nA) are required. The quality of the camera can affect the accuracy of the fitting of pattern shifts to strain, through errors in pattern centre and optical distortions [24]. However for relative (experimental pattern to experimental pattern) HR-EBSD approaches measurement of the pattern centre within 5 pixels [24] and typical optical distortions are reasonable [24].

Once an array of patterns is captured to disk, the processing is conducted offline using software algorithms. An EBSD map is segmented into individual grains containing regions of similar orientation and a reference point, ideally with no strain, but often just of high pattern quality, is selected. The reference point must be less than ~ 12° from the orientation of the test pattern, or there is insufficient overlap between test and reference for a reasonable cross-correlation result [25]. The HR-EBSD technique measures change in strain and lattice rotation with respect to the reference point, and so these may not be variations in absolute strain.

The reference pattern is sub-divided into many 'regions of interest' (ROI), which for a 1,000 x 1,000 pixel image are often 256 x 256 pixels in size [8] and this is repeated for a test pattern, extracted for a point within the same grain. At each ROI, the pattern is extracted and transferred to the Fourier domain for filtering and frequency based cross-correlation. The FFT based cross-correlation enables rapid calculation of the best translation vector between test and reference ROIs. If the translation of four or more well-spaced ROIs are measured well, then with knowledge of the pattern centre, eight components of the deformation tensor can be extracted. For the impact of strain and lattice rotation on the pattern, see figure 1. The ninth of deformation is the hydrostatic strain, which can be calculated only using an out of plane stress traction free boundary condition and Hooke's law.

Reasonable measurement of the translation shift is achieved using sub-pixel precision through up-sampling techniques (i.e., the centre of the cross-correlation peak is determined with sub-pixel accuracy as we assume that the shape of the peak is smooth – see [26] for more information).

Mapping strain and rotation within a map requires capturing of two patterns with the same geometry. Typically, EBSD maps are capturing using 'beam scanning' approaches where the electron beam is deflected and rastered across the sample. If the sample is flat and well aligned in the microscope (i.e., the sample plane normal is well known) then it is possible to construct a simple geometrical model where the beam shift can be corrected for (this is trivially accounted for in commercial software). This correction is important when the beam is shifted by a similar amplitude to the pattern shifts which arise due to strain or lattice rotation. Modern indirect detectors have an effective pixel size of ~20 - 40 μm, and, therefore, uncertainties of in the prediction of the beam shift of 1 μm can result in systematic errors in the calculation of deformation tensors. Alignment of the sample can be assisted greatly through careful insertion into the SEM and presentation with respect to the beam, for instance using special sample holders (for example see page 85, chapter 2.3.3.2 of reference [27]). Note that for GND analysis (on which this paper focusses), beam shift errors have a less significant role as GNDs are calculated using local gradients between neighbour points).

IOP Conf. Series: Materials Science and Engineering **304** (2017) 012003 doi:10.1088/1757-899X/304/1/012003

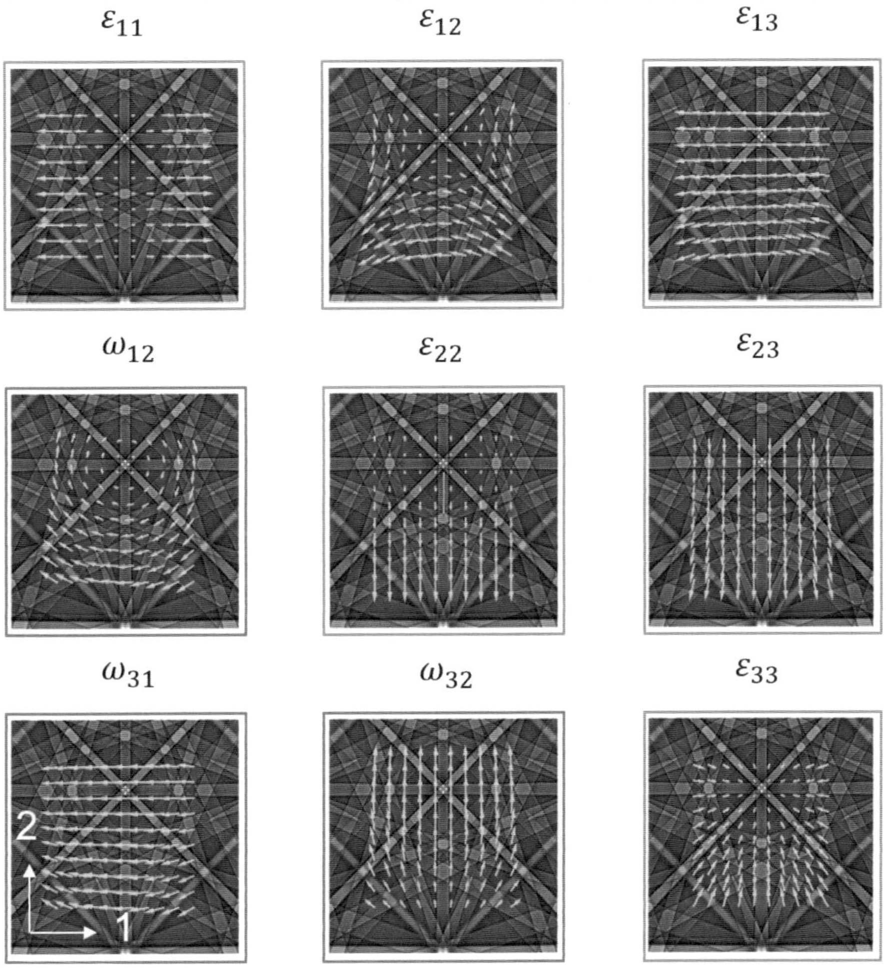

Figure 1. The effect of strain and rotation components on region of interest shifts, for deformations performed in the phosphor frame. Note the presence of the pattern centre, indicated as a red mark. The three normal strain components are not independent and so hydrostatic strains are not accessible with this approach.

Recently, it has been shown by both Maurice *et al.* [28] and Britton and Wilkinson [25] that the presence of a significant rotation gradient between test and reference pattern results in significant inaccuracy in the measurement of elastic strain, e.g., for metallic systems. Therefore, the algorithm has recently been updated to include a remapping step [25], whereby the pattern intensities of one pattern are remapped to better match the orientation of the other prior to extracting the strain components. The total contribution of the remapping and the shift measurements are combined to extract the deformation gradient tensor.

The deformation gradient tensor can be split into strain and rotation using either a finite or infinitesimal decomposition framework. In the case of small strains and small rotations these two are approximately equivalent, and for HR-EBSD, the finite framework is typically used for cases of large lattice rotation differences (as the strain gradients are typically small, due to the elastic limit in most materials).

IOP Conf. Series: Materials Science and Engineering **304** (2017) 012003 doi:10.1088/1757-899X/304/1/012003

The HR-EBSD pattern matching approach provides the best 'guess' to fit a model that maps the intensities of one pattern onto another, and this model is used to calculate the deformation gradient (see figure 2). This approach has been validated extensively using both experimental comparisons, e.g., through comparisons with NIST traceable Raman based stress measurements and AFM in Si [29, 30], comparison with modelling [15, 23, 32], comparison with X-ray based micro-Laue diffraction [32], and algorithm validation using dynamically simulated diffraction patterns [15, 22, 23].

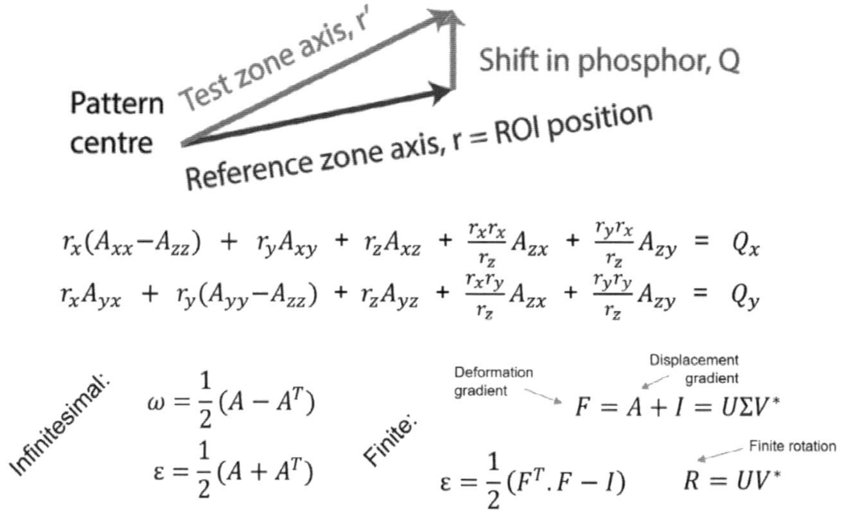

Figure 2. Relationship between shifts in the phosphor and extraction of strain and lattice rotation for HR-EBSD.

One particular aspect of this work has been the generation of 'quality' metrics [26] that indicate when the HR-EBSD method fails. The mean angular error (MAE) describes the difference between the pattern shifts as measured and the predicted pattern shifts per the model, and a high value (i.e., greater than the measurement made) indicates the precision, and not accuracy, obtained. The (normalised) peak height describes the degree of matching for the cross-correlation peak, and it is often formatted as a geometric mean of all ROI peak heights normalised against auto correlation. Measurements using simulations have indicated that when this value falls below 0.3, the measurement is unlikely to be reasonable [32]. These two metrics can be used to rapidly screen data from large maps to remove suspect points, however further care must be taken when reporting stress variations (for example) as for example, not using a remapping step can result in artificially high stresses to be reported which are not physically reasonable [33].

Once the lattice rotations between test and reference are measured, it is possible to utilise Nye's framework [6] to estimate the geometrically necessary dislocation (GND) density and this has proved a very popular aspect of the technique [12, 13, 19, 34, 35]. In short, for each point within the orientation gradient map, a local gradient of the lattice rotation field is calculated and use to determine the lattice curvature (the adaptive kernel we use is described in [36]). As three rotation terms are measured in a 2D map, six components of lattice curvature are calculated. From Nye's formulation, the presence of GNDs can support these lattice curvatures, and it is a 'simple' task of solving the following equation for ρ:

$$\boldsymbol{\xi}(\mathbf{6}\times\mathbf{18})\cdot\boldsymbol{\rho}(\mathbf{18}\times\mathbf{1})=\boldsymbol{\Lambda}(\mathbf{6}\times\mathbf{1}) \quad (1)$$

where $\boldsymbol{\rho}$ are the curvature types, $\boldsymbol{\Lambda}$ are lattice curvatures, and $\boldsymbol{\xi}$ is the outer product of the Burgers vector and line direction for each dislocation type (the origin of this equation and its use for HR-EBSD measurements is explained in Wilkinson and Randman [12]).

For a metallic system, such as for FCC crystals such as Cu, the equation is underdetermined as there are many more potential dislocation types (12 {111} <110> edge, and 6 <110> screw [37]) than constraints from measurement of lattice curvature and therefore following the work of Wilkinson and Randman [12], we have chosen to find a potential solution that reduces the self-energy of the dislocations (accounting for screw and edge types, as well as changes in Burgers vector length) in a so-called "L-1 minimisation" formulation. This solution is one of many solutions to the problem that necessarily supports the lattice curvature, but it may not be unique.

3. Example – Deformed Cu and geometrically necessary dislocations

For this example, we take a map from data reported elsewhere [18, 38] and use this to explore the relative merits of extracting GND content using Hough and HR-EBSD based approaches. There are two important differences between Hough and HR-EBSD analysis: HR-EBSD has a higher angular misorientation precision and measures the misorientation about three independent axes (see figure 2); the Hough-based measurement of disorientation has a greater angular misorientation uncertainty, and as the misorientation angle between two points decreases the axis of misorientation is less well known [39, 40]. This makes the formation of a confident solution to eq. (1) difficult (irrespective of the underdetermined nature of this equation for a metallic system).

The map, taken from a sample of uniaxially deformed Cu, was captured using 1 x 1 binning by Jun Jiang at Oxford University on a JEOL 6500F instrument with 1 x 1 binning and a Digiview II camera and OIM-DC v5 system. HR-EBSD analysis was performed using fifty 256 x 256 pixel ROI.

A comparison of the disorientation fields, using the same reference point between Hough and HR-EBSD techniques, is shown in figure 3. Qualitatively 2D field plots look identical between the techniques, yet a quantitative line scan indicates that the Hough-based approach has a lower signal to noise ratio and the magnitude of these noise fluctuations varies from grain to grain.

These disorientation fields (figure 3) were input into the same GND code and the total GND density was calculated and they are reported in figure 4.

The presence of high frequency noise in the Hough-based disorientation analysis results in an apparent increase in the GND density recovered, this is evident in a simple area fraction analysis (figure 4d). Furthermore, the HR-EBSD analysis is likely to be relatively isotropic in terms of noise floor with regards to crystal orientation, whereas the Hough-based approach seems to show and indicate that there are some grains and some grain boundaries where the apparent GND density is higher. This may be problematic if conclusions regarding the structure and accumulation of GNDs in particular grains or near particular grain boundaries are of interest to the investigator.

4. Example – Indents in interstitial free steel

For this example, we have indented an interstitial free steel sample using a macro-indent and focussed on exploring the GND content surrounding the indenter impression. The indent spanned multiple grains and this study is interesting with respect to an understanding of how deformation patterns around micro-hardness measurements, which are widely used in industry as a measure of microstructure condition and quality assessment.

The sample was prepared as a 10 mm x 10 mm x 0.8 mm token from a rolled interstitial free steel sheet. The sample was heat-treated for 2 hours at 700 °C and quenched in room temperature water to recrystallise the microstructure. The RD-TD surface of the sample was subsequently polished to a 4,000 SiC finish. Final preparation of the sample surface was completed using a Gatan PECSII argon ion polisher using a three-step process, adjusting in voltage and/or angle with each 2 hr step: 8 kV at 5°, 3 kV at 5°, 3 kV at 1°.

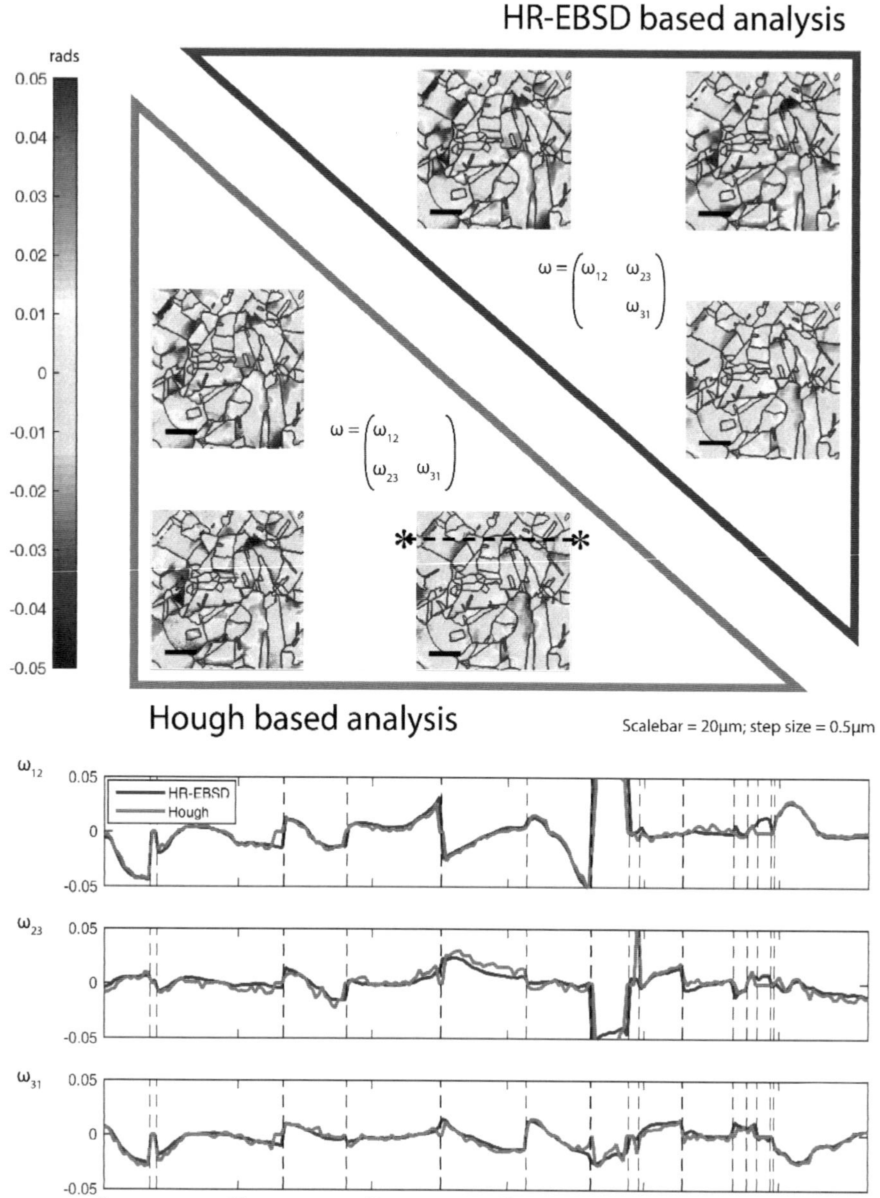

Figure 3. Lattice disorientation maps measured with HR-EBSD and Hough-based methods for copper deformed to 6 % in strain. Vertical dashed lines represent grain boundaries. There is good agreement between both measurements and lattice disorientation varies for points within each grain, and the disorientation is often found to change significantly near grain boundaries. For these maps, the X1 axis points left to right, the X2 axis points down the field. (We thank Dr J. Jiang and Mr C. Zhou for their assistance in generating this map.)

IOP Conf. Series: Materials Science and Engineering **304** (2017) 012003 doi:10.1088/1757-899X/304/1/012003

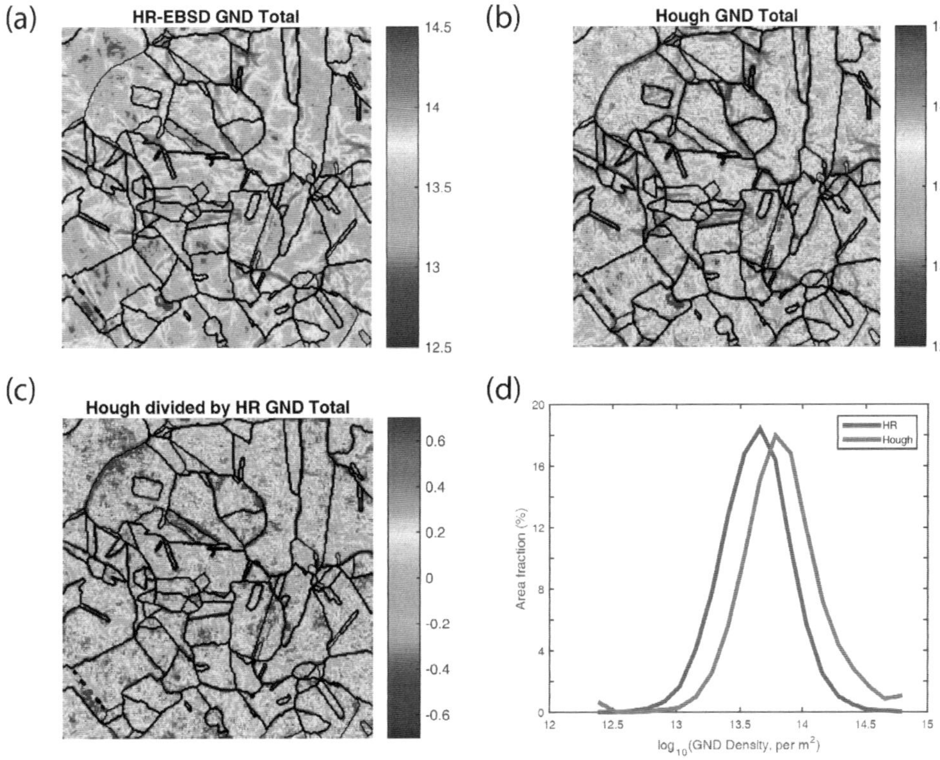

Figure 4. Comparison of GND measurements with Hough and HR-EBSD measurement of lattice curvature. (a and b) GND density distribution measured as $\log_{10}$(dislocations per m^2) for HR-EBSD and Hough based EBSD calculations, respectively; (c) the Hough-derived field divided pixel wise by the HR-EBSD field, presented on a $\log_{10}$ scale; (d) a histogram of the resultant plots. (a-c) represent areas that are 112.5 x 112.5 μm^2. (We thank Dr J. Jiang and Mr C. Zhou for their assistance in generating this map.)

A Vickers hardness indent was made into the surface of the sample within the region that was ion polished. The load applied to the indenter was 10 g for 5 s.

For the EBSD, a FEI Quanta 650g FEG-SEM coupled with a Bruker eFlash HR EBSD detector was used. The working and detector distances were set at 16.4 mm and 15.7 mm respectively. The sample and detector were tilted 70° and 10° respectively. EBSD patterns were captured at full resolution (1,200 x 1,600 pixels) at an exposure time of 0.6 s, probe current of ~ 10 nA, and the step size was 0.3 μm.

HR-EBSD analysis was performed with fifty 256 x 256 pixel ROI, with pattern remapping. GND evaluation was performed using Nye's analysis. As IF steel is ferritic, the <111> {110} slip systems were used (as per Wilkinson and Randman [12]).

The mean angular error (MAE) and cross-correlation peak heights (PHs) maps for the second cross-correlation pass are shown in figure 5. As the indent is broad, patterns were obtained from inside the indentation crater, yet the focus of this work is to explore the GND density surrounding the indenter impression and so these were filtered out of the GND analysis. These quality maps indicate that there is reasonable pattern quality and HR-EBSD analysis for the majority of grains, and a filter of PH > 0.3 and MAE < 5 x10^{-3} was used. We note that there was a reduction in PH specifically in regions that were surrounding the indent and near grain boundaries, most likely to the presence of significant strain gradients and/or pattern overlap which affects diffraction within the interaction volume result in blurring of the diffraction pattern.

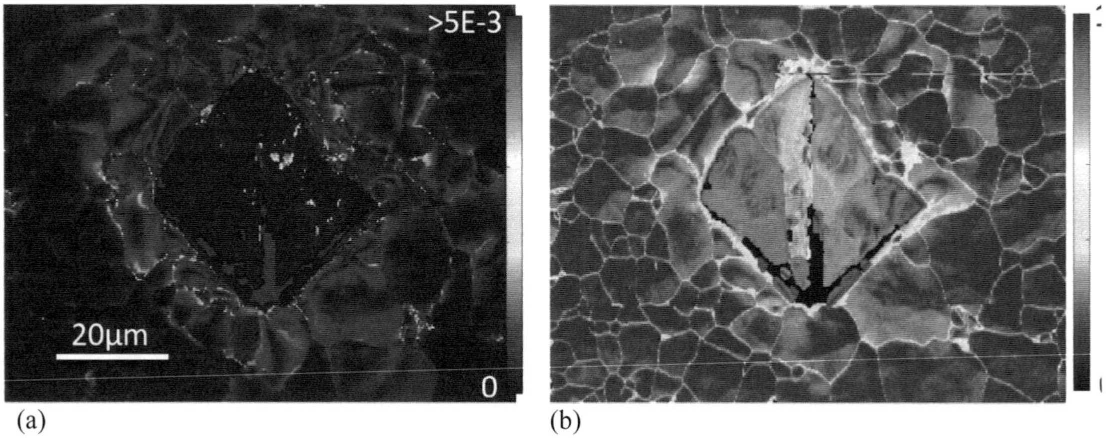

Figure 5. Quality maps for the 2nd pass of the cross-correlation showing: a) mean angular error (in radians); b) Normalised cross-correlation peak heights.

The HR-EBSD rotation gradient fields, combined from passes 1 and 2 of the cross-correlation process, were used to generate a map of the total GND density and this is shown in figure 6.

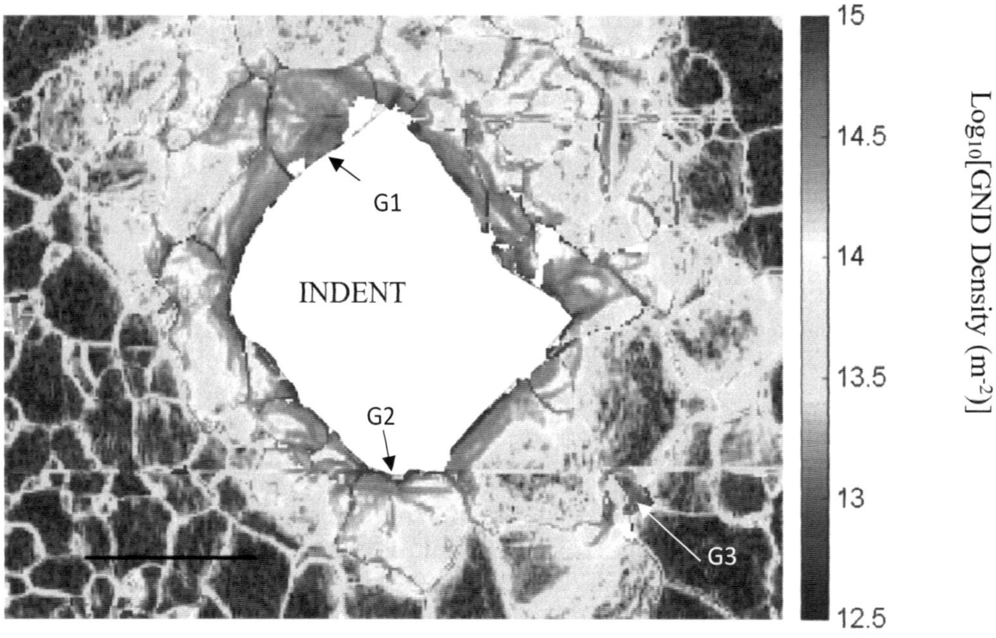

Figure 6. GND field around the indent. The length of the scale-bar is 20 µm.

In this map, we see that there are several notable features that highlight the heterogeneous nature of plastic slip in metallic systems, even when deformation is fiercely localised around an indenter.

A first order evaluation highlights that the GND density extends away from the indenter impression. Evidence of the plastic zone field (of which the GND density is only a part) indicates that plasticity is different within each grain, which is not surprising given the nature of crystal plasticity in a polycrystalline aggregate.

The striking details of this heterogeneous plastic deformation are brought out in this micrograph, for instance there is one grain towards the top of the indent (G1), which has stored a significant amount of GND density in a smooth and continuous field; where as a grain towards the bottom of the indent (G2) has 'plumes' of high GND density extending from the vertex of the indenter and there are many other grains where only half of the grain near the indenter have significant GND density. Furthermore, while most of the high GND density is localised towards the impression edge, there are some regions 'one grain further in' (bottom right of the field, G3) which have significant accumulation of GND density towards the grain boundary region and there is a region clear of GND density nearer the impression.

Finally, compatibility and equilibrium conditions require that grains must 'co-deform' to accommodate strain (or cracks will form) and there is significant evidence of this in the GND field. The vein structure of the GND field is often modulated by the presence of grain boundaries, for instance where the particular stress state imposed during indentation enables one grain to freely plastically deform and store lattice strain gradients, and its neighbour does not. However, there are occasions where the grain boundaries almost seem 'invisible' to the GND structure, as veins of high GND content can extend across the grain boundary region, most likely due to local compatibility requirements.

We also note that there are two horizontal 'flashes' across the map at the bottom and top of the indenter impression which are due to stochastic image drift in the SEM, these are aesthetically displeasing but do not detract from our ability to interpret this data rich micrograph.

Due to the out of plane displacement, associated with pile-up or sink-in around the indent for this example, the exact magnitude of the GND field must be taken with care, as significant changes in surface roughness may impact the spatial field used to evaluate the curvature terms (i.e., we assume that the out of plane displacement is zero in formulating the curvature tensor). The impact of this uncertainty could be addressed using atomic force microscopy (AFM).

5. Discussion

HR-EBSD affords an increase in angular resolution of two orders of magnitude as compared to conventional Hough-based EBSD. This increase in resolution enables probing of lattice strain variations, which can be used to calculate intragranular residual stress variations, and local orientation changes. This is well suited for studies of deformed materials, such as the metallic examples shown here. This manuscript has focussed on GND evaluation, as this is a popular area of study and there have been examples of recent work that focus on the elastic strain measurement aspect of the technique, such as a review in JOM [26].

The HR-EBSD technique is sufficiently mature and robust that there are now multiple groups using the technique in industrial and academic contexts.

In this manuscript, we have outlined two case studies in deformed copper and iron that indicate the potential of the technique to probe the intricate nature of stored dislocation fields in deformed metals.

In the copper example, we highlight that an increased angular resolution, with explicit measurement of the misorientation axis, affords a greater reliability in the calculation of residual deformation content. Often this is more crudely reported through use of the kernel average misorientation (KAM), a step size independent metric of the local orientation gradient, and yet as EBSD provides knowledge of the crystal orientation, it is often possible to provide an educated guess at the potentially contributing slip systems, GND density maps can be easily calculated (and this has been extensively demonstrated by groups worldwide, including excellent work for example by Pantleon and co-workers [39, 40], as well as colleagues at the MPIE in Düsseldorf [43-45]). The absolute different between the two approaches when applied to the same sample is relatively small (a factor of 1.4x), and yet the standard deviation and localised variation is perhaps more significant.

The IF steel indenter impression highlights the potential of this technique to improve and further understand existing metallographic characterisation techniques. Observations of a component of the surface plastic zone indicates that crystal slip remains significantly heterogeneous, even where there is significantly localised plasticity. The presence of different GND density structures within different grains highlights that understanding localised plasticity around indents remains challenging, as there is an interplay of the local crystal orientation, the local grain neighbourhood, and the imposed strain gradient due to the indentation process.

Perhaps of wider interest, broadly we observe similar trends in the dislocation structures in these metallic systems across multiple grains where deformation in polycrystalline aggregates results in fierce localisation of stored dislocation content in some, but not all, grains and near some, but not all, grain boundaries and triple junctions. Understanding how these structures form and where localisation occurs in constrained polycrystalline samples is critical to develop a deeper understanding of deformation and failure of materials, and we hope that HR-EBSD is a valuable tool in the materials characterisation toolbox that will assist us on this journey.

6. Conclusion

We have reviewed the HR-EBSD methodology and outline key steps in direct pattern analysis to capture subtle changes in interplanar angle and lattice disorientation. The two examples presented here highlight the potential for HR-EBSD to add value to the materials characterisation community, affording an extra tool to characterise local structure with high precision within the scanning electron microscopy. The Cu example highlights how the increased angular resolution of HR-EBSD improves our confidence in probing local orientation gradients that can be interpreted in terms of stored GND content. We find that there are systematic variations in the GND density fields recovered, as well as localised regions where the HR-EBSD and Hough-based approaches diverge significantly (such as near certain boundary types). Assessment of the GND content around an indent in IF steel highlights the potential for HR-EBSD to assist our understanding of deformation during metallographic tests and to relate this to the dominant crystal plasticity mechanisms at work.

Acknowledgements

We would like to thank Prof Angus Wilkinson for long standing and productive collaborations on the HR-EBSD technique. We acknowledge that Cu data was obtained by Dr J. Jiang during his PhD work at Oxford. We thank both C. Zhou and Dr J. Jiang for assistance with the initial code used to analyse the Cu GND data. TBB acknowledges funding from the Royal Academy of Engineering and funding from the AVIC-BIAM centre at Imperial College. TBB and JH both acknowledge funding from the Shell UTC in Advanced Interfaces in Materials Science. We would like to thank the reviewers for their useful comments which have improved the quality of this manuscript.

References

[1] Winkelmann A, Trager-Cowan C, Sweeney F, Day A and Parbrook P 2007 Many-beam dynamical simulation of electron backscatter diffraction patterns. *Ultramicroscopy* **107** 414-421
[2] Bragg W 1913 The diffraction of short electromagnetic waves by a crystal. *Proc. Camb. Philol. Soc.* **17** 4
[3] Kikuchi S 1928 Diffraction of cathode rays by mica. *Proc. Imp. Acad.* **4** 271-274
[4] Wilkinson A J and Britton T B 2012 Strains, planes, and EBSD in materials science. *Mater. Today* **15** 366-376
[5] Schwartz A J, Kumar M, Adams B L and Field D P (Eds.) 2010 *Electron backscatter diffraction in materials science.* (New York, NY: Springer)
[6] Nye J F 1953 Some geometrical relations in dislocated crystals. *Acta Met.* **1** 153-162

[7] Wilkinson A J, Meaden G and Dingley D J 2006 High resolution mapping of strains and rotations using electron backscatter diffraction. *Mater. Sci. Technol.* **22** 1271-1278
[8] Wilkinson A J, Meaden G and Dingley D J 2006 High-resolution elastic strain measurement from electron backscatter diffraction patterns: new levels of sensitivity. *Ultramicroscopy* **106** 307-313
[9] Nolze G 2007 Image distortions in SEM and their influences on EBSD measurements. *Ultramicroscopy* **107** 172-183
[10] Humphreys F J 2001 Review - Grain and subgrain characterisation by electron backscatter diffraction. *J. Mater. Sci.* **36** 3833-3854
[11] Brough I, Bate P S and Humphreys F J 2006 Optimising the angular resolution of EBSD. *Mater. Sci. Technol.* **22** 1279-1286
[12] Wilkinson A J and Randman D 2010 Determination of elastic strain fields and geometrically necessary dislocation distributions near nanoindents using electron back scatter diffraction. *Philos. Mag.* **90** 1159-1177
[13] Jiang J, Britton T B and Wilkinson A J 2012 Measurement of geometrically necessary dislocation density with high resolution electron backscatter diffraction: effects of detector binning and step size. *Ultramicroscopy* **125** 1-9
[14] Humphreys F J, Huang Y, Brough I and Harris C 1999 Electron backscatter diffraction of grain and subgrain structures - Resolution considerations. *J. Microsc.* **195** 212-216
[15] Villert S, Maurice C, Wyon C and Fortunier R 2009 Accuracy assessment of elastic strain measurement by EBSD. *J. Microsc.* **233** 290-301
[16] Speller S C, Britton T B, Hughes G M, Krzton-Maziopa A, Pomjakushina E, Conder K, Boothroyd A T and Grovenor C R M 2012 Microstructural analysis of phase separation in iron chalcogenide superconductors. *Supercond. Sci. Technol.* **25** 84023
[17] Speller S C, Britton T B, Hughes G, Lozano-Perez S, Boothroyd A T, Pomjakushina E, Conder K and Grovenor C R M 2011 Analysis of local chemical and structural inhomogeneities in $Fe_ySe_{1-x}Te_x$ single crystals. *Appl. Phys. Lett.* **99** 192504
[18] Jiang F, Britton T B and Wilkinson A J 2013 Evolution of dislocation density distributions in copper during tensile deformation. *Acta Mater.* **61** 7227-7239
[19] Guo Y, Collins D M, Tarleton E, Hofmann F, Tischler J, Liu W, Xu R, Wilkinson A J and Britton T B 2015 Measurements of stress fields near a grain boundary: Exploring blocked arrays of dislocations in 3D. *Acta Mater.* **96** 229-236
[20] Jiang J, Zhang T, Dunne F P E and Britton T B 2016 Deformation compatibility in a single crystalline Ni superalloy. *Proc. Riyal Soc. A Math. Phys. Engng. Sci.* **472** 20150690
[21] Wallis D, Hansen L N, Britton T B and Wilkinson A J 2016 Geometrically necessary dislocation densities in olivine obtained using high-angular resolution electron backscatter diffraction. *Ultramicroscopy* **168** 34-45
[22] Britton T B, Jiang J, Clough R, Tarleton E, Kirkland A I and Wilkinson A J 2013 Assessing the precision of strain measurements using electron backscatter diffraction – Part 1: detector assessment. *Ultramicroscopy* **135** 126-135
[23] Britton T B, Jiang J, Clough R, Tarleton E, Kirkland A I and Wilkinson A J 2013 Assessing the precision of strain measurements using electron backscatter diffraction – Part 2: experimental demonstration. *Ultramicroscopy* **135** 136-141
[24] Britton T B, Maurice C, Fortunier R, Driver J H, Day A P, Meaden G, Dingley D J, Mingard K Wilkinson A J 2010 Factors affecting the accuracy of high resolution electron backscatter diffraction when using simulated patterns. *Ultramicroscopy* **110** 1443-1453
[25] Britton T B and Wilkinson A J 2012 High resolution electron backscatter diffraction measurements of elastic strain variations in the presence of larger lattice rotations. *Ultramicroscopy* **114** 82-95

[26] Britton T B, Jiang J, Karamched P S and Wilkinson A J 2013 Probing deformation and revealing microstructural mechanisms with cross-correlation-based, high-resolution electron backscatter diffraction. *JOM* **65** 1245-1253
[27] Plancher E 2015 Mesures de champs de déformations élastique et totale pour la détermination du comportement mécanique local de matériaux cristallins. *PhD Thesis* (Paris, France: ParisTech)
[28] Maurice C, Driver J H and Fortunier R 2011 On solving the orientation gradient dependency of high angular resolution EBSD. *Ultramicroscopy* **113** 171-181
[29] Vaudin M D, Stan G, Gerbig Y B and Cook R F 2011 High resolution surface morphology measurements using EBSD cross-correlation techniques and AFM. *Ultramicroscopy* **111** 1206-1213
[30] Vaudin M D, Gerbig Y B, Stranick S J and Cook R F 2008 Comparison of nanoscale measurements of strain and stress using electron back scattered diffraction and confocal Raman microscopy. *Appl. Phys. Lett.* **93** 193116
[31] Plancher E, Petit J, Maurice C, Favier V, Saintoyant L, Loisnard D, Rupin N, Marijon J B, Ulrich O, Bornert M, Micha J S, Robach O and Castelnau O 2016 On the accuracy of elastic strain field measurements by Laue microdiffraction and HR EBSD. *Exp. Mech.* **56** 483-492
[32] Britton T B and Wilkinson A J 2011 Measurement of residual elastic strain and lattice rotations with high resolution electron backscatter diffraction. *Ultramicroscopy* **111** 1395-1404
[33] Britton T B 2017 Comment on "An experimental study on evolution of grain-scale stress/strain and geometrical necessary dislocations in advanced TA15 titanium alloy during uniaxial tension deformation". *Adv. Eng. Mater.* **19** 1700051
[34] Jiang J, Britton T B and Wilkinson A J 2015 The orientation and strain dependence of dislocation structure evolution in monotonically deformed polycrystalline copper. *Int. J. Plast.* **69** 102-117
[35] Littlewood P D, Britton T B and Wilkinson A J 2011 Geometrically necessary dislocation density distributions in Ti–6Al–4V deformed in tension. *Acta Mater.* **59** 6489-6500
[36] Britton T B and Wilkinson A J 2012 Stress fields and geometrically necessary dislocation density distributions near the head of a blocked slip band. *Acta Mater.* **60** 5773-5782
[37] Karamched P S and Wilkinson A J 2011 High resolution electron back-scatter diffraction analysis of thermally and mechanically induced strains near carbide inclusions in a superalloy. *Acta Mater.* **59** 263-272
[38] Jiang J, Britton T B and Wilkinson A J 2012 Accumulation of geometrically necessary dislocations near grain boundaries in deformed copper. *Philos. Mag. Lett.* **92** 580-588
[39] Pantleon W, He W, Johansson T P and Gundlach C 2008 Orientation inhomogeneities within individual grains in cold-rolled aluminium resolved by electron backscatter diffraction. *Mater. Sci. Engng. A* **483** 668-671
[40] Prior D J 1999 Problems in determining the misorientation axes, for small angular misorientations, using electron backscatter diffraction in the SEM. *J. Microsc.* **195** 217-225
[41] Wilkinson A J 2001 A new method for determining small misorientations from electron back scatter diffraction patterns. *Scr. Mater.* **44** 2379-2385
[42] Pantleon W 2008 Resolving the geometrically necessary dislocation content by conventional electron backscattering diffraction. *Scr. Mater.* **58** 10-13
[43] Zaafarani N, Raabe D, Roters F and Zaefferer S 2008 On the origin of deformation-induced rotation patterns below nanoindents. *Acta Mater.* **56** 31-42
[44] Zaafarani N, Raabe D, Singh R N, Roters F and Zaefferer S 2006 Three-dimensional investigation of the texture and microstructure below a nanoindent in a Cu single crystal using 3D EBSD and crystal plasticity finite element simulations. *Acta Mater.* **54** 1863-1876
[45] Demir E, Raabe D, Zaafarani N and Zaefferer S 2008 Investigation of the indentation size effect through the measurement of the geometrically necessary dislocations beneath small indents of different depths using EBSD tomography. *Acta Mater.* **57** 559-569

A deterministic model of electron transport for electron probe microanalysis

J Bünger[1], S Richter[2] and M Torrilhon[1]

[1] RWTH Aachen University, Centre for Computational Engineering Science (MathCCES), Schinkelstrasse 2, 52062 Aachen, Germany

[2] RWTH Aachen University, Central Facility for Electron Microscopy (GFE), Ahornstrasse 55, 52074 Aachen, Germany

E-mail: buenger@mathcces.rwth-aachen.de

Abstract. Within the last decades significant improvements in the spatial resolution of electron probe microanalysis (EPMA) were obtained by instrumental enhancements. In contrast, the quantification procedures essentially remained unchanged. As the classical procedures assume either homogeneity or a multi-layered structure of the material, they limit the spatial resolution of EPMA. The possibilities of improving the spatial resolution through more sophisticated quantification procedures are therefore almost untouched. We investigate a new analytical model (M_1-model) for the quantification procedure based on fast and accurate modelling of electron-X-ray-matter interactions in complex materials using a deterministic approach to solve the electron transport equations. We outline the derivation of the model from the Boltzmann equation for electron transport using the method of moments with a minimum entropy closure and present first numerical results for three different test cases (homogeneous, thin film and interface). Taking Monte Carlo as a reference, the results for the three test cases show that the M_1-model is able to reproduce the electron dynamics in EPMA applications very well. Compared to classical analytical models like XPP and PAP, the M_1-model is more accurate and far more flexible, which indicates the potential of deterministic models of electron transport to further increase the spatial resolution of EPMA.

1. Introduction

Electron probe microanalysis (EPMA) provides a popular method to obtain quantitative information about the chemical composition of heterogeneous materials, particularly fine structures and grains in metals and alloys. As an imaging technique, it poses an inverse problem, as the chemical concentrations are not measured directly, but have to be reconstructed from intensity measurements of X-rays.

Usually, the experimental intensities are normalised to so-called k-ratios k^{Exp} and the simulation of a forward model provides computed k-ratios $k^{\mathrm{M}}(c)$ that depend on one or more assumed elemental concentration fields $c(x)$ inside the probe. The inverse problem [1,2] consists of finding the concentration c^* that minimises the error between the experimental and simulated k-ratios k^{M}.

Problem 1 (Inverse problem)
Find c^* such that

$$c^* = \arg\min_c \|k^{\mathrm{M}}(c) - k^{\mathrm{Exp}}\|^2 \tag{1}$$

with some appropriate norm $\|\cdot\|$.

Obviously, the reconstructed concentration $c(x)$ inherits fundamental characteristics like accuracy or assumptions on the material structure from the forward model used to compute $k^{\mathrm{M}}(c)$. Moreover, the forward model substantially determines complexity and computational cost of the inverse problem. Therefore, the choice of the forward model $k^{\mathrm{M}}(c)$ is crucial for both computational speed and accuracy of the reconstruction of concentrations $c(x)$.

Most analytical methods used nowadays use simple forward models, which either assume that the sample is homogeneous within the interaction volume [3, 4], or are based on the calculation of the $\phi(\rho z)$ distribution [5–9], where the latter model the probability of ionisation as a function of depth, expressed in mass thickness. Both types impose strong restrictions on the range of material structures to which they can be applied. The former are only applicable to structures that exceed the interaction volume and the latter require the composition to only vary as a function of depth, like thin films or multi-layers. One approach to improve the spatial resolution of EPMA is to use forward models with less or no restriction on the material structure.

When more complex concentration distributions are of interest, Monte Carlo simulations of the electrons inside the probe typically replace analytical models [10–14]. While these simulations exhibit very good physical accuracy, they are also computationally expensive and suffer from statistical noise. This makes the solution of the inverse problem to recover the chemical composition difficult.

Hence, we believe that more sophisticated forward models that balance computational effort with accuracy have to be developed in order to improve the spatial resolution of EPMA. Attempts on deterministic modelling of electrons inside solids to predict characteristic quantities like backscattered electrons and emitted X-rays used in quantitative EPMA have been made earlier [15] and similar approaches to modelling particle transport are also common in nuclear engineering [16] and electron radiotherapy [17, 18].

This paper investigates a new forward model for the prediction of local ionisation intensity for arbitrary concentration distributions, which conceptually is found between analytical models and Monte Carlo simulations. It uses a deterministic reduction technique to solve the electron transport equation as a partial differential equation. The reduction is based on the minimum entropy approach [19–21], yielding closed moment equations from the Boltzmann equation for electron transport in the continuous slowing down approximation. The simplest case is given by the so-called M_1 model, which computes the local electron energy distribution directly in a deterministic noise-free way. This allows a very efficient computation of the local ionisation intensity and also allows to use modern optimisation techniques for partial differential equations to solve the inverse problem (eq. (1)).

2. Transport theory

2.1. Boltzmann equation for electron transport

The starting point for the derivation of the model is the Boltzmann equation for electron transport, which follows from balancing the physical processes influencing the number of electrons n with velocity v at space point x and time t in a stationary background medium

$$\begin{aligned}\frac{\partial}{\partial t}n(x,v,t) + \underbrace{v\cdot\nabla_x n(x,v,t)}_{\text{in-/outflow}} &= \underbrace{\int_{\mathbb{R}^3} N_V(x)\hat{\sigma}(x,\|v'\|,\|v\|,v'\cdot v)\|v'\|n(x,v',t)dv'}_{\text{in-scattering}} \\ &\quad - \underbrace{\int_{\mathbb{R}^3} N_V(x)\hat{\sigma}(x,\|v\|,\|v'\|,v\cdot v')\|v\|n(x,v,t)dv'}_{\text{out-scattering}}.\end{aligned} \tag{2}$$

The solution $n : \mathbb{R}^3\times\mathbb{R}^3\times\mathbb{R}_{\geq 0} \to \mathbb{R}_{\geq 0}$ is a phase density, such that $n(x, v, t)$ is the probability density of finding an electron at position x with velocity v at time t. The background medium is modelled by $N_V(x)$, the number density of scattering centres at position x, and the scattering cross-section $\hat{\sigma}(x, \|v'\|, \|v\|, v'\cdot v)$, describing the probability that an electron changes its velocity from v' to v when hitting a scattering centre at space point x. Note that we assume the background material to be isotropic, as the scattering cross-section $\hat{\sigma}$ does not depend on the pre- and post-collision velocity vectors v' and v separately but only on the magnitudes and the angle between both vectors.

In the context of EPMA, the time scale of the physical processes is very small compared to the duration of the measurement. Therefore, we can assume that the distribution of electrons inside the material reaches a steady state immediately after switching on the electron beam and thus neglect the time dependency ($\partial/\partial_t \cdot = 0$).

Before deriving the reduced model it is convenient to rewrite eq. (2). First, we perform a transformation to energy and direction by expressing the velocity and scattering cross-section as functions of energy ϵ and direction Ω

$$v(\epsilon,\Omega) = \|v(\epsilon)\|\Omega \tag{3}$$

$$\sigma(x,\epsilon',\epsilon,\Omega'\cdot\Omega) = \hat{\sigma}(x,\|v(\epsilon')\|,\|v(\epsilon)\|,v(\epsilon')\cdot v(\epsilon)). \tag{4}$$

We then split the scattering cross-section into an elastic ($\epsilon' = \epsilon$) and an inelastic ($\epsilon' > \epsilon$) part ($\sigma = \sigma_{\text{el}} + \sigma_{\text{in}}$). Finally we define

$$\psi(x,\epsilon,\Omega) := \|v(\epsilon)\|n(x,v(\epsilon,\Omega)) \tag{5}$$

and obtain

$$\begin{aligned}\Omega\cdot\nabla_x\psi(x,\epsilon,\Omega) &= N_V(x)\int_\epsilon^\infty\int_{S^2}\sigma_{\text{in}}(x,\epsilon',\epsilon,\Omega\cdot\Omega')\psi(x,\epsilon',\Omega')d\Omega'd\epsilon' \\ &\quad + N_V(x)\int_{S^2}\sigma_{\text{el}}(x,\epsilon,\Omega\cdot\Omega')\psi(x,\epsilon,\Omega')d\Omega' \\ &\quad - N_V(x)\sigma_{\text{in}}^{\text{tot}}(x,\epsilon)\psi(x,\epsilon,\Omega) \\ &\quad - N_V(x)\sigma_{\text{el}}^{\text{tot}}(x,\epsilon)\psi(x,\epsilon,\Omega)\end{aligned} \tag{6}$$

with the total scattering cross-sections $\sigma_{\text{el}}^{\text{tot}}(x,\epsilon) = 2\pi\int_{-1}^{1}\sigma_{\text{el}}(x,\epsilon,\mu)d\mu$ and $\sigma_{\text{in}}^{\text{tot}}(x,\epsilon) = 2\pi\int_0^\epsilon\int_{-1}^{1}\sigma_{\text{in}}(x,\epsilon,\epsilon',\mu)d\mu d\epsilon'$ where $\mu = \cos(\theta)$ with θ being the deflection angle. S^2 denotes the sphere with radius one (unit 2-sphere) and accordingly the integral over S^2 in (6) is over all directions Ω'. ψ can be interpreted as an electron fluence in the sense that $\psi(x,\epsilon,\Omega)d\Omega d\epsilon dAdt$ corresponds to the number of electrons that traverse an area dA perpendicular to Ω with a direction within the solid angle $d\Omega$ around Ω at an energy within $[\epsilon, \epsilon + d\epsilon]$ at point x during dt.

2.2. Continuous slowing down approximation

As a first simplification, the continuous slowing down approximation for the energy loss of the electrons is used. The same simplification appears in all analytical models and almost all Monte Carlo models, e.g., [11, 13]. The simplification is based on the assumption that an electron loses a significant amount of energy through a very large number of collisions, with any single collision changing the energy of the electron only slightly. In this case, the inelastic scattering cross-section is peaked about small energy losses and an asymptotic analysis with respect to energy on the inelastic in-scattering term in eq. (6) yields the following approximation

$$\begin{aligned} N_V(x)\int_\epsilon^\infty \int_{S^2} \sigma_{\text{in}}(x,\epsilon',\epsilon,\Omega\cdot\Omega')\psi(x,\epsilon',\Omega')d\Omega' d\epsilon' \\ \approx N_V(x)\int_{S^2} \sigma_{\text{in}}^{CSD}(x,\epsilon,\Omega\cdot\Omega')\psi(x,\epsilon,\Omega')d\Omega' + \frac{\partial}{\partial\epsilon}(S(x,\epsilon)\psi(x,\epsilon,\Omega)) \end{aligned} \tag{7}$$

with

$$\sigma_{\text{in}}^{CSD}(x,\epsilon,\mu) := \int_0^\infty \sigma_{\text{in}}(x,\epsilon,\epsilon',\mu)d\epsilon' \qquad S(x,\epsilon) := 2\pi\int_0^\infty\int_{-1}^1 (\epsilon-\epsilon')\sigma_{\text{in}}(x,\epsilon,\epsilon',\mu)d\mu d\epsilon'.$$

Insertion of the approximation eq. (7) into eq. (6) results in the Boltzmann equation in continuous slowing down approximation (BCSD). The term $-\partial/\partial_\epsilon(S\psi)$ can be interpreted as a transfer from higher to lower energies, with the stopping power S governing the speed of this transfer.

3. Continuum approximation

Knowing the electron fluence ψ we can compute the intensity of a specific characteristic X-ray generated at point x from (cf. [22])

$$I_{0,j}(x) = w_j\ N_{V,j}(x)\int_0^\infty \sigma_{\text{ion},j}(x,\epsilon)\int_{S^2}\psi(x,\epsilon,\Omega)d\Omega d\epsilon \tag{8}$$

where w_j and $\sigma_{\text{ion},j}$ are the fluorescence yield and ionisation cross-section of some chemical element identified by j. Note that detailed information about the angular distribution of the electron fluence ψ is not necessary to evaluate eq. (8), as only the average in angle of ψ is required. This, and the fact that solving the BCSD is computationally very expensive due to the high dimensionality of ψ, motivates further simplification with respect to the direction variable Ω.

3.1. Method of moments

The method of moments is a mathematical concept to reduce the dimension of the phase space in differential equations by integrating out selected independent variables of the phase space. Its objective is to derive equations for weighted integrals of the unknown, the so-called moments, and solve for these instead. As we want to eliminate the direction Ω, our moments of interest are integrals of ψ over the unit 2-sphere weighted by polynomials of the direction Ω

$$\psi^{(n)}(x,\epsilon) := \int_{S^2} \underbrace{\Omega\otimes\ldots\otimes\Omega}_{\text{n times}}\,\psi(x,\epsilon,\Omega)d\Omega. \tag{9}$$

The moment equations follow from multiplying the BCSD equation with monomials in Ω, integrating over the unit 2-sphere and identifying the moments. Here, we choose the simplest case of only the first two moments and therefore consider moments of up to first order ($n = 0, 1$) for which we get the following moment equation system

$$-\frac{\partial}{\partial \epsilon}(S(x,\epsilon)\psi^{(0)}(x,\epsilon)) + \nabla_x \psi^{(1)} = 0 \tag{10}$$

$$-\frac{\partial}{\partial \epsilon}(S(x,\epsilon)\psi^{(1)}(x,\epsilon)) + \nabla_x \psi^{(2)} = -T\psi^{(1)} \tag{11}$$

with

$$T(x,\epsilon) := 2\pi N_V(x) \int_{-1}^{1} (1-\mu) \left(\sigma_{\text{el}}(x,\epsilon,\mu) \int_0^\infty \sigma_{\text{in}}(x,\epsilon',\epsilon,\mu) d\epsilon' \right) d\mu. \tag{12}$$

Note that the zeroth moment $\psi^{(0)}$ corresponds to the integral of ψ in eq. (8). Hence, the theory computes essentially only the quantity of interest. Due to the divergence-term in the BCSD equation, the system of moment equations contains higher order moments and is thereby underdetermined. In order to obtain a closed system of partial differential equations, a closure relation that expresses the highest order moments as a function of the moments of interest

$$\psi^{(2)} = \psi^{(2)}(\psi^{(0)}, \psi^{(1)}) \tag{13}$$

has to be defined. The closure is crucial for the accuracy and mathematical properties of the moment equation system and should be chosen carefully.

3.2. Minimum entropy closure

The minimum entropy closure is physically motivated. It can be thought of as choosing the electron fluence ψ such that it minimises the Boltzmann entropy for a one particle system while generating a given set of moments $\psi^{(i)}$ $i = 0, \ldots, n$ and using it to close the system of moment equations by simply evaluating the integral expression of the highest moment.

Problem 2 (Minimum entropy closure)

1. Find ψ_{ME} so that

$$\psi_{\text{ME}} = \arg\min_{\psi} \left(- \int_{S^2} \psi \log(\psi) d\Omega \right) \quad \text{such that} \quad \int_{S^2} \Omega^i \psi_{\text{ME}} d\Omega = \psi^{(i)} \quad 0 \le i \le n \tag{14}$$

2. Compute $\psi^{(n+1)}$ from ψ_{ME}

$$\psi_{ME}^{(n+1)}(x,\epsilon) = \int_{S^2} \Omega^{(n+1)} \psi_{\text{ME}}(x,\epsilon,\Omega) d\Omega. \tag{15}$$

This procedure yields the so-called M_1-model for electron transport [19–21]. Figure 1(a) visualises a histogram for the direction of travel of electrons in a homogeneous sample generated during a Monte Carlo simulation. The sample is subdivided into rectangular cells with each cell containing a group of lines. The lines visualise the direction of travel distribution in the cell by pointing in the corresponding direction (taking the cell centre as starting point) and having a length proportional to the number of electrons moving in this direction. For a better orientation, the mean energy of electrons is plotted in the background. From fig. 1(a), it is clear that the electrons are in a well structured configuration, enforced by the electron beam, when entering the sample. As the electrons penetrate the material they are deflected through collisions and the direction of travel distribution is more and more spread over all directions. Minimisation of the mathematical entropy corresponds to maximising the disorder of the electrons with respect

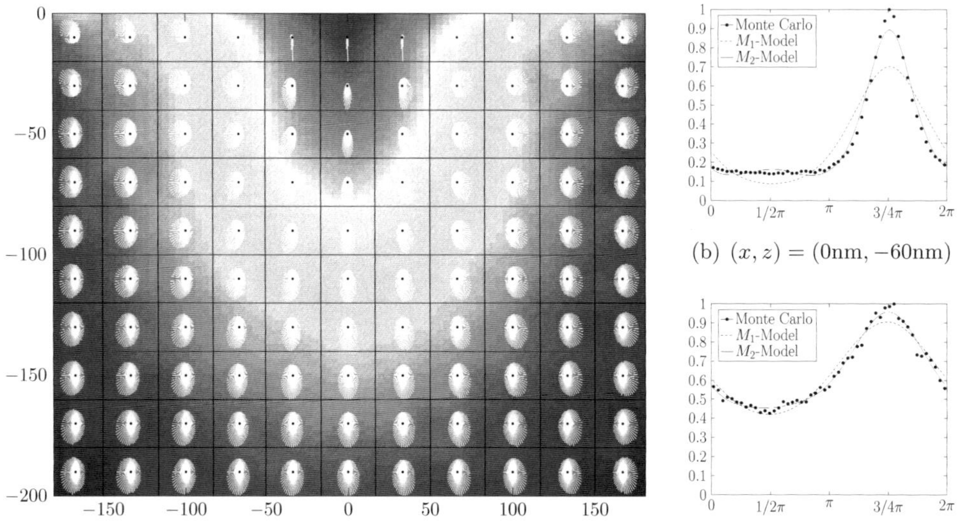

(a) Cellwise normalized direction of travel distribution

(b) $(x, z) = (0\text{nm}, -60\text{nm})$

(c) $(x, z) = (0\text{nm}, -175\text{nm})$

Figure 1. (a) Normalised direction of travel distribution of electrons in copper with $\epsilon \in [5\,\text{keV}, 10\,\text{keV}]$ obtained from a Monte Carlo simulation in DTSA-II [13,23]. (b) and (c) Angular distribution at two different depths in the Monte Carlo simulation from subfigure (a) compared to the M_1 reconstruction ψ_{M_1} and M_2 reconstruction ψ_{M_2}.

to the direction of travel for a given set of moments. This assumption of maximal disorder is not accurate over the whole range of energies. Nevertheless, the angular distribution of electrons is not rich in detail, having only one dominant direction of travel in each cell. We will show in section 4 that the M_1-model is able to reproduce the electron transport in the context of EPMA well. Figures 1(b) and 1(c) show the direction of travel distribution at two different space points, one close to the surface and another at deeper into the sample, as well as the M_1 and M_2 distribution reconstructed from the moments. As expected the minimum entropy reconstruction is better far inside the material than close to the surface. Figures 1(b) and 1(c) furthermore illustrate the improvement of the reconstruction when increasing the number of considered moments.

3.3. M_1-model

For the case of taking only the first two moments the minimum entropy closure problem for the second order moment $\psi^{(2)}$ can be solved analytically and we find

$$\psi^{(2)}_{\text{ME}} = \psi^{(0)} \left(\frac{1 - \chi(\|\alpha\|)}{2} I + \frac{3\chi(\|\alpha\|) - 1}{2} \frac{\alpha}{\|\alpha\|} \otimes \frac{\alpha}{\|\alpha\|} \right) \tag{16}$$

where χ is the Eddington factor which depends implicitly on the Euclidean norm of the anisotropy parameter $\alpha = \psi^{(1)}/\psi^{(0)}$ [19–21, 24]. The M_1-model now follows from closing the moment eqs. (10) and (11) with the minimum entropy closure relation for $\psi^{(2)}$ given by eq. (16). We define a variable vector

$$U(x, \epsilon) := (\psi^{(0)}(x, \epsilon), \psi^{(1)}(x, \epsilon)) \tag{17}$$

and write the M_1 equations in a compact form to point out the similarity to classical conservation laws with relaxation [25]

$$-\frac{\partial}{\partial\epsilon}(S(x,\epsilon)U(x,\epsilon)) + \nabla_x \cdot F(U(x,\epsilon)) = T(x,\epsilon)U(x,\epsilon) \tag{18}$$

with

$$F(U) := \begin{pmatrix} U_1^T \\ \psi_{\mathrm{ME}}^{(2)}(U_0,\ U_1) \end{pmatrix} \qquad T(x,\epsilon) := \begin{pmatrix} 0 & 0 \\ 0 & -T(x,\epsilon)I \end{pmatrix}.$$

Compared to classical evolution equations, $\partial/\partial_\epsilon(SU)$ takes the part of a time derivative describing an evolution of U in energy space from high to low energies. The highest possible energy is the beam energy ϵ_b, hence the initial condition is given by $U(x,\epsilon_b)$, which we approximate by Gaussian distributions centred at the point where the beam electrons enter the material. For the solution of such initial value problems, a variety of well-established numerical methods is available [25]. We decided to employ the discontinuous Galerkin (DG) method for the discretisation in space with a Runge-Kutta (RK) time stepping scheme for the integration in time [26–28]. The results presented in the following section were obtained using a second order discretisation.

At this point, our focus is to study the accuracy of the M_1-model and its numerical solutions. We therefore developed our current M_1-solver with respect to flexibility using the finite element library deal.II [29, 30]. Yet the long-term objective is to develop a M_1-solver tailored to EPMA application to efficiently solve the inverse problem.

4. Results and discussion

Although the M_1-model is designed for complex geometries, we first assess simple samples to study its validity, as more data is available for these geometries. We take results obtained from Monte Carlo (MC) simulations as reference solutions since this method can model the electron transport most accurately. The reference solutions presented in this section were generated using the DTSA-II MC software [23], which also implies the CSD approximation. A more precise physical model is implemented in the Monte Carlo code PENELOPE [14], which we plan to use for the generation of additional reference solutions in the future. In order for the comparison between the M_1- and MC-model to be as fair as possible, we use the same data for material parameters (elastic scattering cross-section, shell ionisation cross-section, etc.) in both models and therefore favour data available for both models over more elaborated data available only in DTSA-II. Table 1 gives an overview of the material models used in the simulation results presented in this section. The setups of the three test cases we will look at in this section are shown in table 2. In all test cases, we approximate the initial electron distribution as a Gaussian distribution with a standard deviation of 10 nm, corresponding to a full width at half maximum (FWHM) of approximately 23.5 nm.

The test cases are two dimensional in the sense that the systems are homogeneous in the third direction. Note that the electrons still move in three dimensions in space but their phase space distribution function is constant along one spatial direction.

At this point we study only the most essential processes, namely the electron-transport and the related generation of X-rays in the material, and leave modelling of X-ray transport and associated fluorescence effects to the future.

Table 1. Material models used in M_1 and MC simulations.

	M_1-model	MC-model
Elastic scattering cross-section	Screened Rutherford [31] (McKinley-Feshbach formula [33])	Screened Rutherford [31] (NIST Mott [32])
Stopping power	Bethe formula [31]	Bethe formula [31]
Mean ionisation potential	Sternheimer64 [34]	Sternheimer64 [34]
Absolute ionisation cross-section	Casnati82 [35]	Casnati82 [35]
X-Ray transition probability	ENDLIB-97 [36]	ENDLIB-97 [36]

Table 2. Setup of three test cases (homogeneous, thin film and interface).

Setup of test case 1 (homogeneous material)

Beam energy ϵ_B:	12, 20 keV
Beam diameter D_B:	10 nm
Material (density) El_1:	Cu ($8.96 \cdot 10^3$ kgm^{-3})

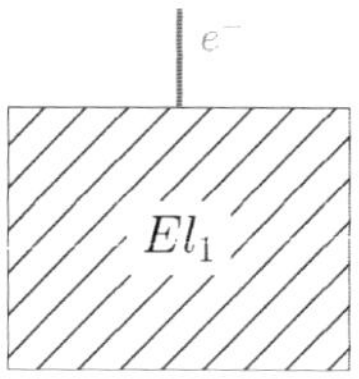

Setup of test case 2 (thin film)

Beam energy ϵ_B:	10 keV
Beam diameter D_B:	10 nm
Film thickness L:	100 nm
Material (density) El_1:	Cr ($7.2 \cdot 10^3$ kgm^{-3})
	Al ($2.7 \cdot 10^3$ kgm^{-3})
Material (Density) El_2:	Ni ($8.9 \cdot 10^3$ kgm^{-3})

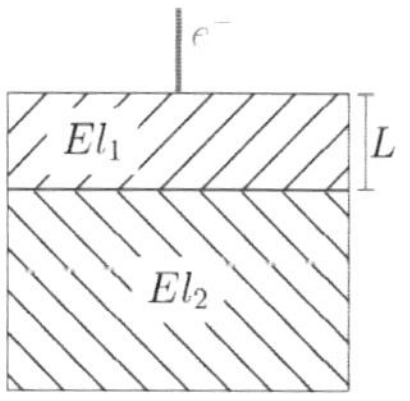

Setup of Test Case 3 (Interface)

Beam energy ϵ_B:	10 keV
Beam diameter D_B:	10 nm
Material (density) El_1:	Cr ($7.2 \cdot 10^3$ kgm^{-3})
Material (density) El_2:	Ni ($8.9 \cdot 10^3$ kgm^{-3})

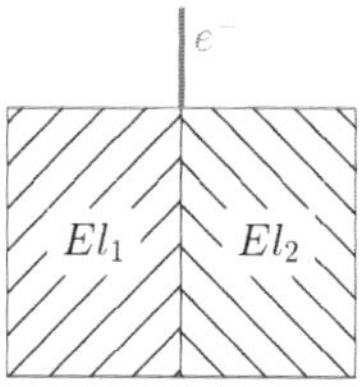

4.1. Homogeneous material

In order to start with the simplest material structure possible, we consider a homogeneous bulk of copper excited by a narrow Gaussian electron beam for different energies. As most methods for computing the k-ratios of homogeneous samples are based on the calculation of the depth distribution of radiation ($\phi(\rho z)$ curve), we use it as basis for the validation of the M_1-model here. Figure 2 shows the depth distributions of Cu-$K\alpha$ radiation computed from the M_1- and MC-model.

In addition to the M_1 and MC solutions, we added the depth distributions obtained with two common analytical models, the XPP [9] and the PAP [4] model. For a better comparability, the M_1 and MC results are scaled such that the maximum value of the depth distribution is equal to the maximum value of the PAP depth distribution. Note that the scaling corresponds to changing the number of simulated electrons and thereby does not have an influence on normalised quantities like k-ratios.

For all beam energies, the depth distributions obtained from the M_1-model match the MC

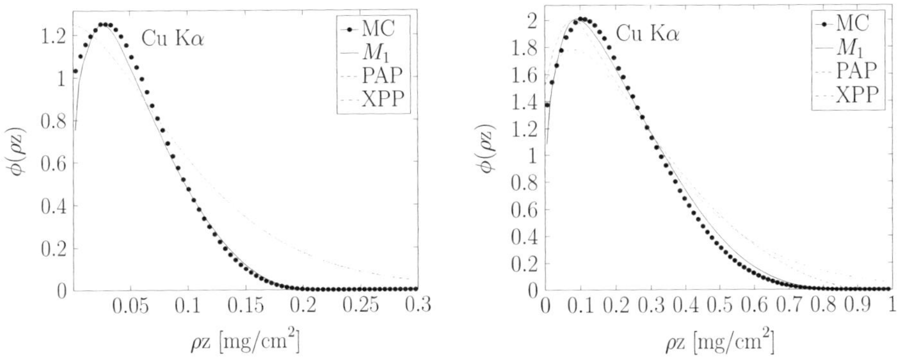

(a) 12 keV (b) 20 keV

Figure 2. Normalised depth distribution of Cu-Kα radiation ($\phi(\rho z)$) in copper at different beam energies. Comparison of the M_1 moment model to Monte Carlo and two common analytical models (XPP and PAP).

results very well. The results for XPP and PAP model approach the MC results for increasing beam energy but nevertheless clearly deviate from the MC results. The deviation is most evident near the sample surface for the lower beam energy of 12 keV where the graphs deviate also qualitatively, which reflects the limited applicability of the PAP and XPP model to cases in which the beam energy is close to the minimal excitation energy of the shell of interest [4].

4.2. Thin film

A common sample structure of interest are multi-layered samples. This motivates the material structure of our second test case, a thin film on a substrate. Both the substrate and the thin film consist of only one chemical element as visualised in the schematic drawing shown in Table 2. We look at two material combinations with nickel as substrate in both cases. We chose the material of the thin film to be chromium and aluminium in order to have, respectively, a moderate and a strong difference of the atomic number between both materials. These correspond respectively to the cases of a moderate and a strong change in material properties at the interface.

Analogous to the homogeneous test case, we validate the M_1-model based on the depth distributions of $K\alpha$-radiation, which are plotted in figs. 3(a) and 3(b). We computed the depth distribution of $K\alpha$-radiation neglecting the fact that each element is present only in either the thin film or the substrate. This is not physically correct but provides more information for comparison and can easily be corrected by ignoring the first or second part of the graph. As for the homogeneous test case, we scaled the M_1 and MC results such that the maximum Cr/Al $K\alpha$-radiation equals the respective maximum computed using the PAP model. The dashed vertical lines in figs. 3(a) and 3(b) indicate the position of the interface between the thin film and the substrate.

For the moderate case of chromium on nickel (fig. 3(a)), our observations are similar to the homogeneous case. We observe very good agreement between the M_1 and the MC results whereas the XPP and PAP results deviate significantly from the MC solution. The same applies to the depth distributions of Ni-$K\alpha$ radiation for the case of a thin aluminium film (fig. 3(b)). The M_1 results for the depth distribution of aluminium $K\alpha$-radiation start to deviate slightly from the MC results at a mass depth of around 0.1 mg/cm^2. Comparing the deviation with the deviations between the results for different elastic scattering cross-sections (screened Rutherford and Mott) we can see that the deviation is in the range of uncertainties due to material parameters. Even though the XPP and PAP results for the depth distribution of Al-$K\alpha$ radiation are relatively

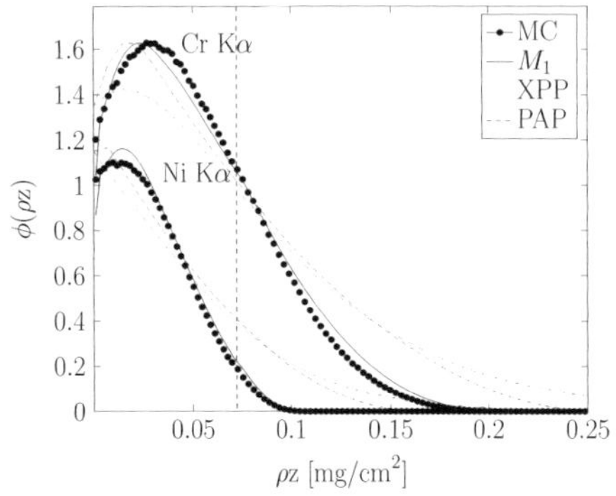

(a) Cr-Ni

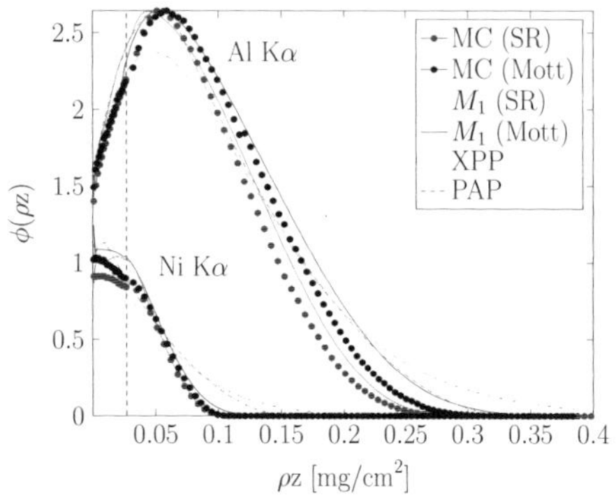

(b) Al-Ni

Figure 3. Depth distribution of Kα-radiation for two one layer thin film systems ((a) 100 nm chromium on nickel; (b) 100 nm aluminium on nickel) at 10 keV. Comparison of the M_1 moment model to Monte Carlo a two common analytical models (XPP and PAP).

good, the M_1 results are clearly better.

4.3. Interface setup

The third test case is also a binary structure. Two one-element materials separated vertically, with the interface aligned to the electron beam as visualised below. As the chemical composition of this sample is not a function of depth, classical analytical models like XPP and PAP are not applicable. We will, therefore, compare the M_1 solution with MC results only.

We will value the M_1 solution based on the spatial distribution of the electron number density n. A reference solution n_{ref} was calculated from a weighted histogram for the position and energy of electrons in a MC simulation. We used 100 bins to discretise the energy interval $[0, \epsilon_B]$ and a 100×100 equidistant grid to discretise the sample in space. A comparison of the electron number densities at different energies is presented in fig. 4. Comparing the M_1 with the Monte Carlo solution, we see that the M_1-model is able to reproduce the dynamics of the electrons very well. For both models, the electrons penetrate the chromium region ($x < 0$) about the same amount further then the nickel region ($x > 0$). The area occupied by electrons and the location of the maximum electron density for the different energies match remarkably well too. Furthermore, the M_1-model reproduces the sharp increase in density at the bottom. The main difference between both solutions is the earlier smoothing out of the electron density distribution towards the centre of the occupied region in the MC solution. This effect has to be investigated further.

Also noticeable in fig. 4 is the stochastic character of the Monte Carlo solution, resulting in a noisy solution even for a fairly high number of 200,000 simulated electrons. The distortion of the solution by the stochastic noise increases with decreasing energy (fig. 4), as the records in the histogram are distributed over more bins, which increases the uncertainty per bin. This effect makes the calculation of gradients, necessary for solving the inverse problem, difficult. As a deterministic model, the M_1 solution always gives smooth solutions, which are well suited for computing gradients.

5. Conclusion

Taking Monte Carlo as a reference, the results for the three test cases presented in the previous section show that the M_1-model is able to reproduce the electron dynamics in EPMA applications very well. Compared to classical analytical models like XPP and PAP, the M_1-model is more

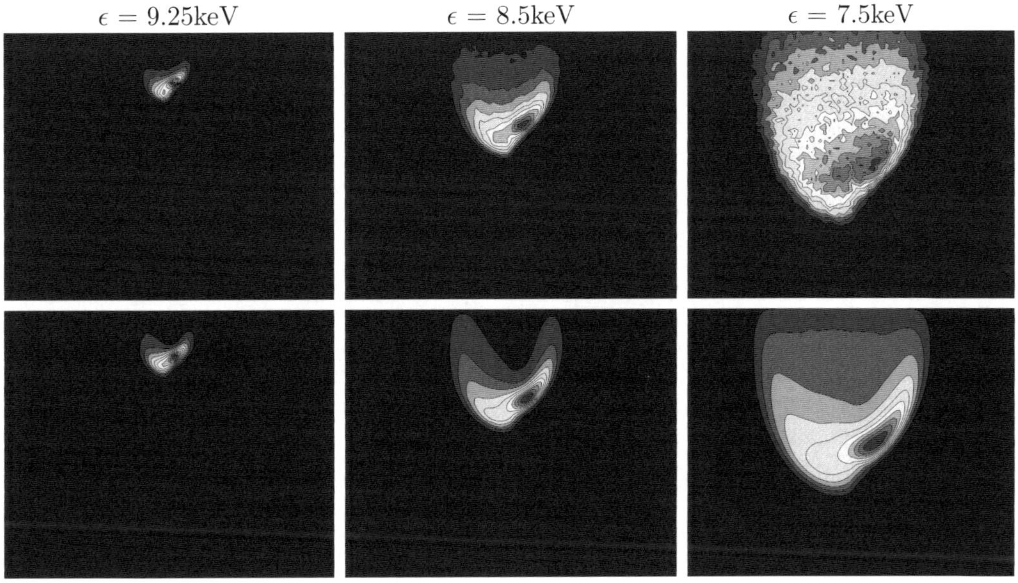

Figure 4. Spatial electron number density distribution in a binary structure (chromium (left) and nickel (right)) at different energies for a beam energy of $\epsilon_B = 10\,\text{keV}$. Comparison of the M_1 solution n_{M_1} (bottom) to Monte Carlo n_{ref} (top).

accurate and far more flexible. Nevertheless, the M_1-model is still an approximation to the BCSD equation and hence it is natural that M_1 solutions usually deviate slightly from MC solutions. Further investigations are necessary to better understand the cause of the deviations, the impact of such deviations on the accuracy of k-ratios and possible improvement through higher order moment models.

In summary, we are convinced that a minimum entropy moment model of the BCSD together with a sophisticated method for computing gradients of PDE constrained optimization problems (i.e., adjoint state method) is an efficient, flexible and accurate concept for solving the inverse problem of reconstructing the chemical material composition in EPMA.

References

[1] Tarantola A 2005 *Inverse problem theory and methods for model parameter estimation.* (Philadelphia, PA: Society for Industrial and Applied Mathematics)
[2] Aster R C, Borchers B and Thurber C H 2013 *Parameter estimation and inverse problems (Second Edition).* (Boston, MA: Academic Press)
[3] Bastin G F and Heijligers H J M 1991 Quantitative electron probe microanalysis of ultra-light elements (boron–oxygen). *Electron Probe Quantitation* (Heinrich K F J and Newbury D E, Eds.) (New York, NY: Plenum Press) 145–161
[4] Pouchou J L and Pichoir F 1991 Quantitative analysis of homogeneous or stratified microvolumes applying the model "PAP". *Electron Probe Quantitation* (Heinrich K F J and Newbury D E, Eds.) (New York, NY: Plenum Press) 31–75
[5] Bastin G F, Dijkstra J M and Heijligers H J M 1998 *X-ray Spectrom.* **27** 3–10
[6] Packwood R H and Brown J D 1981 *X-ray Spectrom.* **10** 138–146
[7] Philibert J 1963 A method for calculating the absorption correction in electron probe microanalysis. *X-ray optics and X-ray microanalysis* (Pattee H H, Coslett V E and Engström A, Eds.) (New York, NY: Springer) 379–392
[8] Love G, Sewell D A and Scott V D 1984 *J. Phys. Colloques* **45** 21–24
[9] Pouchou J L and Pichoir F 1984 *Rech. Aérospat.* **3** 167–192
[10] Ammann N and Karduck P 1990 A further developed Monte Carlo model for the quantitative EPMA of

complex samples. *Microbeam Analysis* (Michael J and Ingram P, Eds.) (San Francisco, CA: San Francisco Press) 150–154
[11] Gauvin R, Lifshin E, Demers H, Horny P and Campbell H 2006 *Microsc. Microanal.* **12** 49–64
[12] Joy D C 1991 *Scanning Microscopy* **5** 329–337
[13] Ritchie N W M 2005 *Surf. Interface Anal.* **37** 1006–1011
[14] Salvat F 2015 *PENELOPE-2014: a code system for Monte Carlo simulation of electron and photon transport.* (Issy-Les-Moulinex, France: OECD-NEA)
[15] Fathers D and Rez P 1979 Transport equation theory of electron backscattering. *Scanning Electron Microscopy* (Johari O, Ed.) (AMF O'Hare, IL: SEM, Inc.) pp 55–66
[16] Case K and Zweifel P 1967 *Linear transport theory* (Boston, MA: Addison-Wesley)
[17] Frank M, Herty M and Sandjo A N 2010 *Math. Models Methods Appl. Sci.* **20** 661–678
[18] Küpper K 2016 *Models, numerical methods, and uncertainty quantification for radiation therapy.* PhD-thesis (Aachen, Germany: RWTH Aachen University)
[19] Duclous R, Dubroca B and Frank M 2010 *Phys. Med. Biol.* **55** 3843–3857
[20] Pichard T 2016 *Mathematical modelling for dose depositon in photon therapy.* PhD-thesis (Bordeaux, France: Université de Bordeaux)
[21] Olbrant E 2012 *Models and numerical methods for time- and energy-dependent particle transport.* PhD-thesis (Aachen, Germany: RWTH Aachen University)
[22] Ammann N 1989 *Monte Carlo simulation der röntgenstrahlungs - Erzeugung durch elektronenbeschuß in schicht - Substrat - Systemen.* MSc-thesis (Aachen, Germany: RWTH Aachen University)
[23] Ritchie N W M 2009 *Micros. Microanal.* **15** 454–522
[24] Mevenkamp N 2013 *Inverse modeling in electron probe microanalysis based on deterministic transport equations.* MSc-thesis (Aachen, Germany: RWTH Aachen University)
[25] LeVeque R J 1992 *Numerical methods for conservation laws* 2nd ed (Basel, Switzerland: Birkhäuser)
[26] Cockburn B and Shu C W 1989 *Math. Comput.* **52** 411–435
[27] Cockburn B, Lin S Y and Shu C W 1989 *J. Comput. Phys.* **84** 90–113
[28] Cockburn B, Hou S and Shu C W 1990 *Math. Comput.* **54** 545–581
[29] Bangerth W, Davydov D, Heister T, Heltai L, Kanschat G, Kronbichler M, Maier M, Turcksin B and Wells D 2016 *J. Numer. Math.* **24** 135–141
[30] Bangerth W, Hartmann R and Kanschat G 2007 *ACM Trans. Math. Softw.* **33** 24/1–24/27
[31] Reimer L 1998 *Scanning electron microscopy* 2nd ed (Berlin: Springer)
[32] Powell C, Salvat F and Jablonski A 2007 *NIST electron elastic-scattering cross-section database* (Gaithersburg, MD: National Institute of Standards and Technology)
[33] McKinley W A and Feshbach H 1948 *Phys. Rev.* **74** 1759–1769
[34] Berger M J and Seltzer S M 1964 *NASA Special Publication* **3012**
[35] Casnati E, Tartari A and Baraldi C 1982 *J. Phys. B: At. Mol. Opt. Phys.* **16** 505
[36] Cullen D E, Hubbell J H and Kissel L 1997 *EPDL97 The Evaluated Data Library '97 Version.* (Livermore, CA: Lawrence Livermore National Laboratory)

IOP Conf. Series: Materials Science and Engineering **304** (2017) 012005 doi:10.1088/1757-899X/304/1/012005

Benefits from bremsstrahlung distribution evaluation to get unknown information from specimen in SEM and TEM

F Eggert, P P Camus, M Schleifer and F Reinauer

EDAX Inc., AMETEK Materials Analysis Division, 91 McKee Drive, Mahwah, NJ 07430, U.S.A.

E-mail: frank.eggert@ametek.com

Abstract. The energy-dispersive X-ray spectrometer (EDS or EDX) is a commonly used device to characterise the composition of investigated material in scanning and transmission electron microscopes (SEM and TEM). One major benefit compared to wavelength-dispersive X-ray spectrometers (WDS) is that EDS systems collect the entire spectrum simultaneously. Therefore, not only are all emitted characteristic X-ray lines in the spectrum, but also the complete bremsstrahlung distribution is included. It is possible to get information about the specimen even from this radiation, which is usually perceived more as a disturbing background. This is possible by using theoretical model knowledge about bremsstrahlung excitation and absorption in the specimen in comparison to the actual measured spectrum. The core aim of this investigation is to present a method for better bremsstrahlung fitting in unknown geometry cases by variation of the geometry parameters and to utilise this knowledge also for characteristic radiation evaluation. A method is described, which allows the parameterisation of the true X-ray absorption conditions during spectrum acquisition. An 'effective tilt' angle parameter is determined by evaluation of the bremsstrahlung shape of the measured SEM spectra. It is useful for bremsstrahlung background approximation, with exact calculations of the absorption edges below the characteristic peaks, required for P/B-ZAF model based quantification methods. It can even be used for ZAF based quantification models as a variable input parameter. The analytical results are then much more reliable for the different absorption effects from irregular specimen surfaces because the unknown absorption dependency is considered. Finally, the method is also applied for evaluation of TEM spectra. In this case, the real physical parameter optimisation is with sample thickness (mass thickness), which is influencing the emitted and measured spectrum due to different absorption with TEM measurements. The effects are in the very low energy part of the spectrum, and are much more visible with most recent windowless TEM detectors. The thickness of the sample can be determined in this way from the measured bremsstrahlung spectrum shape.

1. Introduction

The energy-dispersive X-ray spectrometer (EDS) spectrum depends on both the material elements and the measurement geometry of the excited specimen area. The emission intensity of low energy X-rays is influenced significantly both by self-absorption of the generated X-rays inside the specimen and by the surface topography at the analysed point. For example, self-absorption exceeds 70 % for aluminium X-rays in the high absorption case versus 35 % with low absorption (example almandine, electron excitation energy 15 keV, scanning electron microscope (SEM)). It is well known that bremsstrahlung X-rays are similarly affected. This is utilised in peak-to-background (P/B) quantitative methods to analyse particles and specimens with undefined surfaces. Nevertheless, a good background approximation is necessary for accurate determination of the P/B values for inputs with the P/B-ZAF approach (ZAF is correction for atomic number **Z**, absorption **A**, fluorescence **F**). The goal of this investigation was to develop a method for better bremsstrahlung fitting in unknown geometry cases by variation of the geometry parameters. It is possible to determine unknown absorption by evaluating the measured bremsstrahlung background shape. This reduces the result uncertainties for specimens with irregular surfaces. The optimised physical parameter is the 'effective tilt', which describes the unknown absorption of the sample at the measured surface point [1]. It parameterizes the unknown geometry with an 'equivalent value' for a hypothetical flat sample of the same composition, which would have similar absorption to the investigated sample surface. The theory of this method has been previously reported in [1]. This study provides verification of the method with real measurements.

The method is also applied to transmission electron microscope (TEM) measurements for the first time. With TEM the absorption influences are primarily due sample thickness. Because the typical sample thicknesses are not much higher than 100 nm it affects only the very low energy region of measured X-rays (< 1 keV). The recent innovations with windowless detectors for TEM [2] have triggered a revision of older bremsstrahlung background models, because much more X-rays are now visible in the low energy part of the spectrum [3]. This provides then also more detailed information of the low energy region bremsstrahlung distribution, which is significantly affected by sample thickness. Therefore, the idea is to also apply the absorption evaluation from bremsstrahlung shape to TEM-spectra, but now with variation of the sample thickness.

The used bremsstrahlung background calculation is briefly described in the next section and is based on an earlier publication [6]. There were improvements of the basis method, published in [8], and the dedicated application for TEM spectra was investigated and published in [3]. However, the described method should also be applicable for other bremsstrahlung calculation models.

2. Method

The calculation of the bremsstrahlung distribution for background approximations follows Kramer's law [4], using the modification by Lifshin [5], and was substituted by a quick calculation of the absorption terms $\varepsilon_{E\nu}$ and $\boldsymbol{A}_{E\nu}$ [6]:

$$N_{E_\nu}^{br} = \varepsilon_{E_\nu} A_{E_\nu} \left[a \frac{E_0 - E_\nu}{E_\nu} + b \frac{(E_0 - E_\nu)^2}{E_\nu} \right] \tag{1}$$

with:

$\boldsymbol{N_{E\nu}}^{br}$ number of bremsstrahlung counts measured in spectrum-channel with related energy $\boldsymbol{E_\nu}$
$\boldsymbol{\varepsilon}_{E\nu}$ detector efficiency with photon energy $\boldsymbol{E_\nu}$ (the fraction which is detected)
$\boldsymbol{A}_{E\nu}$ absorption correction with photons of energy $\boldsymbol{E_\nu}$ (the fraction which is not absorbed)
a, b empirical coefficients for spectra fit (if $\boldsymbol{b}$= 0, then $\boldsymbol{a}$ represents absolute adjustment only)
$\boldsymbol{E}_0$ primary electron energy.

It is possible to use as many fit-regions as desired to get the best background fit. The coefficients $\boldsymbol{a}$ and $\boldsymbol{b}$ are determined on the real acquired spectrum and are constant during the calculation between two fit regions. The coefficient $\boldsymbol{b}$ is simply zero between the first spectrum channel and the first fit region. The same is between the last fit region and the last spectrum channel. Therefore, there are F+1 segments with individual application of (1), where F is the number of the used fit regions. Each segment is always separately approached by using both boundary fit regions (or only one with the first and last section). Finally, F-1 pairs of $\boldsymbol{a}$ and $\boldsymbol{b}$ are determined with (1) plus 2 times $\boldsymbol{a}$ only. The measured bremsstrahlung counts for mean energy of a background fit region are determined from the average of the channel contents. If two measurement points are used, it is possible to determine 2 unknown parameters (***a*** and ***b***). With one measurement point only (first and last section) then ***b*** is required to be set to zero.

The fit regions are automatically selected in peak-free spectrum regions, with minimum distances predefined, based on a top-hat filter peak-search. The used fit-regions are changeable by the operator, based on a visual inspection of the bremsstrahlung fit compared to the measured spectrum.

The energy dependent self-absorption model is different for SEM- and TEM-spectra.

For **SEM spectra** evaluation, it is possible to describe the absorption part of (1) with $\varepsilon_{Ev} A_{Ev}$ with only a single absorption formula [6]:

$$\varepsilon_{E_v} A_{E_v} = e^{(k_a \, d_{Be,E_v}^{equiv} / E_v^{3.162})} \tag{2}$$

The absorption for all energies is computable as a function of the energy E_v with the same exponential absorption formula at each energy (k_a is a constant). A virtual Be material thickness $d_{Be,Ev}^{equiv}$ is used to describe the total material, which is along the absorption path for all absorbing sample and detector window elements. The basic idea is quite simple. All absorption calculations (mass absorption coefficients (MAC)) have the same characteristic in energy dependence. The only serious differences in energy dependence between the elements are the absorption jumps, which are characteristic for each element. That is why the mass absorption coefficient calculation for beryllium is applicable using an equivalent beryllium thickness. The value $d_{Be,Ev}^{equiv}$ is constant between absorption edges and is incremented at each edge energy E_i (start beginning from low energies towards the high energy end of the spectrum).

$$d_{Be,E_v}^{equiv} = \sum_{E_i<E_v} d_{Be,E_i}^{equiv} \tag{3}$$

The increment steps must consider all edges, due both to detector entrance window materials (ε_{Ev}), and all elements inside the specimen (A_{Ev}). E_i is the critical excitation energy of the electron shell (jump energy). The value of the needed $d_{Be,Ei}^{equiv}$ increments depends on edge energy. They are calculated before the bremsstrahlung computation process. In the case of specimen element edges the formula is [6]:

$$d_{Be,E_i}^{equiv} = k_b \left(\tfrac{\mu}{\rho}\right)_{E_i}^{diff} L \,(E_0^2 - E_i^2)\, c_i \, E_i^{3.162} \tag{4}$$

where L is a numerical value which considers geometry (ratio between absorption path and X-ray emission depth), E_0 is the primary electron energy, k_b is a constant, c_i is the mass fraction of the element in specimen and $(\mu/\rho)^{diff}$ is the difference between mass absorption coefficient after and before the regarded edge energy E_i (all values in keV). Described differently, eq. (4) describes the thickness of an additional virtual beryllium layer which would have the same additional absorption effect (additional contribution to all effects) as the shell-excitation jump of the element at energy E_i.

The calculation of the entire bremsstrahlung deviation (the bremsstrahlung spectrum) starts with an initial energy and value of $d_{Be,0}^{equiv}$. This initial value is the fictive beryllium thickness, which would have same absorption (at the starting energy) with all absorption effects, from the detector efficiency and from all elements in the specimen, considered together, at the relevant excitation conditions.

Following the model derivations in [6], a basic estimation of a mean generation depth of photons in the specimen was made, based on the Thomson-Whiddington law [7]. The self-absorption is then calculated based on the assumption of a mean emission depth of X-rays with specific energies. Later investigations have shown this simplification limits the accuracy of the calculation for low energy bremsstrahlung of high atomic number elements [8]. There was no way to fit the calculated background to measured spectra with one exponential mass absorption function by modification of the parameters. This means, a depth distribution is required also for a bremsstrahlung calculation to consider the X-rays that come from different sample depths, which is mathematically a superposition (addition) of several exponential absorption functions.

Figure 1 shows the bremsstrahlung model after modification with a depth distribution consideration. The background fit is improved both in the low energy region and after the absorption edges. The model introduction considering N-shell edges was the final improvement. This is not usual in common published mass absorption coefficient calculation formulae.

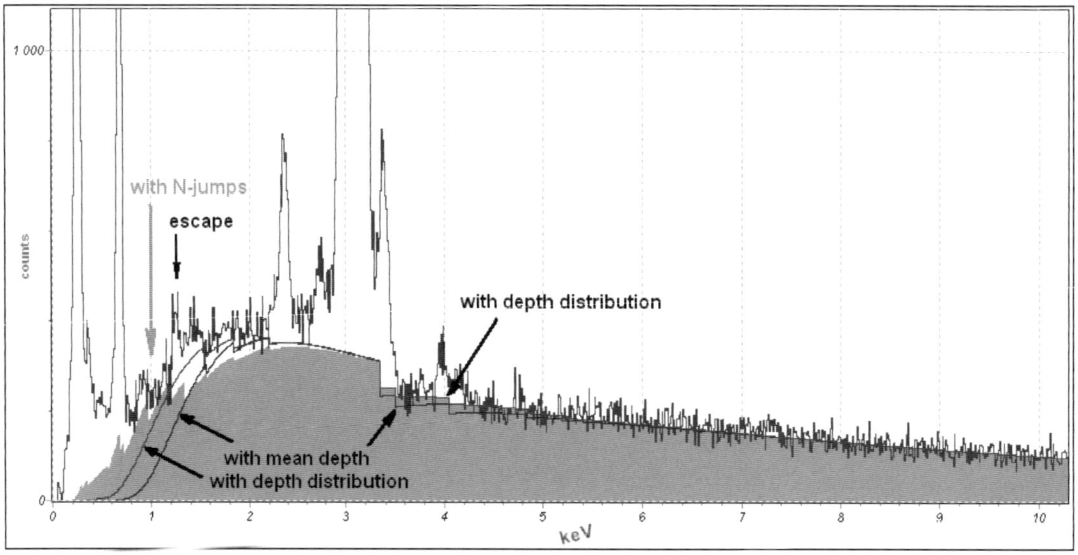

Figure 1. Comparison between measured pure element spectrum of Thorium (E_0 = 20 keV) and bremsstrahlung background calculation (one single fit region only at 5 keV). The low energy absorption is overestimated if the depth distribution is not accounted for. The model is improved with consideration of a depth distribution of X-rays for bremsstrahlung emission and with including N-jumps into the model (filled line) [8].

For **TEM spectra**, a different X-ray excitation and absorption model is required. Because the TEM specimens are very thin, the absorption calculation $A_{E\nu}$ depends primarily on sample thickness and density, and follows a simple X-ray absorption law using a linear approach based on an equal generation of X-rays at all depths. If the same background calculation basic model is used, equation (4) is replaced by (5) for TEM spectra evaluation:

$$d_{Be,E_i}^{equiv} = \rho t_{sample} \frac{(\frac{\mu}{\rho})_{E_i}^{diff}}{\rho_{Be}(\frac{\mu}{\rho})_{E_i}^{Be}} L \; c_i$$

$$d_{Be,E_i}^{equiv} = t_{sample} \frac{\rho_{sample}}{\rho_{Be}} k_a (\frac{\mu}{\rho})_{E_i}^{diff} L \; c_i \; E_i^{3.162} \tag{5}$$

where $\boldsymbol{t}$ is the sample thickness and $\boldsymbol{\rho_{sample}}$ the density of sample. $\boldsymbol{k_a}$ is the same constant as in eq. (2).

3. Examples for SEM

Figure 2 compares the absorption effects on almandine spectra simulated for different tilt angles. Different absorptions not only affect the characteristic X-rays, but also the bremsstrahlung background shape. This is also illustrated in figure 3 which shows two different linear intensity scalings to better illustrate the similar absorption effects on both the characteristic lines (a) and on the bremsstrahlung emission (b).

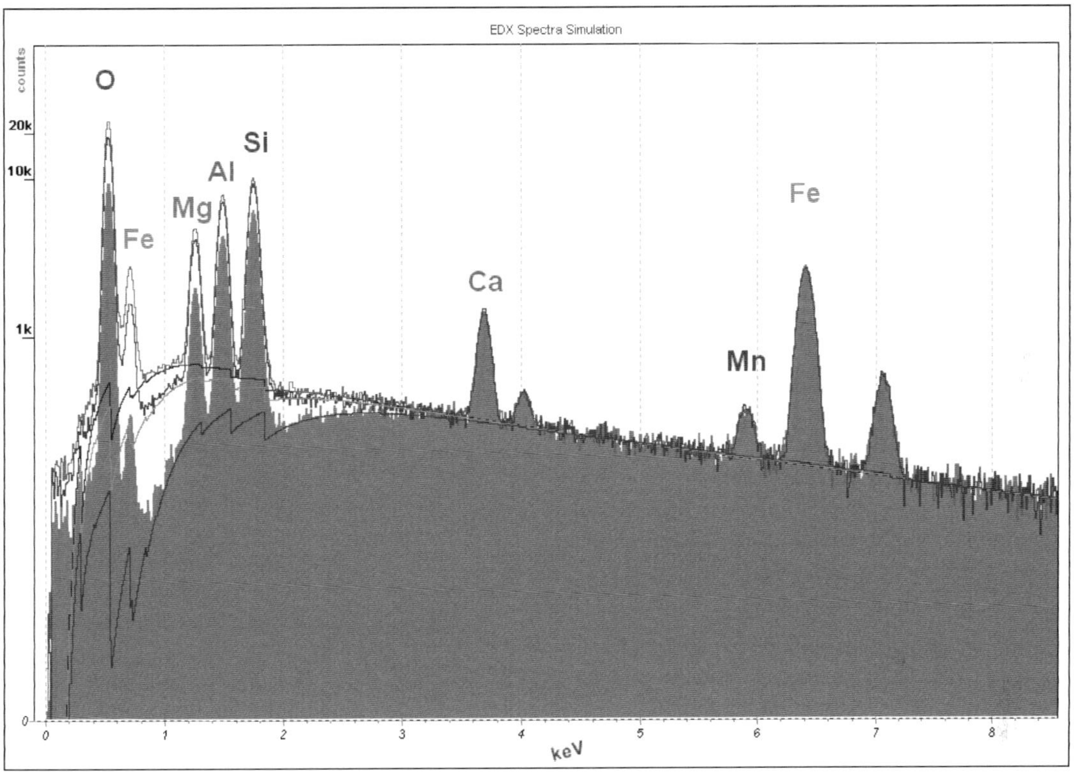

Figure 2. Simulation of almandine spectra for a 15 keV primary electron energy at -24° tilt (red bars), 0 tilt (magenta line) and 48° tilt (green line), and the related bremsstrahlung backgrounds (logarithmic y-axis). (Simulation method described in [13]).

It is possible to account for the effect of variation of the tilt-angle by using eqs. (2) and (4). If physical parameter L in eq. (4) is used as an unknown value for the background fitting, then the most affected low energy part of the spectrum can be better adjusted for the non-linear absorption effects than it would be with the two linear parameters $\boldsymbol{a}$ and $\boldsymbol{b}$. Therefore, this is the method of choice to fit the background shape in cases where the specimen surface is not well known, e.g., with unpolished samples.

Although the P/B-ZAF approach is already nearly insensitive for surface geometries [9, 10], application of the physical model based geometry optimisation is required to get reliable measured P/B values, which is always an interpolation and needs as real as possible modelling of the large absorption edges (which depends on specimen composition and excitation). Therefore, iteration is required anyway, which includes the background approximation with perhaps changing weight fraction results until convergence is obtained. With the presented procedure, an optimised approach of

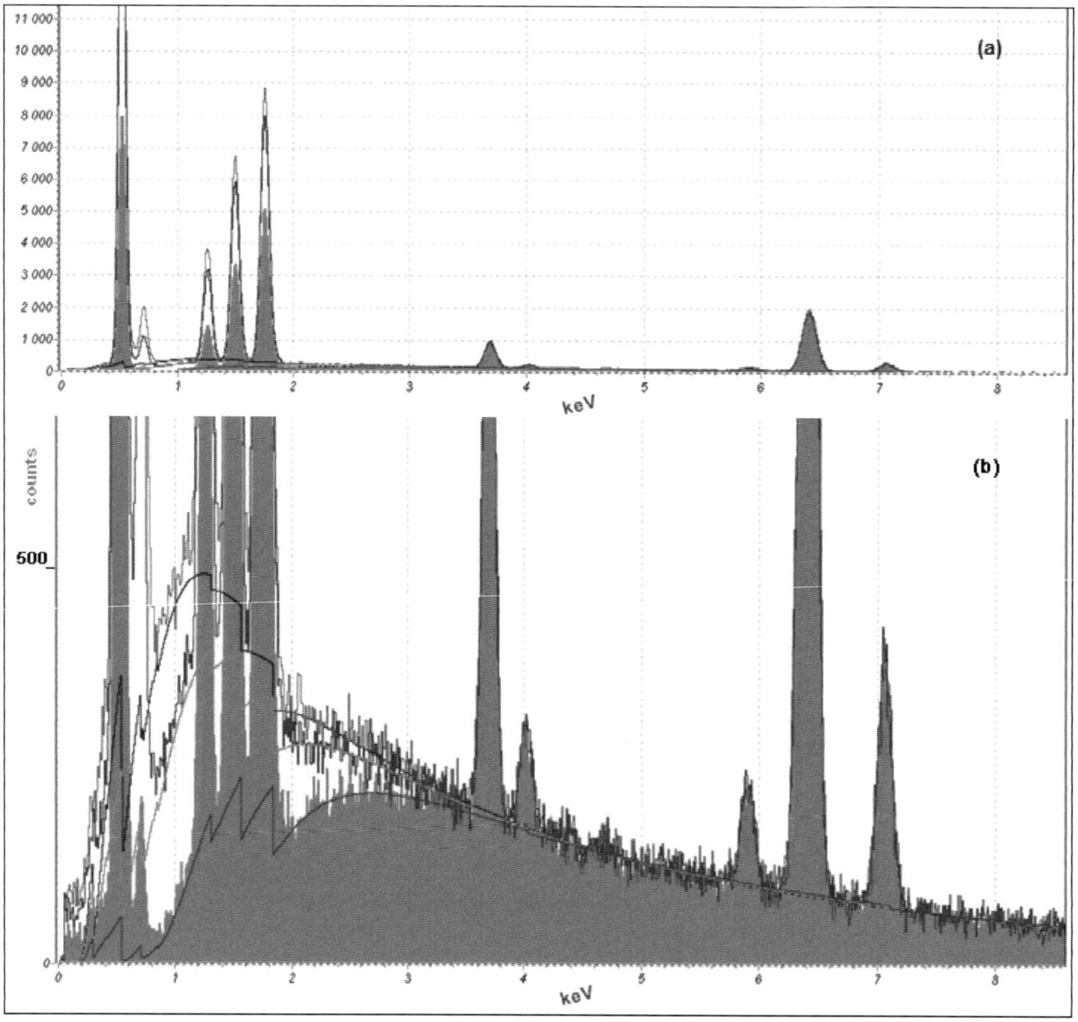

Figure 3. Same as with figure 2 but linear y-scale to show better the absorption influences with characteristic lines (a) compared to background shape (b).

background calculation is also possible with acquired spectra, which are from rough specimen regions and particles with different unknown absorption situations. The calculation is additionally optimised with absorption modelling by fitting the low energy part of background approximation. This is performed by optimising a real physics meaning parameter, the 'effective tilt', which will be optimised additionally to mathematical parameters in equation (1).

The proper ‘effective tilt’ is searched by variation of absorption with different tilt estimations (setting the azimuth angle to zero) from the most possible negative tilt (high absorption case) towards high positive tilt cases (low absorption). It is a numerical solution by repeated tilt angle variation calculations, searching for the best fitting background in regard of three numerical test criteria within the checked spectrum region between 550 eV (above the oxygen absorption edge at 536 eV) and 4 keV:

- Minimal sum of the deviation squares (least squares) in peak free regions.
- Least squares deviations of all calculated background channels which are above the measured spectrum.

- Statistical criteria, the calculated background should be never higher than acquired counts, all channels, plus 3 times the square root of channel content.

The best fitting background is found per a decision tree which combines the three criteria. This tilt angle is then used as the found result for the 'effective tilt'. The described procedure provides both the background spectrum shape with all important low energy edges in irregular surface cases, and provides the best fit angle with beneficial use in quantitative model calculations for final concentration determination with characteristic radiation evaluation.

The search for unknown absorption is with an additional free parameter with additional result uncertainties. Therefore, this should be performed only if the specimen surface is not perfectly flat. It is always better to avoid this with well-polished and positioned specimen, but the daily praxis in SEM labs is often not with this ideal case.

4. Results with SEM measurements

First some simulated 'almandine' spectra were evaluated to test the method in regarding reliability. The absorption is known and can be compared with determined values. This allows a good assessment about abilities of the method.

Figure 4 presents the calculated tilts with bremsstrahlung background fit versus known values of the simulation between -30° and +48°. The uncertainty in determination of the tilt angle by bremsstrahlung background fitting (taken as unknown) is high for low absorption cases (high positive tilts), because it is then less sensitive. Fortunately, where the tilt determination has higher uncertainties the influences of those tilt uncertainties to the quantitative results are also less (also low absorption case). But for the much more important high absorption case (low and negative tilts) all determined results are reasonably aligned with the real values. So, this is all in good agreement with the needs. The determination of the 'effective tilt angle' itself is not the final goal, but reduction of the entire result uncertainty.

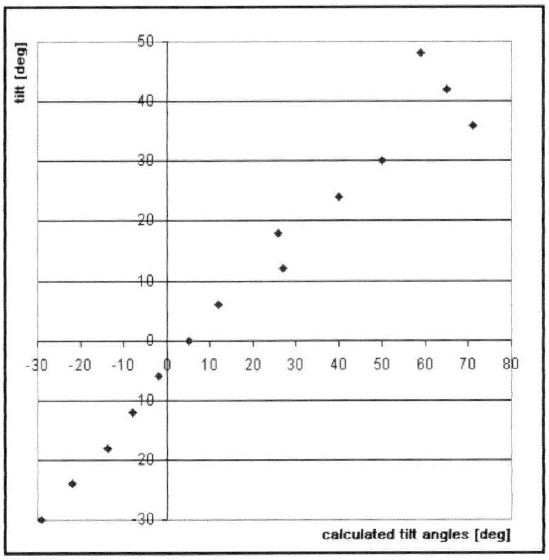

Figure 4. Comparison of original tilt angles (y-axis) with results, determined by background optimisation (x-axis) evaluation of simulated spectra.

The determined tilt angle for an unknown geometry is even beneficial for a classical net-counts evaluating ZAF correction, which it is usually only possible to apply for polished specimens. But figure 5 demonstrates the possibility of the evaluation of irregular surfaces even with such a classical ZAF correction. Because the low X-ray energy elements are much more absorbed in high absorption cases, the results contain an underestimation if not considered (figure 5a). This provides too low element concentrations at negative tilts (means high absorption case in a rough surface). The low absorption influenced elements like Iron show the opposite behaviour, due to normalisation effects. The individual quantitative results differ up to 100 % relative (Fe result variation between 20 - 43 %) depending on the absorption situation.

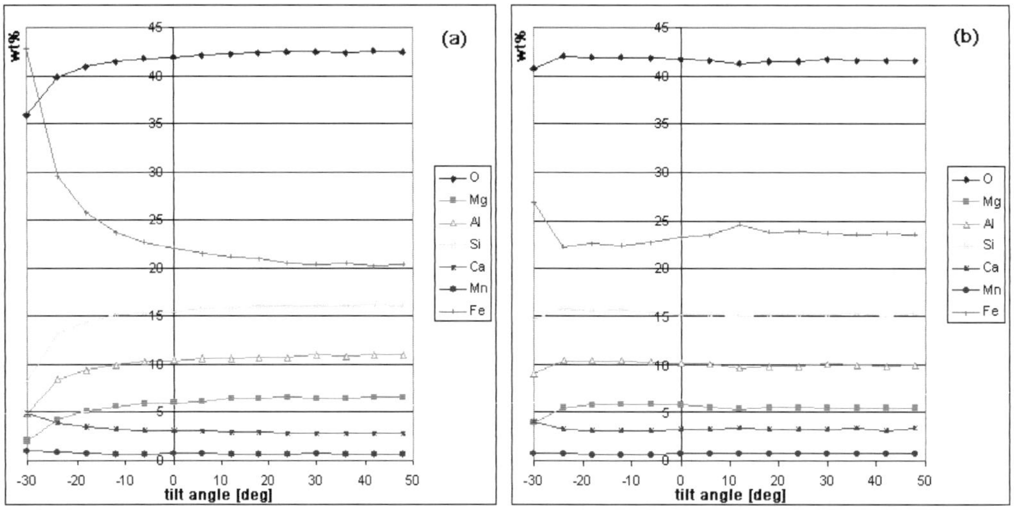

Figure 5. ZAF based evaluation [14] of almandine simulated spectra with -30° to +48° specimen tilt (normalised results, oxygen by stoichiometry); a) without considering the different self-absorption in ZAF model; b) with considering the 'effective tilt' angle which was determined by bremsstrahlung background fit.

By using the additional knowledge of the already calculated 'effective tilt angle' the results are much more stable (see figure 5b). Therefore, even the quantitative results of ZAF-based evaluations depend less on the actual absorption case, if the absorption parameter is used which was almost independently ascertained. The usually small mutual dependence from specimen composition is solved by iteration.

With the second step, the method was tested with measured spectra of a Nb/Ni alloy specimen (figure 6). To make an assessment possible, again a flat surface specimen was used to compare the evaluation results of calculated tilt with a known measurement tilt. If this succeeds, then it is certain also improving the results in real irregular surface cases.

All quantitative evaluations were performed by evaluating the L-line series for Nb and the K-lines for all other elements. Figure 7 presents the determined tilts vs. known. In some cases there are two different result values for each angle, because due to iteration with slightly different quantitative results, the best fit angles can differ slightly between the P/B-ZAF based and ZAF-based standardless analysis. These also contain higher uncertainties in small absorption cases. These calculated values for tilt also vary little with selection of background fit regions.

The results for P/B-ZAF based evaluation [11-13] are presented with figure 8 and the net-count based ZAF results [14] are within figure 9. In both cases the final quantitative result changes in the right- hand parts of the figure are much reduced with the optimised tilt, compared to the left-hand part where all was calculated with the assumption of zero tilt.

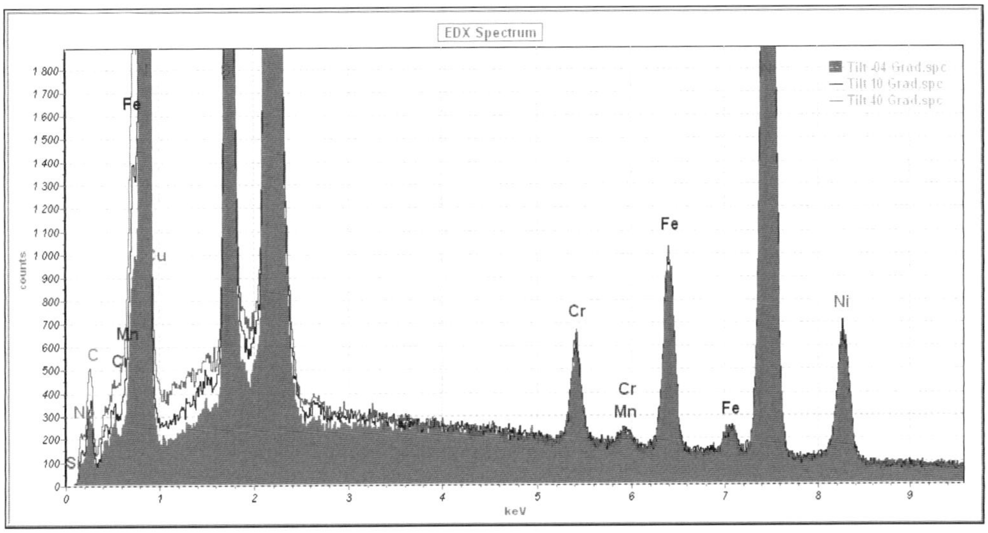

Figure 6. Comparison of spectra from a Nb/Ni alloy measured at three tilt angles: -4°, +10°, and +40°.

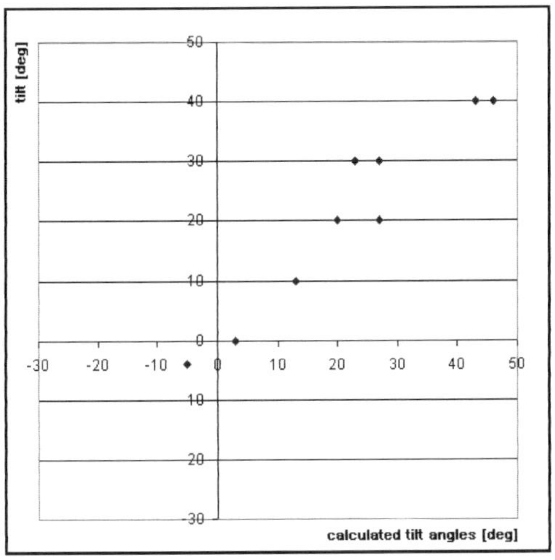

Figure 7. Comparison of real tilt angles (y-axis) with results, determined by background optimisation (x-axis) for measured Nb/Ni alloy spectra.

It is apparent that even the P/B method needs background optimisation by varying the absorption. This is demonstrated well with figure 10, where the background calculation performed for zero tilt is compared to calculation using the optimised tilt. The background shape differs most below Si-K and Nb-L X-ray line series. This is what influences the P/B value determination and it does not matter that the final P/B-model correction does not depend much on absorption. The difficulties are already with determination of the measured P/B values.

But as expected, the net-count based results fluctuate more due to changing absorption cases. But it can be shown for the first time even those are possible to improve if irregular surfaces are measured and the proposed method is used to determine the unknown specimen self-absorption. The resultant fluctuation was reduced both for the P/B method (Si 27% → 7%; Nb 15% → 5%) and for ZAF based quantification (Si 33% → 14%; Nb 25% → 8%). Nevertheless, the P/B method remains the better choice for irregular surfaces.

IOP Conf. Series: Materials Science and Engineering **304** (2017) 012005 doi:10.1088/1757-899X/304/1/012005

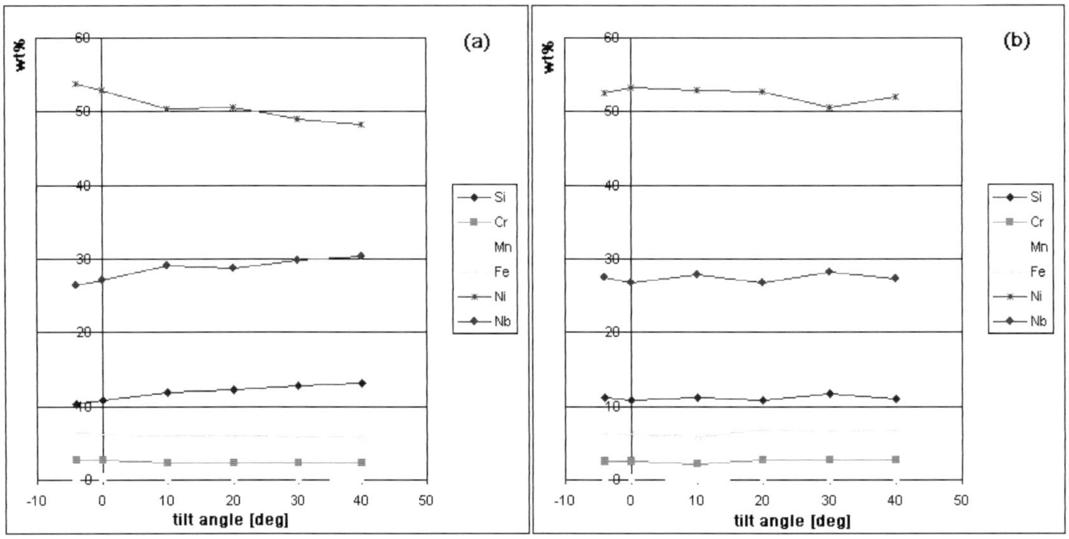

Figure 8. P/B-ZAF based evaluation of the measured alloy spectra with -4° to +40° specimen tilt (normalised results); a) without considering the different self-absorption for background calculation; b) with performing the 'effective tilt' angle determination by bremsstrahlung background fit.

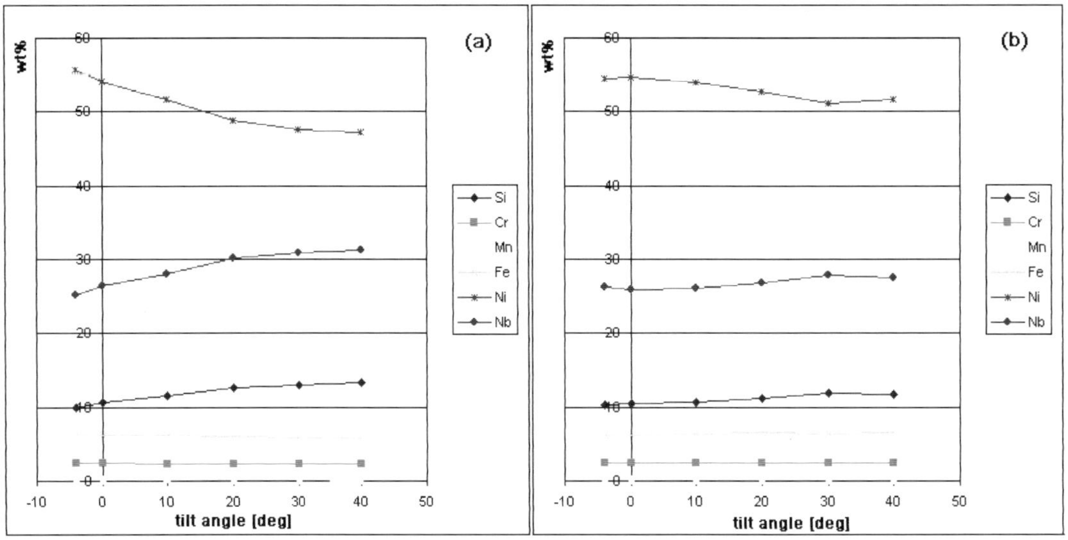

Figure 9. Same as with figure 8, but all results are net-count based ZAF quantifications; a) without considering the different self-absorption in ZAF model; b) with considering the 'effective tilt' angle which was determined by bremsstrahlung background fit.

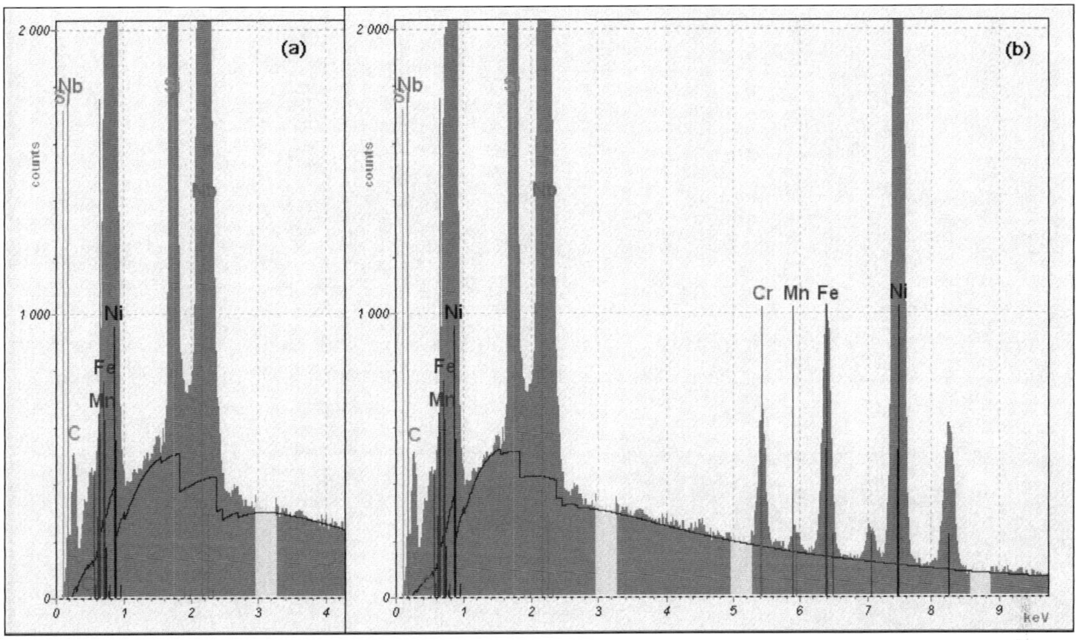

Figure 10. Measured alloy spectrum with +40° tilt (automatically determined fit regions for background approximation are visible); a) with using a zero degree tilt for background fit; b) the background fit is performed with the described additional 'effective tilt angle' optimisation.

5. Examples for TEM

It is also possible to beneficially apply the same method for TEM-EDS spectra evaluation. The thickness of the TEM sample is almost always unknown at the location where the analysis is performed. The improved ability to evaluate low-energy X-rays with new types of detectors [2] now makes the small absorption effects visible. Following the proven method with SEM-EDS spectra, it is possible to determine the unknown specimen thickness (the mass thickness) automatically from the measured TEM-EDS bremsstrahlung shape, not the absorption by tilt variation (as with SEM, eq. (4)) but by variation of a specimen thickness parameter for TEM samples (eq. (5)).

The effect of different thicknesses on the bremsstrahlung background shape is demonstrated in figure 11. Because the very thin samples thickness affects only the low energy X-ray part of the spectrum, the parameter optimisation is typically performed by fitting spectrum energy range below 2 keV.

6. Results with TEM measurements

TEM measurements with FIB lamella were performed along a chain of points (figure 12) with determination of thickness by automatic fitting of the low energy spectrum part of the bremsstrahlung distribution by variation of the specimen thickness parameter. The expectations were thicknesses between about 40 and 100 nm. The results with automatic thickness determination have met the expectation (figure 13).

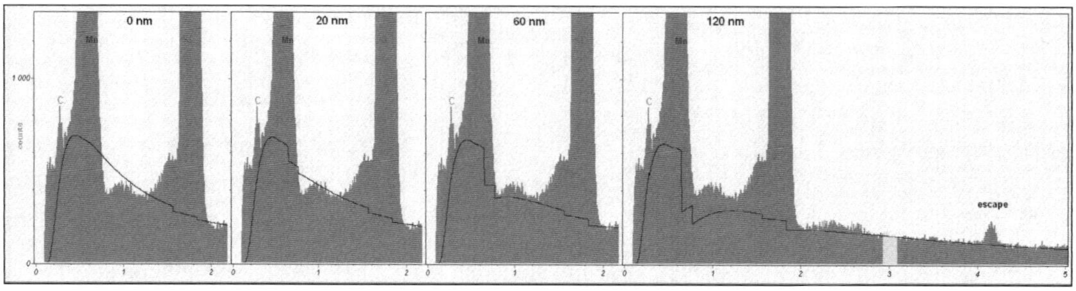

Figure 11. $MnSi_2$ TEM spectrum background calculation always with first fit region at 3 keV but with different sample thickness parameter assumptions, from 0 nm until 120 nm (density 3 g/cm^3).

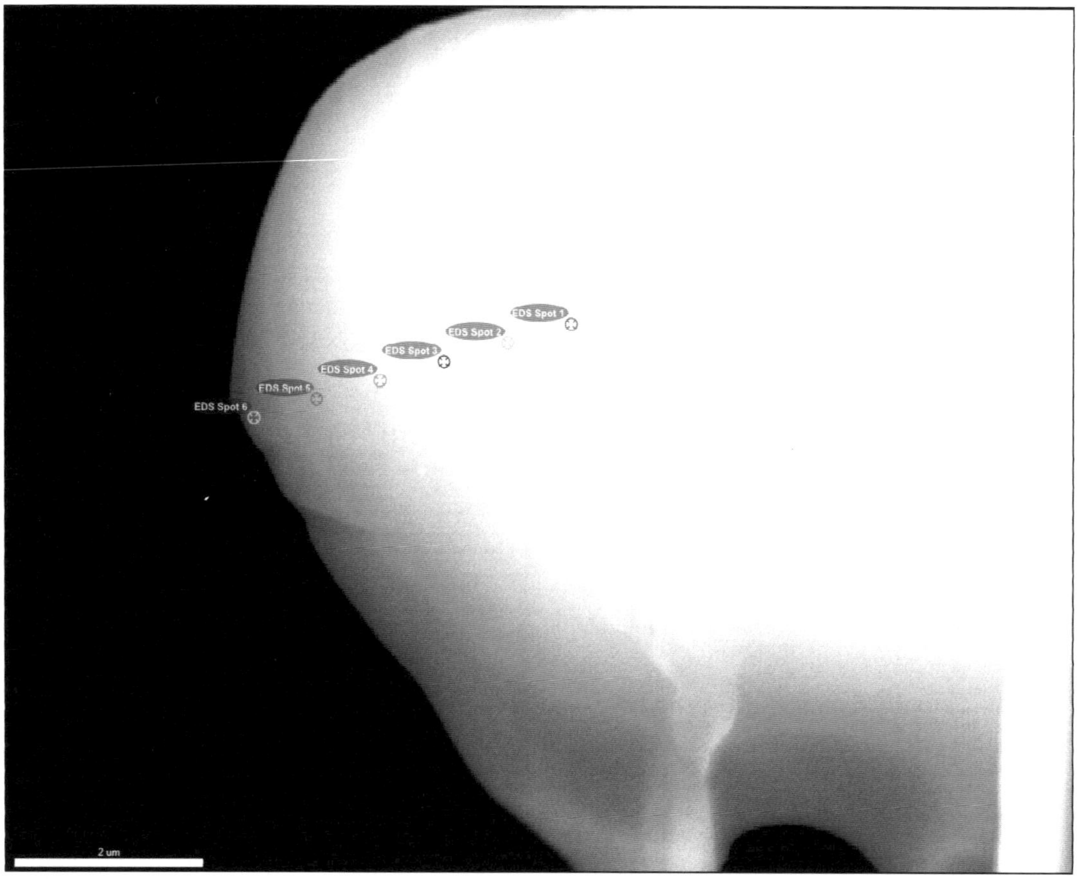

Figure 12. TEM image of FIB lamella with a chain of measurement points for EDS spectra.

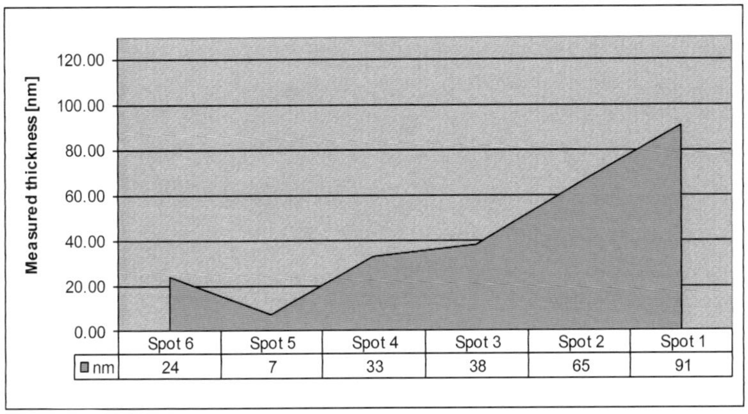

Figure 13. Plot of the calculated thicknesses (with assumed density of 4.23 g/cm^3) by evaluation of bremsstrahlung spectra for different measurement spots marked in figure 12.

The accuracy of the thickness determination is expected to be in the range ±15 to 20 nm. However, the goal is not primarily to calculate thickness, but to automatically determine values for better background fitting for improved automated qualitative analysis of TEM-spectra [3] with EXpertID [15] (automatic element identification). Finally, this optimised spectrum evaluation with thickness determination should also be beneficial for reduction of uncertainties with quantitative results, as it is with SEM spectra evaluation. Figure 14 gives an example of spectrum background evaluation for 'EDS Spot1' in figure 12.

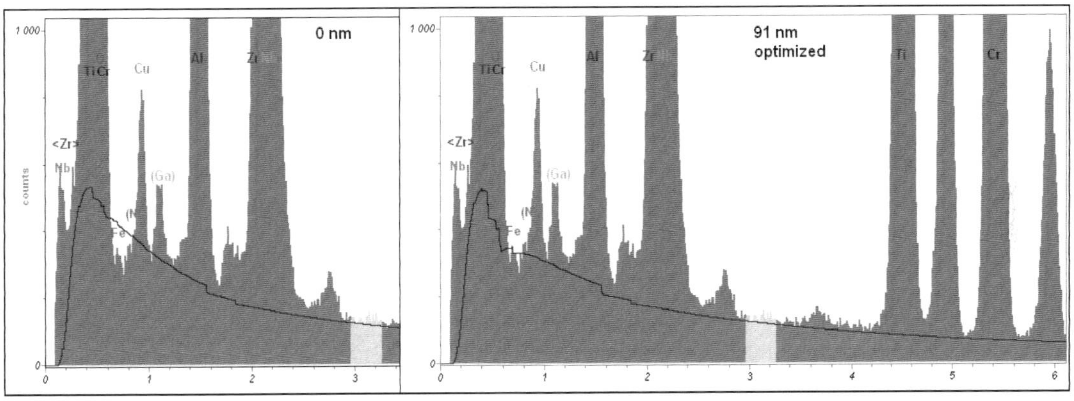

Figure 14. TEM spectrum from 'EDS Spot1' in figure 12 evaluation with optimisation of bremsstrahlung spectrum shape by automatic variation and optimisation of TEM sample thickness (with assumed density of 4.23 g/cm^3).

7. Conclusions

This work demonstrates the advantages, reasonability and the justifications to calculate the bremsstrahlung distribution with evaluation of EDS spectra. The used bremsstrahlung model was outlined for SEM and for the first time also for TEM applications. It was then demonstrated that the bremsstrahlung distribution (shape) is usable to ascertain the real specimen absorption situation during EDS spectrum acquisition, for both SEM and TEM spectra. This is beneficial and applicable for the exact background approximation for non-flat sample surfaces (SEM) and unknown TEM sample thicknesses. It was shown that it is also possible to use as an additional knowledge for ZAF correction because the final analytical results become much more reliable. The application for TEM with included automatic sample thickness determination was presented for the first time. The thickness results were also in the expected range and provide extra knowledge. However, the primary goal with TEM sample thickness determination was to optimise and improve the spectrum background fit to make all following spectra evaluation steps more reliable and with less uncertainties.

References

[1] Eggert F 2012 *Microsc. Microanal.* **18** (Suppl 2) 1744-1745
[2] EDAX 2014 *Technical Note June 2014*
[3] Eggert F, Camus P P, Sandborg A and Zaluzec N J 2015 *Microsc. Microanal.* **21** (Suppl 3) 825-826
[4] Kramers H A 1923 *Phil. Mag.* **46** 836
[5] Lifshin E 1974 in: *Proc. 9th Ann. Conf. Microbeam Analysis Soc.* 53
[6] Eggert F 1985 *Experim. Techn. Physik* **33** 441-448
[7] Whiddington R 1912 *Proc. Royal Soc. London* **A86** 360
[8] Eggert F 2007 in: *EMAS 2007 Workshop, Book of Tutorials and Abstracts.* (Karlsruhe, Germany: European Microbeam Analysis Society) 255-256
[9] Small J A, Heinrich K F J, Newbury D E, Myklebust R L and Fiori C E 1978 *Nat. Bur. Stand. Spec. Publ.* **533** 29
[10] Statham P J 1979 *Mikrochim. Acta* **Suppl. 8** 229
[11] Heckel J and Jugelt P 1984 *X-ray Spectrom.* **13** 159
[12] Eggert F and Heckel J 1986 *Experim. Techn. Physik* **34** 201-211
[13] Eggert F 2006 *Microchim. Acta* **155** 129-136
[14] EDAX TEAM software, eZAF correction procedure
[15] Eggert F 2010 *IOP Conf. Ser.: Mater. Sci. Engng.* **7** 012007

IOP Conf. Series: Materials Science and Engineering **304** (2017) 012006 doi:10.1088/1757-899X/304/1/012006

Development of a compact FE-SEM and X-ray microscope with a carbon nanotube electron source

M Irita[1,2], S Yamazaki[1], H Nakahara[1] and Y Saito[1]

[1] Nagoya University, Department of Quantum Engineering, 464-8603 Nagoya, Japan
[2] Nagoya University, Venture Business Laboratory, 464-8603 Nagoya, Japan

E-mail: masaru.irita@surf.nuqe.nagoya-u.ac.jp; ysaito@nagoya-u.jp

Abstract. A carbon nanotube (CNT) possesses various benefits as a field electron emitter. Multi-walled CNT (MWNT) have exhibited high brightness and the field emission (FE) current is stable even under poor high-vacuum condition of 10^{-6} - 10^{-7} Pa without ion pumping. In this study, we have experimentally manufactured a compact composite field emission scanning electron microscope (FE-SEM) and X-ray microscope (XRM) with a single isolated MWNT electron source. An electron beam from a MWNT was focussed and accelerated by a Butler electrostatic lens. Stable SEM images were observed at a Butler voltage V_{but} = 1.5 kV and accelerating voltage V_{acc} ≃ 15 kV with a focussed probe current I_{foc} = 61 pA. The spatial resolution of the SEM, which was estimated from 80 - 20 % edge profile of polystyrene latex spheres, was about 9 nm. XRM images were acquired by 30-min exposure at V_{but} = 1.5 kV and V_{acc} ≃ 17 kV. A spatial resolution of about 200 nm was obtained for XRM. The present results prove the high performance of the compact FE-SEM and XRM using a single isolated MWNT electron source at 10^{-7} Pa.

1. Introduction

Carbon nanotube (CNT) possesses various benefits as a field electron emitter, e.g., needle shape with a sharp tip, high chemical stability, mechanical strength and electrical conductivity. The sharp tip enables not only low voltage operation but also realisation of high brightness and restriction of emission area, which are advantageous for forming a finely focussed electron beam with a simple lens system. Chemical inertness and low atomic diffusion of a CNT surface make an electron emission stable even under moderate ultrahigh vacuum, which alleviates the need for a massive vacuum pumping system. By employing CNT as an electron source, therefore, a field emission scanning electron microscope (FE-SEM) can be miniaturised whilst keeping high performance.

FE from CNT have been studied for over 25 years [1]. Multi-walled CNT (MWNT) have exhibited high brightness of 10^9 - 10^{10} A/(cm^2•sr) as an electron source for FE-SEM [2]. Furthermore, the FE current is stable even under poor high-vacuum condition of 10^{-6} - 10^{-7} Pa without ion pumping [3]. In previously demonstrated CNT-based FE-SEM [4, 5], the residual gas pressure in their electron gun chambers was at a level of 10^{-3} Pa, being inadequate to maintain stable FE even for CNT. Thus, a ballast resistance of 10 - 20 MΩ had to be added in series to the emitter in order to stabilise the emission current. [4, 5]. In the scene of an archaeological site excavation and planetary exploration, for example, mobile microscopes are necessary to investigate samples on site.

In the present study, we have experimentally manufactured a compact dual-mode microscope of FE-SEM and X-ray microscope (XRM) with a single isolated MWNT electron source. The electron source and electrostatic lenses for extraction and focussing (Butler lens, indicated in figure 1b) are installed in an ultra-high vacuum chamber. A stable electron emission without any ballast resistance is demonstrated, and the performance of the SEM and XRM is reported.

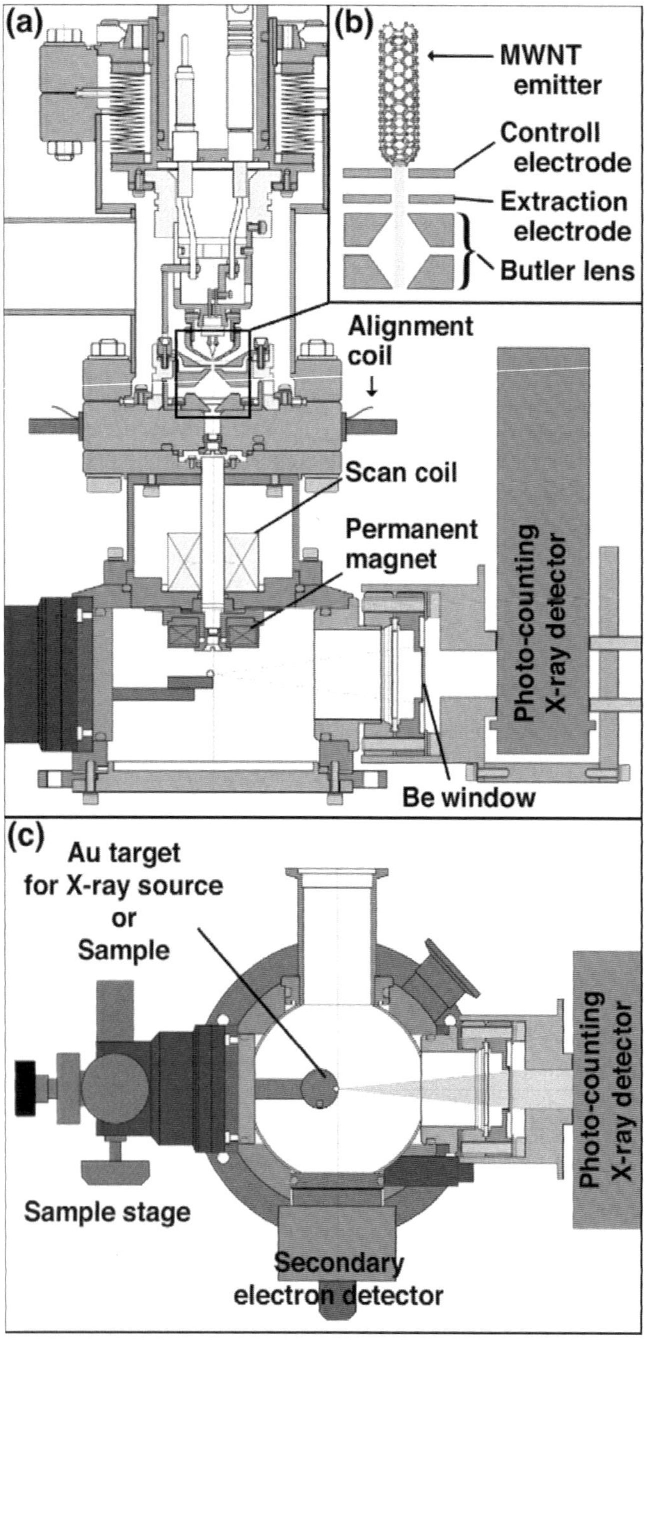

Figure 1. (a) Structure of the CNT-FE-SEM/XRM. (b) Electrostatic lens system. (c) Top view of sample chamber design.

2. Experimental

A drawing of the dual microscope equipped with a CNT FE emitter is shown in figure 1a. The CNT electron source and a set of electrostatic lenses for extracting and condensing an electron beam are installed in an ultra-high vacuum gun chamber, which can be baked and is evacuated by a turbo-molecular pump system. The gun/condenser lens chamber is connected, via an ICF 114 flange, with a column of a commercially-available compact SEM (Tiny-SEM, Technex Lab Co) in which scanning coils and an objective lens made of permanent magnets are installed. Figure 1c shows a schematic of the sample chamber. The Tiny-SEM sample-chamber is equipped with a secondary electron detector and is also employed. The electron source chamber and the sample chamber were differentially pumped with two turbo-molecular pumps. The ultimate pressure of the electron source chamber was 10^{-7} Pa. For measuring X-ray micrographs, a target made of Au is placed at the position where the electron beam is focussed and a sample is set between the X-ray source and the X-ray detector (XRI-UNO, XRAY IMATEK Co) as shown in figure 1c. The X-ray detector is a two-dimensional imaging device of 14.1 × 14.1 mm^2 with 256 × 256 pixels. The detector is based on a silicon sensor being effective for X-ray photons with energy in a range between 4 and 25 keV. A projection-type XRM based on the CNT-FE-SEM was constructed with the Au target for the X-ray source. The total length of the column from the top of the gun to the bottom of the sample chamber is 317 mm.

An electrostatic lens system, consisting of a control electrode, an extraction electrode and Butler lens, was specially designed for the CNT emitter as shown in figure 1b. The control electrode, covering the CNT emitter, was set at the same electric potential as the CNT emitter. The geometrical sizes and potentials of the electrostatic electrodes were optimised by paraxial approximation and by electron trajectories simulations (ELFIN). At an acceleration voltage of between 15 and 17 kV, an electron beam from a CNT emitter was focussed at an object plane of the magnetic objective lens, and then the focussed beam is projected on a sample plane. The CNT used was a MWNT produced by an arc discharge method. A single isolated MWNT, which was attached to the tip of a tungsten (W) needle using micromanipulators under an ordinary SEM observation [3], was used as the FE electron source. Figure 2 shows the MWNT electron sources used for SEM and XRM operation, respectively. The length of MWNTs was 150 and 30 nm extending from the W needle apex as shown in figures 2a and 2b, respectively.

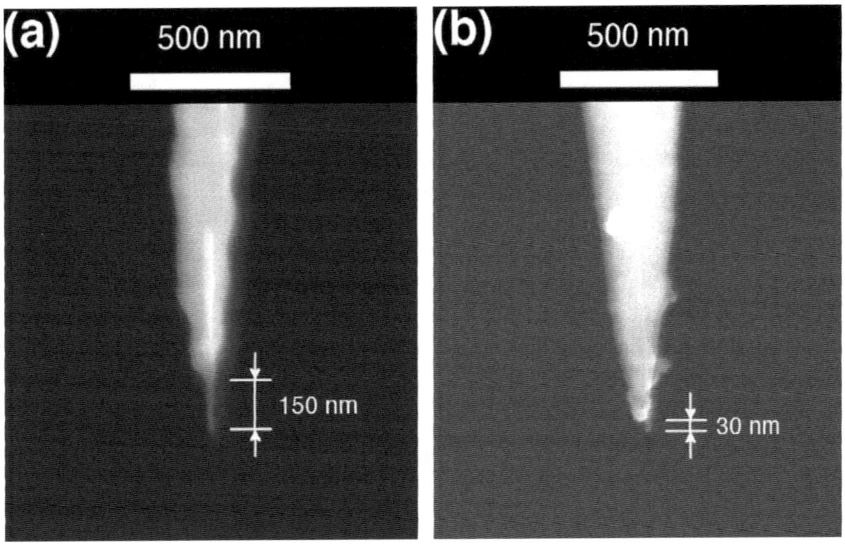

Figure 2. SEM images of CNT emitters used for a) SEM, and b) XRM operation, respectively.

IOP Conf. Series: Materials Science and Engineering **304** (2017) 012006 doi:10.1088/1757-899X/304/1/012006

3. Results and discussion

The MWNT electron source and Butler electrostatic lens were installed in the gun chamber, which was evacuated with a turbo-molecular pump down to 10^{-7} Pa. The CNT emitter operated stably without any ballast resistance. Insertion of a ballast resistance is not desirable, because the electric potential at the emitter tip fluctuates due to a voltage drop in the ballast resistance. Fluctuation of the emitter potential broadens the energy distribution of emitted electrons, which deteriorates the resolution of SEM images. In the present FE-SEM, such fluctuation of the emitter potential due to a ballast resistance is avoided. FE from a single isolated MWNT (figure 2a) was focussed and accelerated by a Butler electrostatic lens. FE current I_{emi} and focussed beam current I_{foc} (at Butler voltage V_{but} = 1.5 kV and accelerating voltage V_{acc} ≃ 15 kV) as a function of the extraction voltage V_{ext} are presented in figures 3a and 3b, respectively. The focussed beam current was measured by a Faraday cup. It has been experimentally observed that I_{foc} / I_{emi} was less than 0.01 %. High current electron beam is not necessary for achieving nano-resolution SEM images. The beam diameter becomes large when a high beam current is used. Thus, the resolution of SEM images depended on the current of the focussed beam. The most stable SEM images were observed at I_{foc} = 61 pA in our microscope.

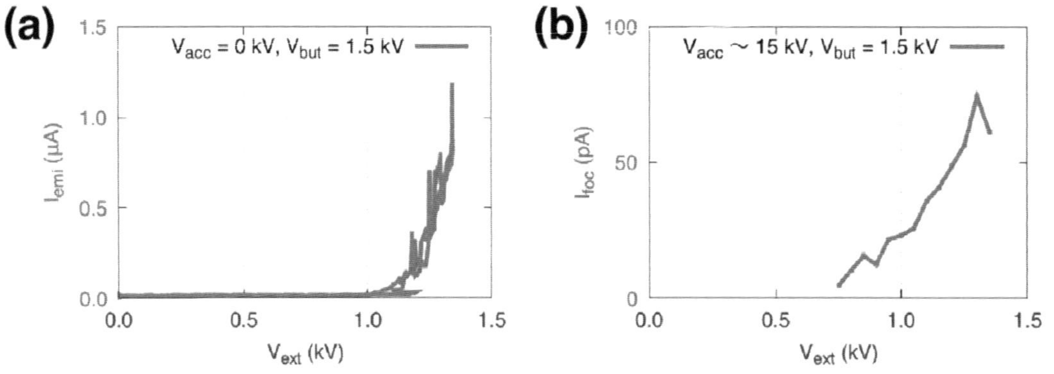

Figure 3. a) FE current, and b) beam current versus the extraction voltage characteristics.

An example of an SEM image acquired using I_{foc} = 61 pA at V_{but} = 1.5 kV and V_{acc} ≃ 15 kV is shown in figure 4a. The sample is Au-Pd coated polystyrene latex spheres (PLSs), which are 200 or 500 nm in diameter. Figure 4c shows the line profile across an edge of a PLS. The spatial resolution, which was estimated from the width at 80 - 20 % height of a fitting profile across the SEM image, was 9 nm, suggesting that the beam diameter was approximately 9 nm.

It should be noted that a jaggy-like noise is contained in the SEM image as shown in figure 4a. The period of noise was 60 Hz, indicating the noise is due to a power source. If the noise is removed, the resolution would be improved.

Next, we observed XRM images using the MWNT electron source shown in figure 2b. A typical XRM image is shown in figure 4b, which was acquired by 30-min exposure at V_{but} = 1.5 kV and V_{acc} ≃ 17 kV. The sample is 400-mesh Au grid. In contrast to the SEM image, the XRM image shows that the regions of high transmission appear bright. In the XRM image, the mesh hole appears bright and the bar appears dark. Figure 4d shows a line profile across the bar. The spatial resolution is estimated to be about 200 nm from the width at 80 - 20 % of an intensity change at the edge.

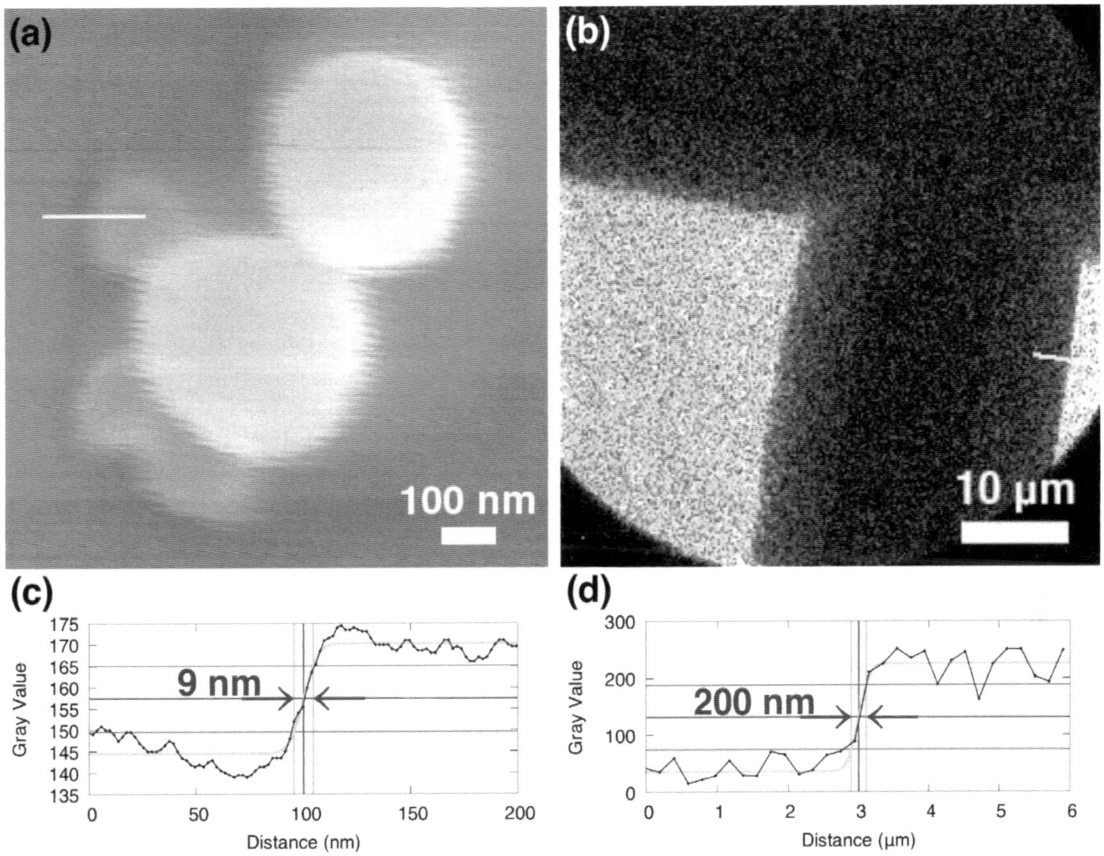

Figure 4. a) SEM image of Au-Pd coated PLS. b) XRM image of Au mesh. c) and d) Line profiles of the SEM (a) and XRM (b) images obtained at lines indicated in respective images. The solid green lines are the result of fitting using the cumulative distribution function of the standard normal distribution.

In our microscope, the electron beam spot size δ_e is considered to be approximately equal to the resolution of SEM, i.e., ~ 9 nm. The X-ray source size δ_X is estimated to be 220 nm for the Au target at 15 keV by using Castaing's equation [6]. Smearing due to Fresnel diffraction is expressed as $\delta_F \simeq \sqrt{a\lambda}$, where a is the distance between source and sample, a = 0.48 nm and λ is wavelength of X-ray. Using λ = 0.13 nm, which is the characteristic X-ray of Au-Lα_1, $\delta_F \simeq \sqrt{a\lambda} = 200$ nm. Therefore, the theoretical spatial resolution is $\delta = \sqrt{\delta_e^2 + \delta_X^2 + \delta_F^2} \sim 300\,nm$. The theoretical resolution is a little bit worse (1.5 times) than the experimental one, which may be due to the rough estimation of δ_X.

4. Conclusion

The present study has demonstrated the high performance of FE-SEM and XRM observation equipped with a single MWNT emitter operated at 10^{-7} Pa. The microscope has been miniaturised to a compact size that it can be placed on a desk. The present study shows a potential that the CNT-based SEM and XRM are applied to the scene of an archaeological site excavation and future planet search.

Acknowledgments
The authors also thank H. Murata, Meijo University for paraxial trajectory calculations and T. Ohno, Technex Lab Co. for valuable advice on modification of the Tiny SEM. The authors acknowledge the financial support from the Ministry of Education, Culture, Sports, Science and Technology of Japan (Grant-in-Aid for Scientific Research (B), No. 25286024).

References
[1] Saito Y, *et al.* 1997 Conical beams from open nanotubes. *Nature* **389** 554-555
[2] Nakahara H, *et al.* 2009 Evaluations of carbon nanotube field emitters for electron microscopy. *Appl. Surf. Sci.* **256** 1214-1217
[3] Nakahara H, *et al.* 2011 Carbon nanotube electron source for field emission scanning electron microscopy. *e-J. Surf. Sci. Nanotech.* **9** 400-403
[4] Suga H, *et al.* 2006 Stable multiwalled carbon nanotube electron emitter operating in low vacuum. *Surf. Interface Anal.* **38** 1763-1767
[5] Yabushita R, *et al.* 2007 Development of compact field emission scanning electron microscope equipped with multiwalled carbon nanotube bundle cathode. *J. Vac. Sci. Technol.* B **25** 640-642
[6] Castaing R 1951 Application of electron probes to local chemical and crystallographic analysis. PhD Thesis (Paris, France: University of Paris)

Focussed ion beam thin sample microanalysis using a field emission gun electron probe microanalyser

Y Kubo

Sumitomo Electric Industries, Ltd., Analysis Technology Research Center, 1-1-3, Shimaya, Konohana-ku Osaka-shi, 554-0024 Osaka, Japan

E-mail: kubo-yugo@sei.co.jp; ykubo2733@gmail.com

Abstract. Field emission gun electron probe microanalysis (FEG-EPMA) in conjunction with wavelength-dispersive X-ray spectrometry using a low acceleration voltage (V_{acc}) allows elemental analysis with sub-micrometre lateral spatial resolution (SR). However, this degree of SR does not necessarily meet the requirements associated with increasingly miniaturised devices. Another challenge related to performing FEG-EPMA with a low V_{acc} is that the accuracy of quantitative analyses is adversely affected, primarily because low energy X-ray lines such as the L- and M-lines must be employed and due to the potential of line interference. One promising means of obtaining high SR with FEG-EPMA is to use thin samples together with high V_{acc} values. This mini-review covers the basic principles of thin-sample FEG-EPMA and describes an application of this technique to the analysis of optical fibres. Outstanding issues related to this technique that must be addressed are also discussed, which include the potential for electron beam damage during analysis of insulating materials and the development of methods to use thin samples for quantitative analysis.

1. Introduction

In the last decade, field emission gun electron probe microanalysis (FEG-EPMA) has been rapidly adopted by laboratories worldwide [1]. An FEG provides superior spatial resolution (SR) for X-ray imaging compared with conventional W- or LaB_6-units because it allows for a smaller electron beam spot size, even if a low acceleration voltage (V_{acc}) is used [1-7]. In this review, the SR is defined as the lateral width of the X-ray generation region. However, the SR obtained when using a FEG-EPMA under typical measurement conditions (meaning the V_{acc} and probe current (I_{prob})) applied in an actual laboratory is typically in the range of approximately 150 to 200 nm [4]. This does not necessarily meet the recent demands for materials design in various fields of engineering, in which the sizes of internal structures of devices are being increasingly reduced. A related topic is the Er-doped optical fibre amplifier (EDF) [8]. As shown in figures 1a and 1b, an EDF generally consists of a core composed of SiO_2 glass doped with impurities surrounded by a cladding of pure SiO_2 glass [8]. The refractive index of the core relative to that of the cladding is increased by doping with Ge at a level of approximately 1 - 5 % (by mass), to confine the optical signals to the core (figure 1a). Approximately 1 - 5 % (by mass) Al is also added to the core to assist in controlling the deviation in refractive gain, one of most important product characteristics [8]. Figures 1c-e show backscattered electron (BSE) images of the EDF cross-section ((c: perspective, (d: around the core, and (e: around the centre of the core). The core is the bright circular area with a diameter of about 2 μm in figure 1d. A local dark

IOP Conf. Series: Materials Science and Engineering **304** (2017) 012007 doi:10.1088/1757-899X/304/1/012007

area with a diameter of 100 nm is observed in the centre of the core (figure 1e), which indicates the fine distribution of the dopants. Analysis of the Al and Ge levels in these cores requires a minimum detection limit (MDL) of less than 5,000 ppm in conjunction with an SR below 50 nm. In addition to the aforementioned insufficient SR, another challenge associated with FEG-EPMA is the low V_{acc} values that are often employed, which can adversely affect the accuracy of quantitative analyses [9]. When operating FEG-EPMA instrumentation at a low V_{acc}, it is typical to use low-energy X-ray lines such as the L- and M-lines. However, these lines are often affected by spectroscopic effects such as peak shifts and overlaps [9]. The fluorescence yields for the L- and M-lines are also lower than that for conventional K-lines [9].

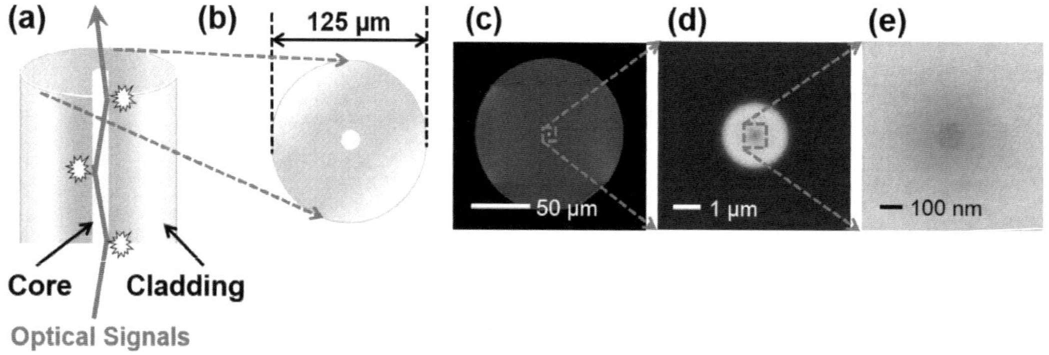

Figure 1. Diagrams of an EDF consisting of a core (SiO_2 glass doped with some impurities) surrounded by a cladding (pure SiO_2 glass). a) Side view, and b) cross-section. BSE images of an EDF cross-section showing c) the entire section, d) the region around the core, and e) the centre of the core. (Modified with permission from Ref. [14], copyright © Microscopy Society of America 2015.)

One approach to achieve a high SR while employing low energy X-rays is to thin the sample and also apply a high V_{acc} [10-14]. Figure 2 presents a schematic diagram showing SR improvement by sample thinning. As shown in the left diagram, the SR obtainable with conventional FEG-EPMA is poor when analysing bulk samples, since the area within which X-rays are generated is overly large. The SR is at best 150 - 200 nm under these conditions, even when using a low V_{acc} [4]. Thinned samples will increase the MDL because the characteristic X-ray intensity decreases as the sample is made thinner. Therefore, it is necessary to optimise both the measurement conditions and the sample thickness so as to balance the SR and MDL.

The general approach to the production of thin samples for electron beam microanalysis using energy-dispersive X-ray spectrometry (EDS) or wavelength-dispersive X-ray spectrometry (WDS) microanalysis, and the inherent advantages of this process with regard to data quantification have been known for quite some time. The basic concept appears to have been first suggested by Cliff and Lorimer's group in the 1970s [10, 15]. Subsequently, Joy *et al.* investigated the minimum detectable mass and minimum mass fraction for thin-sample X-ray analysis [16]. In the 1980s, Horita *et al.* proposed a method that does not require the sample thickness to be known for X-ray absorption corrections to be applied [17]. In addition, Watanabe's group reported the determination of the minimum detectable mass for electron-beam based EDS of thin samples at high acceleration voltages [18]. Thus, we consider that sample thinning will offer greater improvements in conjunction with FEG-EPMA than with conventional EPMA because the diameter of the electron beam can be made very small in the case of the former technique. Despite this, sample thinning as a preparation for FEG-EPMA has seldom been employed. This may be partly the result of the generally accepted view

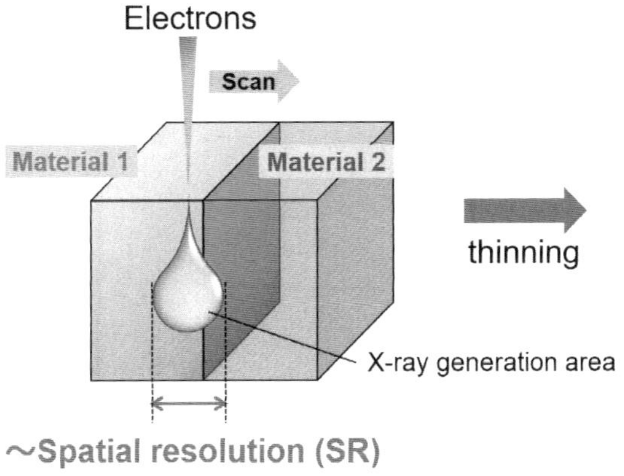

Figure 2. Schematic diagram of SR improvement by sample thinning. (Modified with permission from Ref. [13], copyright © 2015 Elsevier B.V. All rights reserved.)

that sample thinning significantly reduces the characteristic X-ray signals and, therefore, degrades the detection sensitivity. However, to the best of our knowledge, there have been no systematic investigations of the SR and MDL values associated with thin-sample FEG-EPMA. To address this issue, our own research group has been conducting studies related to thin-sample FEG-EPMA [12-14]. The purposes of this mini-review are to present a detailed examination of the potential SR and MDL values that can be obtained using this technique, to describe an example application to the analysis of an EDF, and to discuss remaining challenges. The present review also provides SR and MDL values determined for bulk samples when using FEG-EPMA and for thin samples via scanning transmission electron microscopy in conjunction with EDS (STEM-EDS). The latter technique is considered as an example of a more conventional microanalytical technique. In the work presented herein, sample thinning was performed using a Ga focussed ion beam (Ga-FIB) because this method allowed the fabrication of test specimens with positional accuracy on the nanometre scale. This technique was therefore well-suited to analysis of the EDF core at the centre of the optical fibre cross-section. The combination of Ga-FIB and EPMA has been previously demonstrated by Richter *et al.* [19-21]. They described a depth profile analysis of the structure of a multilayer system consisting of Al/SiO_2/GeSbTe on a SiO_2 substrate, during which they performed a line scan of a sample cross-section at an oblique angle prepared by FIB [19]. In another study [20], a similar technique was employed to examine the interfacial reactions within a multilayered NiAl-Hf-hBN system on a sapphire fibre. More recently, this same group described further improvement of the proposed bevel method by reducing the level of FIB artefacts and using microprobes with enhanced focussing optics [21].

2. Experimental methods

2.1. Sample preparation

InGaP films were grown on GaAs substrates using metal-organic vapour phase epitaxy. These InGaP/GaAs samples were subsequently thinned perpendicular to the layer structure, as shown in figure 2. STEM images confirmed that atomic diffusion across the InGaP/GaAs interface occurred over a distance of only several nanometres, and therefore these samples were suitable for evaluation of the SR.

2.2. *Ga-FIB*

The samples were thinned using Ga-FIB to thicknesses (T) of approximately 100, 130, 210, 310, and 430 nm, as confirmed by secondary electron images of the cross-sections [12, 13]. The sample mounting method has been described in detail in our previous report [12]. The cleaved cross-section of a bulk InGaP/GaAs sample was also analysed as a reference.

2.3. *FEG-EPMA*

X-ray line profiles and energy scans were obtained using an FEG-EPMA instrument (JEOL JXA-8530F). In each measurement, the working distance was approximately 11 mm, the take-off angle was 40° and In-Lα X-rays (critical excitation voltage: 3.73 keV) were employed. The radius of the Rowland circle was 100 mm (an H-type spectrometer with a JEOL microprobe) and the analysing crystal was made of pentaerythritol (PETH). The X-rays were collimated through a 0.5 mm wide slit and then detected using a sealed proportional counter filled with xenon. The measurement conditions included a probe current (I_{prob}) in the range of 5 - 200 nA, V_{acc} values of 30 kV (for thin samples), 5.5 or 10 kV (bulk samples), and a sampling time of 1 s per point. The electron beam was focussed. To assess the SR that could be obtained during FEG-EPMA, an In-Lα line profile was measured across the InGaP/GaAs interface. The electron beam was perpendicular to the plane of the thin samples and the line scan direction was perpendicular to the interface (figure 2). It should also be noted that the scan direction was set parallel to the PETH analysing crystal for all line scans so as to exclude X-ray defocusing effects.

2.4. *STEM-EDS*

In Lα X-ray line profiles across the InGaP/GaAs samples were acquired using a Si (Li) solid state detector (Genesis, EDAX) in conjunction with STEM (Hitachi, HD-2700). The sample was situated 2.43 mm below the objective lens and a take-off angle of 25.7° was employed. The distance between the sample and detector was 9.08 mm and the solid angle was 0.31 sr. The measurement conditions applied during STEM-EDS included an I_{prob} of 0.2 nA, a V_{acc} of 200 kV, a sampling time of 1 s per point, and a beam size of approximately 0.2 nm.

2.5. *Evaluation of the SR*

The SR values obtained with FEG-EPMA and STEM-EDS were defined as the distance over which the intensity of the In-Lα signal dropped from 84 % to 16 % of the difference between the average value in the InGaP layer and that in the GaAs substrate (figure 3a) [3, 22].

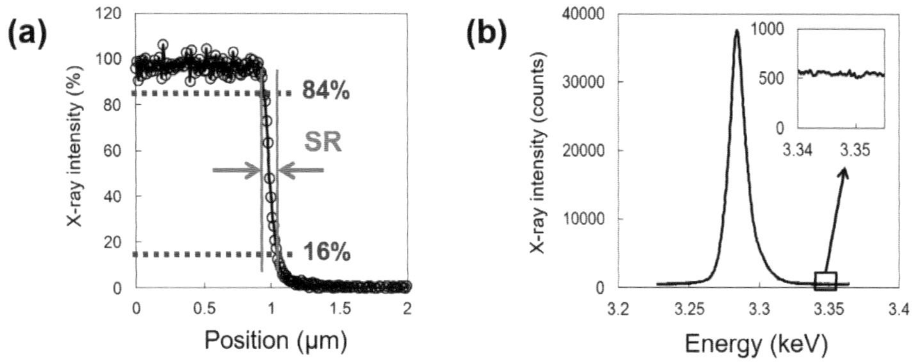

Figure 3. a) SR evaluated as the distance over which the intensity of the In-Lα signal drops from 84 % to 16 % of the difference between the average value in the InGaP layer and in the GaAs substrate. b) Energy scan of the In-Lα X-ray intensity for the InGaP layer. To evaluate the MDL values for thin-sample FEG-EPMA, the standard deviation (inset) in the area associated with the peak was calculated. (Modified with permission from Ref. [13], copyright © 2015 Elsevier B.V. All rights reserved.)

2.6. Evaluation of the MDL

The MDL values obtained from FEG-EPMA and STEM-EDS of In were evaluated based on the results obtained from energy scans of the InGaP layer using In-Lα X-rays. Firstly, the standard deviation *S*, for the background count associated with the peak [23] was calculated as depicted in figure 3b using the equation:

$$S = \left(\sum \frac{(Yi - Yave)^2}{n}\right)^{0.5}, \tag{1}$$

where *Y* is the X-ray signal at a given energy position in the energy scan and n is the number of data points. The *S*-values were converted from X-ray intensity (in counts) to concentration (in ppm (by mass)), where the conversion factor was determined as the ratio of the In-Lα peak intensity to the In concentration (known), assuming a linear relationship between the X-ray intensity and concentration. The MDL was then defined as equal to 3*S* (in ppm (by mass)). In the case of STEM-EDS, the background derived from continuous X-rays was subtracted prior to calculating *S*.

In addition, an EDF preform was prepared by conventional chemical vapour deposition [24] and formed into fibres that were subsequently used for SiO_2-based fibre fabrication. The EDF was cleaved and a flat, smooth cross-section was confirmed using an optical microscope. Thinning of the EDF sample was performed using Ga-FIB (Hitachi, FB2100) to a *T* of approximately 200 nm. The details of this preparation process have been fully described in our previous report [14].

3. Results and discussion

3.1. Variation of SR with V_{acc} in thin and bulk FEG-EPMA

Figures 4a and 4b show the In-Lα line profiles obtained from bulk and thin (*T* = 100 nm) InGaP/GaAs samples, respectively, at V_{acc} = 5.5, 10, 15, 20, 25, and 30 kV, and at an I_{prob} of 50 nA. Figure 4a indicates that the change in the In-Lα intensity across the InGaP/GaAs interface in the bulk sample is more gradual at higher V_{acc}. In contrast, the intensity change is sharper at higher V_{acc} in the case of the thin sample, as shown in figure 4b. Furthermore, at V_{acc} = 5.5 kV, the sharpness of the intensity change for the bulk sample is similar to that for the thin sample.

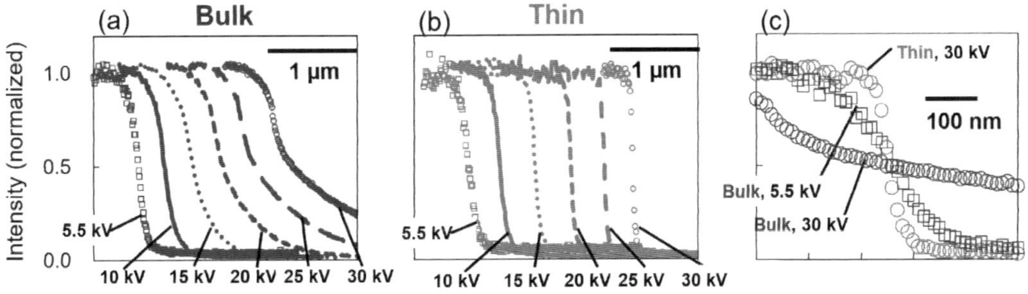

Figure 4. In-Lα line profiles across the InGaP/GaAs interface for a) bulk, and b) thin samples, obtained at V_{acc} of 5.5, 10, 15, 20, 25, and 30 kV at an I_{prob} of 50 nA. c) Results for a thin sample (V_{acc} = 30 kV), and bulk samples (V_{acc} = 30 kV and 5.5 kV). (Modified with permission from Ref. [12], copyright © 2013 Elsevier B.V. All rights reserved.)

To explain this contrast between the results for bulk and thin samples, the relationship among V_{acc}, the electron scattering area, and the SR is summarized in figure 5. The shape of the electron scattering area in the sample changes from hemispherical to droplet-shaped with an increase in V_{acc}, as depicted in the upper right panel of figure 5. This has been reported by many authors [25] and can be briefly confirmed by Monte Carlo simulation. Therefore, the lateral width of the electron scattering in the bulk sample is closely related to the SR, such that a higher V_{acc} results in a reduced SR. When working with thin samples, the SR is largely determined by the lateral size of the scattering area at the bottom of the thin film, as demonstrated in the lower right panel of figure 5. As a result, a higher V_{acc} value improves the SR. Thus, we can make the important conclusion that a high SR value is obtained when a higher V_{acc} is employed in the case of thin-sample FEG-EPMA, whereas a lower V_{acc} is necessary for bulk-sample FEG-EPMA. In addition, the use of a lower V_{acc} limits the detectable characteristic X-ray lines. Furthermore, at lower V_{acc}, the SR in thin-sample FEG-EPMA is similar to that in bulk-sample FEG-EPMA. This is because the width of the electron scattering area becomes as small as the sample thickness in thin-sample FEG-EPMA at lower V_{acc}, as shown in figure 5b.

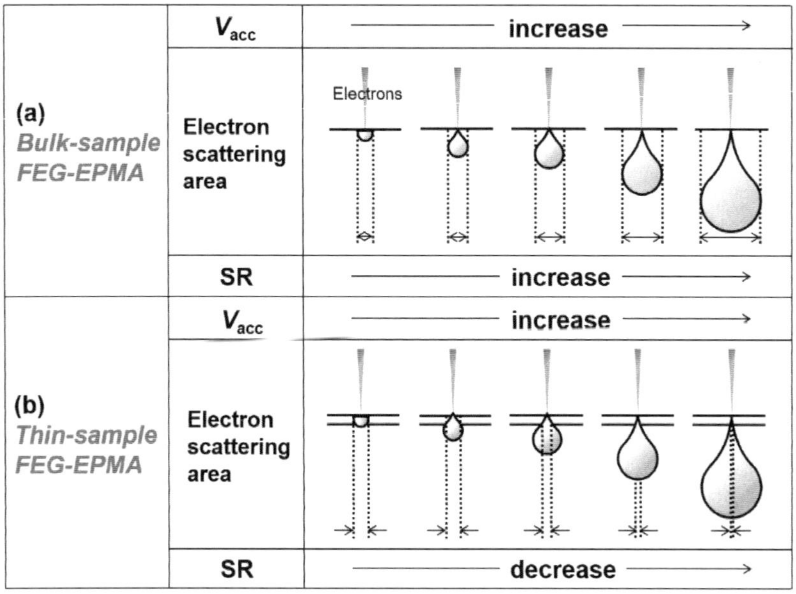

Figure 5. The V_{acc}, electron scattering area, and SR values for a) bulk- and b) thin-sample FEG-EPMA. (Modified with permission from Ref. [12], copyright © 2013 Elsevier B.V. All rights reserved.)

Figure 4c shows the line profiles for (1) a thin sample (V_{acc} = 30 kV), and (2) a bulk sample at V_{acc} = 30 and 5.5 kV. A comparison of (1) and (2) shows that the variation in the In-content across the InGaP/GaAs interface of the thin sample is significantly sharper than that in the bulk samples, which demonstrates enhancement of the FEG-EPMA SR as a result of sample thinning. In bulk-sample FEG-EPMA, the use of a lower V_{acc} enables higher resolution analyses; however, even bulk-sample FEG-EPMA at V_{acc} = 5.5 kV is inferior to thin-sample FE-EPMA at 30 kV in terms of the SR, as evident from a comparison of (1) and (3).

3.2. SR and MDL values for bulk- and thin-sample FEG-EPMA

Figure 6 summarizes the relationships between the SR and MDL for bulk- and thin-sample FEG-EPMA and STEM-EDS analyses of In within the InGaP layer of InGaP/GaAs samples. Based on the discussion in Section 3.1, V_{acc} was set to 30 kV for thin-sample FEG-EPMA in conjunction with T values of 100, 130, 210, 310, and 430 nm, while V_{acc} was 5.5 or 10 kV for bulk-sample FEG-EPMA. In all cases, I_{prob} values of 5, 20, 50, 100, or 200 nA were used. The results presented in figure 4 indicate that thin-sample FEG-EPMA simultaneously results in better SR and higher sensitivity, i.e., a lower MDL, compared to bulk-sample FEG-EPMA and STEM-EDS. The SR values obtained from thin-sample FEG-EPMA range from 40 to 350 nm and the MDL values from 13,000 to 600 ppm under the present measurement conditions. In addition, it is evident that both the SR and MDL can be tuned by varying *T*, and that *T* should be less than approximately 300 nm, because the advantages of thin-sample FEG-EPMA are lost as the samples become thicker. The SR and MDL values required for EDF analysis (50 nm and 5,000 ppm) are indicated by the region labelled "Requirements" in figure 6. Thin-sample FEG-EPMA partly met these requirements, while bulk-sample FEG-EPMA and STEM-EDS did not with an I_{prob} of 50 - 100 nA and *T* of 100 nm. Here we also note that it is important to consider the difference in density between InGaP/GaAs and impurity-doped SiO_2, and that the SR in the SiO_2 can be estimated via a Monte Carlo simulation in conjunction with the measured SR for InGaP/GaAs [12-14]. In Ref. [13], the SR was estimated and the MDL was measured for trace Cl-doped SiO_2 thinned to 140 nm. The SR values of Cl were 31, 33, 41, 55, and 113 nm for I_{prob} of 5, 20, 50, 100, and 200 nA, whereas the MDL values were 655, 491, 446, 392, and 224 ppm (by mass), respectively. These combinations of SR and MDL are plotted as "Thin-sample FEG-EPMA (Cl-doped SiO_2)" in figure 6.

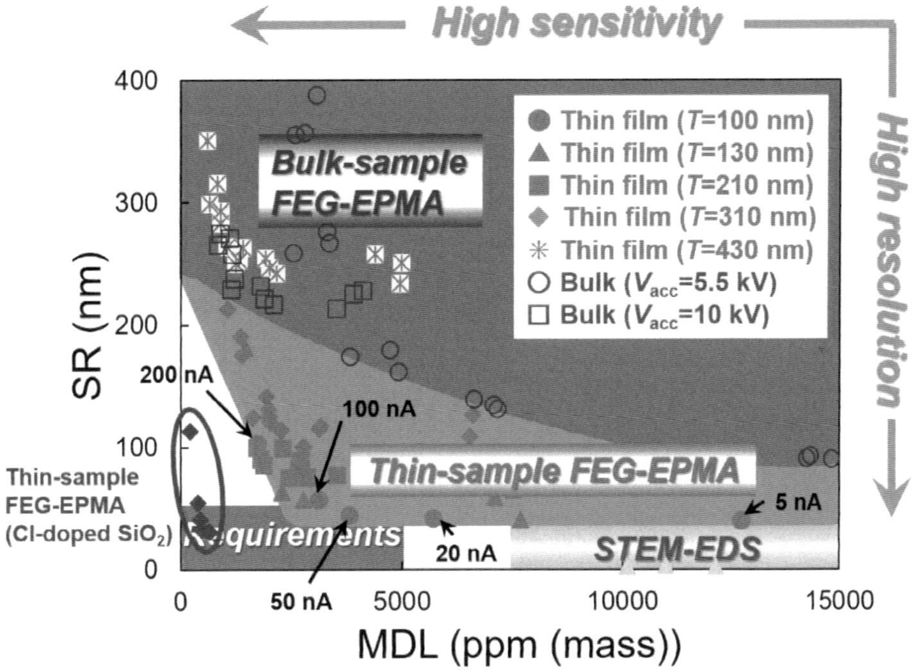

Figure 6. Relationships between SR and MDL for thin- and bulk-sample FEG-EPMA along with that for STEM-EDS. For thin-sample FEG-EPMA, V_{acc} was set to 30 kV and the *T* value ranged from 100 to 430 nm. For bulk-sample FEG-EPMA, V_{acc} was set to 5.5 or 10 kV. I_{prob} values were set to 5, 20, 50, 100, and 200 nA. For thin film ($T = 100$ nm), the I_{prob} indications were added in the figure. These data show that thin-sample FEG-EPMA allows combinations of SR and MDL that bulk-sample EPMA and STEM-EDS do not. (Modified with permission from Ref. [13], copyright © 2015 Elsevier B.V. All rights reserved.)

IOP Conf. Series: Materials Science and Engineering **304** (2017) 012007 doi:10.1088/1757-899X/304/1/012007

3.3. Variation of SR and MDL with T

At this point, we discuss two issues connected with thin-sample FEG-EPMA: the reason why a high SR can be obtained without significant degradation of the MDL and why the advantages of this method are not evident for samples thicker than 300 nm. The white circles, triangles, and squares in figure 7a show the relationship between *T* and the measured SR values for thin-sample FEG-EPMA (V_{acc} = 30 kV, I_{prob} = 5 - 50 nA). Monte Carlo simulations were also performed to examine the validity of the experimental SR values, as follows [13]. The SR is mainly determined by the sum of the diameter of the electron beam on the sample surface and the lateral width of the X-ray generation region in the sample [11]. We first calculated the lateral width of the area containing 70 % of the number of collisions between incident electrons and atomic electrons using Monte Carlo simulation for sample thicknesses of 100, 130, 210, 310, and 430 nm. Here, these calculated lateral widths are defined as *R*. For the sample thickness of 100 nm, the beam diameter on the sample surface was determined such that the sum of the beam diameter and the calculated *R* was equal to the measured SR. The beam diameter on the sample surface was determined to be independent of the sample thickness. In figure 7a, we show the sum of *R* for the respective sample thicknesses and the beam diameters together with the measured SR as solid circles, triangles, and squares. These data indicate that these sums are larger than the measured SR in the case of thicker samples. We attribute this result to the smaller lateral width of the characteristic X-ray generation area compared with that of the electron scattering area. In addition, both the measured and simulated SRs exhibit approximately linear correlations with *T* up to 210 nm, although this correlation drops off at 310 nm.

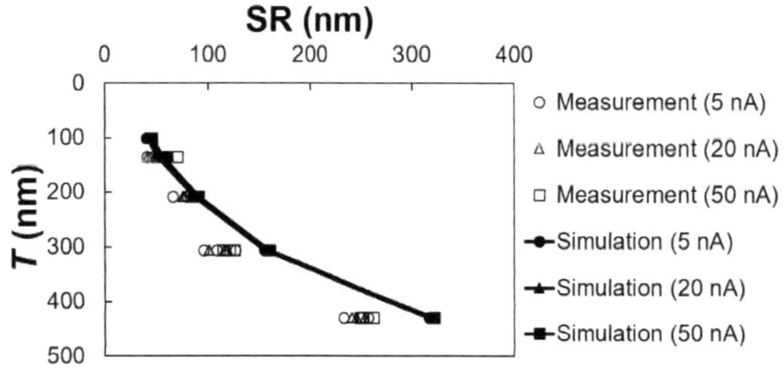

Figure 7. Relationship between *T* and SR for thin-sample FEG-EPMA. (Modified with permission from Ref. [13], copyright © 2015 Elsevier B.V. All rights reserved.)

To better explain these findings, figure 8 presents diagrams of the electron scattering regions in bulk and thin samples. The scattering area is droplet-like in the bulk sample, as depicted in figure 2. At depths below approximately 300 nm, the lateral width of the scattering area increases. For thin-sample FEG-EPMA, the lateral size of the scattering area at the bottom of the thin sample primarily determines the SR. Therefore, the degradation in SR with increasing *T* exceeds that expected from a linear relationship at large *T* (figure 8b).

Figure 9a plots the linear relationship between the X-ray intensity and *T* for I_{prob} values of 5 to 200 nA, based on the average In-Lα X-ray intensity along the line profile in the InGaP layer. Nomura *et al.* previously reported the relationship between V_{acc} and X-ray intensity for thin-sample FE-SEM-EDS analysis with various *T* [26]. A close examination of their plots reveals that the X-ray intensity was proportional to *T*, although this was not specifically identified by the authors. In the present study, figure 9b plots the relationship between the MDL and the reciprocal of the square root of the X-ray intensity for different combinations of *T* and I_{prob}. The data can be well fitted using the linear function y = 137836x + 172.89, i.e., the MDL is approximately proportional to the reciprocal of the square root of the X-ray intensity [27]. Considering the results in both figures 9a and 9b, it can be concluded that the MDL is also proportional to the reciprocal of the square root of *T*. The results of

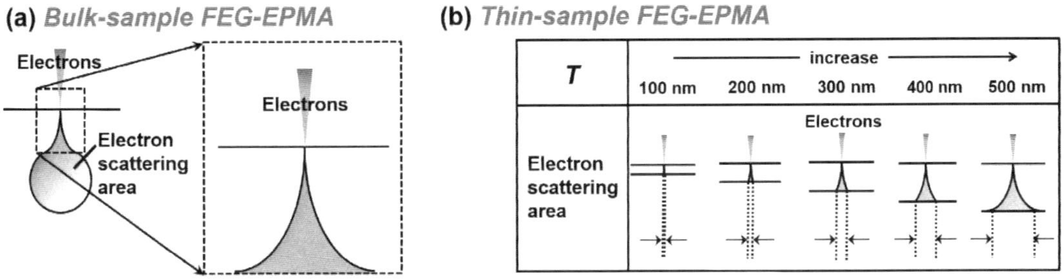

Figure 8. Schematic diagrams of the electron scattering areas for bulk- and thin-sample FEG-EPMA. a) Droplet-like scattering area for bulk-sample FEG-EPMA. b) Electron scattering area for thin-sample FEG-EPMA, which varies with *T*. At depths greater than approximately 300 nm, the lateral width of the scattering area increases more strongly. In the case of thin-sample FEG-EPMA, SR is largely determined by the lateral width of the scattering area at the bottom of the thin sample. Therefore, the degradation in SR with increasing T exceeds that predicted by a linear relationship. (Modified with permission from Ref. [13], copyright © 2015 Elsevier B.V. All rights reserved.)

this detailed investigation into the effect of *T* on the SR and MDL lead to several conclusions. Firstly, the SR value exhibits a linear dependence on *T*, whereas the MDL is inversely proportional to the square root of *T*. Therefore, thinning of a sample for FEG-EPMA allows the SR to be improved without any significant degradation of the MDL. In addition, the lateral width of the scattering area increases at depths greater than about 300 nm, such that an increase in *T* beyond 300 nm significantly reduces the SR without any appreciable degradation of the MDL. Based on the first point, thinning of the sample to less than 100 nm would be expected to significantly improve the SR without a significant increase in the MDL. However, reproducible fabrication of samples thinner than 100 nm is difficult, and raises concerns regarding sample deformation and non-uniform *T*. Furthermore, any damage to the sample is likely to have an effect on the FEG-EPMA measurement results for very thin samples.

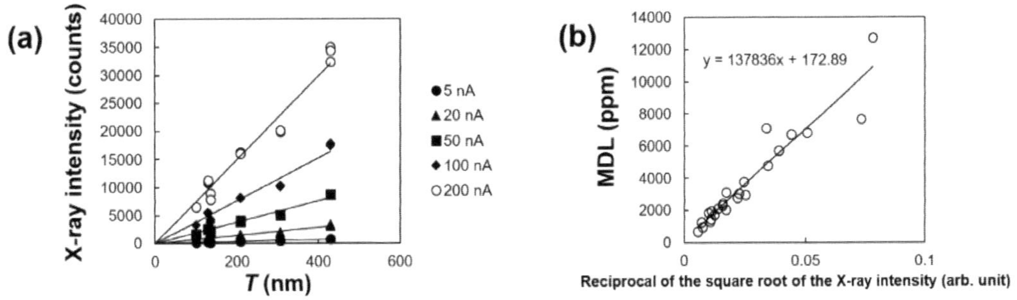

Figure 9. Relationship between a) X-ray intensity and *T* for thin-sample FEG-EPMA, and b) MDL as a function of the reciprocal of the square root of the X-ray intensity for various *T* and I_{prob}. (Modified with permission from Ref. [13], copyright © 2015 Elsevier B.V. All rights reserved.)

3.4. Thin-sample FEG-EPMA of EDF

Figure 10 shows line profiles obtained using a) bulk- and b) thin-sample FEG-EPMA around the EDF core, along with diagrams depicting the correlation between the size of the electron scattering area and the actual elemental distribution around the EDF core. Reductions at the centre of the core are clearly evident for both the Al and Ge intensities. The details of the sample preparation procedure and measurement conditions for the analysis of the EDF core are described in Ref. [14]. Prior to the FEG-EPMA measurements, the EDF samples were assessed using STEM-EDS to determine the distribution of dopants obtained via conventional microsampling. However, our experimental work demonstrates that the analysis of an EDF using conventional STEM-EDS is difficult due to both insufficient MDL values and potential sample damage [14].

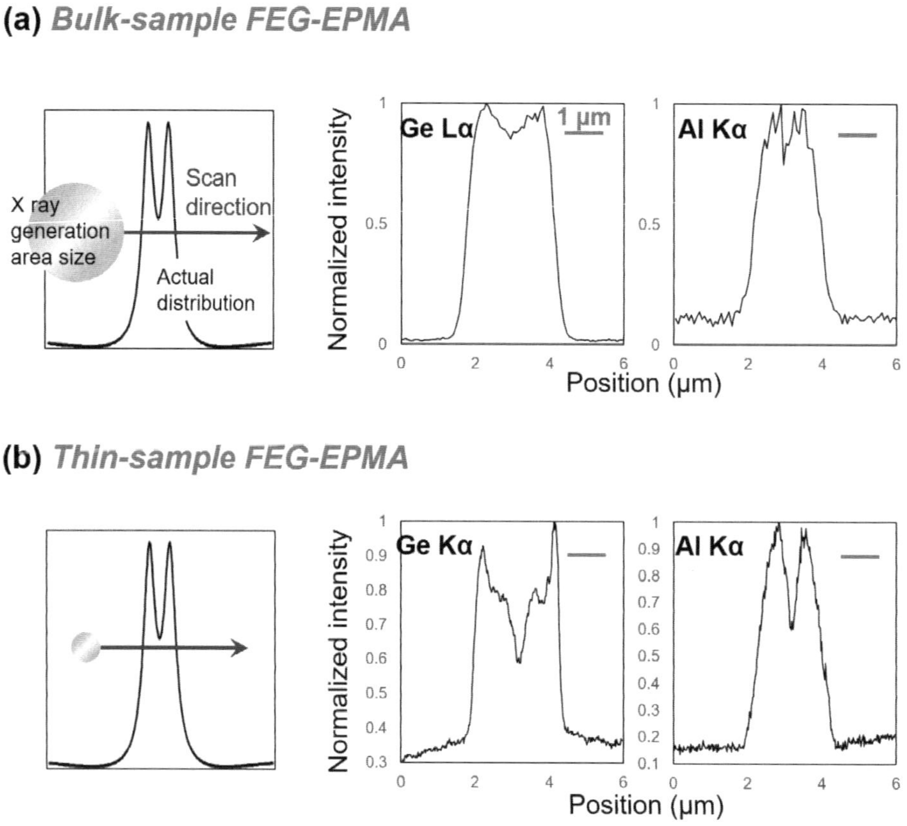

Figure 10. Comparison of line profiles for a) bulk-, and b) thin-sample FEG-EPMA analysis around the EDF core containing Er. The Y-axis is the X-ray intensity normalised with respect to the maximum. (Modified with permission from Ref. [14], copyright © Microscopy Society of America 2015.)

Finally, another advantage of thin-sample FEG-EPMA is discussed along with two important associated issues that require further thorough study. This additional advantage is related to the X-ray lines that are employed in the analysis. One of the most serious challenges in bulk-sample FEG-EPMA is the use of low V_{acc} during high resolution analyses, because a low V_{acc} can adversely affect the accuracy of quantitative work. When operating FEG-EPMA instrumentation at low V_{acc}, it is typical to use low-energy X-ray lines such as the L- or M-lines. In addition, the fluorescence yields for the L- and M-lines are lower than those for the conventional K-line. In contrast, the thin-sample FEG-EPMA technique described herein employs high V_{acc} that enhances the SR. As a result, it is possible to use high-energy X-ray lines for data quantification [9].

The first issue for further consideration is the potential for electron beam damage during the analysis of insulating samples [12]. Under typical measurement conditions, the doses applied in FEG-EPMA are lower than those used during STEM-EDS. The electron probe doses at the sample surface were calculated for both FEG-EPMA and STEM-EDS analysis and were estimated to be 6.37×10^5 C/cm^2 (STEM), 7.08×10^4 C/cm^2 (EPMA, I_{prob} = 5 nA) and 7.08×10^3 C/cm^2 (EPMA, I_{prob} = 200 nA), assuming respective probe sizes of 0.2, 3 and 60 nm. On the basis of these simple estimations, it can be assumed that (1) sample damage due to electron beam radiation will be considerably smaller in the case of FEG-EPMA than for STEM-EDS based on the relative doses, and (2) a larger I_{prob} results in less sample damage during FEG-EPMA. Nevertheless, electron beam damage can become a serious problem because much higher V_{acc} and I_{prob} are required to maximise the advantages of thin-sample FEG-EPMA. In our previous experiments [14], application of V_{acc} at 30 kV and I_{prob} at 200 nA resulted in serious plastic deformation of the sample due to a local rise in temperature by electron beam irradiation. Therefore, it is apparent that novel sample preparation techniques will be required to limit sample damage during thin-sample FE-EPMA. We have reported one such technique in a previous paper [14], applied to the analysis of the EDF core described in this review.

The second challenge is related to the development of methods to use thin samples for quantitative analysis, just as techniques for the quantitative microanalysis of EDS and WDS samples have been widely researched for several decades [28, 29]. In the case of thin-sample FEG-EPMA, the quantification techniques used for bulk samples cannot be directly transferred, because the matrix effect is dependent on T, as with the chemical composition. If calibration standards can be prepared with the same T as the sample to be analysed, then quantitative analysis can be readily performed, otherwise such analysis is difficult. Despite these two challenges, we consider that thin-sample FEG-EPMA is a very promising technique for the analysis of various materials because these two issues have been addressed in a number of individual laboratory studies.

It should be noted that we have omitted the contributions of Monte Carlo simulations to this field of research in the present review for reasons of space. However, such simulations were performed to a significant extent and were found to be extremely helpful to confirm the validity of the experimental results. The interested reader is referred to our previous reports [13, 14] for details of these simulations.

4. Conclusions

This mini-review described the SR and MDL values obtained when employing FEG-EPMA. InGaP samples were thinned to approximately 100, 130, 210, 310, and 430 nm to allow for effective thin-sample FEG-EPMA in the resolution range of 40 - 350 nm and the MDL range of 13,000 - 600 ppm (by mass). It is clear that thin-sample FEG-EPMA can achieve both lower MDL and better SR values than are possible with bulk-sample FEG-EPMA or STEM-EDS. This is because the majority of the X-rays that determine the MDL are generated in the surface region of the sample to depths of about 300 nm during FEG-EPMA. The SR and MDL can also be tuned by variation of the I_{prob} and T in accordance with the analytical purpose. Finally, thin-sample FEG–EPMA was applied to the analysis of dopant distributions in an EDF core. Using this technique, the Al and Ge distributions in an EDF could be observed. Thus, the present technique is suitable as a laboratory technique for microstructural analysis.

Acknowledgements
The author wishes to thank the following individuals for fruitful discussions: Prof. J. Fujita and Prof. T. Kizuka of the University of Tsukuba, Dr. H. Takahashi and Mr. N. Mori of JEOL, Ltd., and Mr. K. Hamada, Dr. K. Yamaguchi, Dr. J. Iihara and Mr. S. Matsukawa of Sumitomo Electric Industries, Ltd. The author also thanks the following individuals for technical support during this work: Mr. K. Sugawara of the Foundation for the Promotion of Material Science and Technology of Japan, and Dr. K. Kuramochi, Mr. T. Haruna, Mr. K. Sasaki, and Mr. K. Hase of Sumitomo Electric Industries, Ltd.

References
[1] Rinaldi R and Llovet X 2015 *Microsc. Microanal.* **21** 1053
[2] Kimura T, Nishida K and Tanuma S 2006 *Mikrochim. Acta* **155** 175
[3] Berger D and Nissen J 2014 *IOP. Conf. Ser.: Mater. Sci. Engng.* **55** 012002
[4] McSwiggen P 2014 *IOP Conf. Ser.: Mater. Sci. Engng.* **55** 012009
[5] Pinard P T and Richter S 2014 *IOP Conf. Ser.: Mater. Sci. Engng.* **55** 012016
[6] Fournelle J, Cathey H, Pinard P T and Richter S 2016 *IOP Conf. Ser.: Mater. Sci. Engng.* **109** 012003
[7] Kubo Y and Yonezawa K 2017 *Anal. Chem.* **89** 8772
[8] Haruna T, Iihara J, Yamaguchi K, Saito Y, Ishikawa S, Ohnishi M and Murata T 2006 *Opt. Express* **14** 11036
[9] Merlet C and Llovet X 2012 *IOP Conf. Ser.: Mater. Sci. Engng.* **32** 012016
[10] Lorimer G W, Cliff G and Clark J N 1975 in: *Developments in electron microscopy and analysis. EMAG 75.* (Venables J A ; ed.) (London: Academic Press) 153
[11] Goldstein J I, Costley J L, Lorimer G W and Reed S J B 1977 *Proc. of the Workshop on Analytical Electron Microscopy, SEM 1977,* vol. 1 (Chicago: IIT Research Institute) 315
[12] Kubo Y, Hamada K and Urano A 2013 *Ultramicroscopy* **135** 64
[13] Kubo Y and Hamada K 2015 *Ultramicroscopy* **157** 48
[14] Kubo Y and Kuramochi K 2015 *Microsc. Microanal.* **21** 1398
[15] Champness P E, Cliff G and Lorimer G W 1982 *Ultramicroscopy* **8** 121
[16] Joy D C and Maher D M 1977 *Proc. of the Workshop on Analytical Electron Microscopy, SEM 1977,* vol. 1. (Chicago: IIT Research Institute) 325
[17] Horita Z, Sano T and Nemoto M 1987 *Ultramicroscopy* **21** 271
[18] Watanabe M and Williams D B 1999 *Ultramicroscopy* **78** 89
[19] Richter S, Bückins M, Aretz A, Kyrsta S, Spähn M and Mayer J 2004 *Microchim. Acta* **145** 187
[20] Richter S, Kyrsta S, Schneider J, Hajas D and Mayer J 2006 *Microchim. Acta* **155** 257
[21] Richter S and Pinard P T 2016 *IOP. Conf. Ser.: Mater. Sci. Engng.* **109** 012014
[22] Barkshire I, Karduck P, Rehbach W P and Richter S 2000 *Mikrochim. Acta* **132** 113
[23] Japanese Industrial Standards Committee 2012 JIS K0114:2012, General Rules for Gas Chromatography (Tokyo: Japanese Standards Association)
[24] French W G, Pearson A D, Tasker G W and Macchesney J B 1973 *Appl. Phys. Lett.* **23** 338
[25] For example, Duncumb P and Shields P K 1963 *Br. J. Appl. Phys.* **14** 617
[26] Nomura S, Todokoro H and Komada T 1975 in: *Scanning electron microscopy 1.* (Johari O and Corvin I; eds.) (Chicago: IIT Research Institute) 103
[27] Jenkins R, Gould R W and Gedcke D 1981 *Quantitative X-ray spectrometry.* (New York: Marcel Dekker, Inc.)
[28] Cliff G and Lolimer G W 1975 *J. Microsc.* **103** 203
[29] Watanabe M and Williams D B 2006 *J. Microsc.* **221** 89

IOP Conf. Series: Materials Science and Engineering **304** (2017) 012008 doi:10.1088/1757-899X/304/1/012008

Microscopic stress characterisation of functional iron-based alloys by white X-ray microbeam diffraction

E P Kwon[1], S Sato[2], S Fujieda[3], K Shinoda[3], K Kajiwara[4], M Sato[4] and S Suzuki[3]

[1] Korea Institute of Industrial Technology, Carbon & Light Materials Application R&D Group, 54853 Jeonju, Republic of Korea
[2] Ibaraki University, Graduate School of Science and Engineering, 316-8511 Hitachi, Japan
[3] Tohoku University, Institute of Multidisciplinary Research for Advanced Materials, 980-8577 Sendai, Japan
[4] Japan Synchrotron Radiation Research Institute, SPring-8, 679-5198 Hyogo, Japan

E-mail: ssuzuki@tagen.tohoku.ac.jp

Abstract. Microscopic residual stress evolution in an austenite (γ) grain during a shape-memory process in an Fe–Mn–Si–Cr alloy was investigated using the white X-ray microbeam diffraction technique. The stresses were measured on a coarse grain, which had an orientation near <144>, parallel to the tensile loading direction with a high Schmid factor for a martensitic transformation. The magnitude of the residual stresses in a grain of the sample, which was subjected to a 23 % tensile strain and subsequent shape-recovery heating, was found to be very small and comparable to that prior to tensile deformation. Measurements of the recovery strain and microstructural analyses using electron backscatter diffraction suggested that the low residual stresses could be attributed to the significant shape recovery caused by a highly reversible martensitic transformation in the grain with a particular orientation.

1. Introduction

Iron-based alloys are utilised not only for structural materials but also for functional materials. These functional materials include Fe–Mn–Si-based shape-memory alloys [1], in which the shape-memory effect appears through a stress-induced martensitic transformation by deformation and a reverse transformation by subsequent heating. It is known that the shape-memory effect depends on the geometrical relationship between the crystallographic orientation of grains and the external loading direction [2]. The orientation dependence of the shape-memory effect is observed in both single crystals and polycrystalline alloys [1, 2].

Recent studies [3-5] have suggested that residual stress developed during a stress-induced martensitic transformation is an important factor contributing to the improvement of the shape-memory effect. Tomota *et al.* [6] insisted that the effects of solid-solution hardening and training treatments in shape-memory alloys can be understood from the viewpoint of back stress, i.e., internal stress built up by a stress-induced γ (fcc) → ε (hcp) martensitic transformation. According to a study by Tomota and Maki [4], the back stress forms in the γ phase around the growing ε martensite plate tips, and it serves an important role in promoting the reversible motion of the partial dislocation in the shape-recovery process on heating, thereby improving the recovery strain of the alloys. The evolution of internal stress would depend on the grain orientation because of the orientation dependence of the martensitic transformation. Although the macroscopic strain change has been commonly examined to evaluate the shape-memory effect, almost no systematic studies on the evolution of microscopic stress in iron-based shape-memory alloys have been performed thus far. For a more comprehensive understanding of the relationship between the residual stress and shape-memory effect, the evolution of residual stress during the shape-memory deformation process must be characterised.

Recently, white X-ray microbeam diffraction has emerged as a powerful tool for measuring the microscopic stress within grains [7-9]. The recently developed microstructure visualisation technique, coupled with energy spectra measurements with a solid-state detector at the BL28B2 beamline of SPring-8, enables the measurement of strain/stress on the local area of interest in materials [10]. In our previous study, this technique was applied to investigate the stress in individual grains in a Cu-Al-Mn superelastic alloy, and the results showed an inhomogeneous stress distribution, both at the granular and intragranular scales [11]. This technique was also applied to analyse the stress distribution in Fe–Mn–C twinning-induced plasticity steels [12].

In the present study, we investigated the microscopic stress evolution within a grain during the shape-memory behaviour of an Fe–Mn–Si–Cr shape-memory alloy using the white X-ray microbeam diffraction technique. The diffraction experiments were performed on a grain with a specific crystallographic orientation undergoing a highly reversible martensitic transformation.

2. Experimental procedures

2.1. Materials

The material used in this study was a polycrystalline Fe-28%Mn-6%Si-5%Cr (mass%) shape-memory alloy produced by centrifugal casting [13]. The microstructure of the alloy consisted of large columnar grains. The alloy was annealed at 1,000 °C for 8 h to coarsen the grains. Small tensile samples were cut from the cast along a specific direction so that the gauge part contained a coarse columnar grain with a specific crystallographic orientation. The tensile samples were additionally annealed at the austenite temperature range (900 °C) for 30 min to eliminate any deformed microstructures formed during sample preparation. To use the samples in white X-ray microbeam diffraction experiments, the thickness (t) was reduced to about 200 μm by electropolishing with a solution of acetic acid and 20 vol% perchloric acid. The diffraction experiments were carried out on a measurement area of 1.5 mm × 0.5 mm in the sample gauge region (3.0 mm × 0.6 mm), as shown in figure 1a.

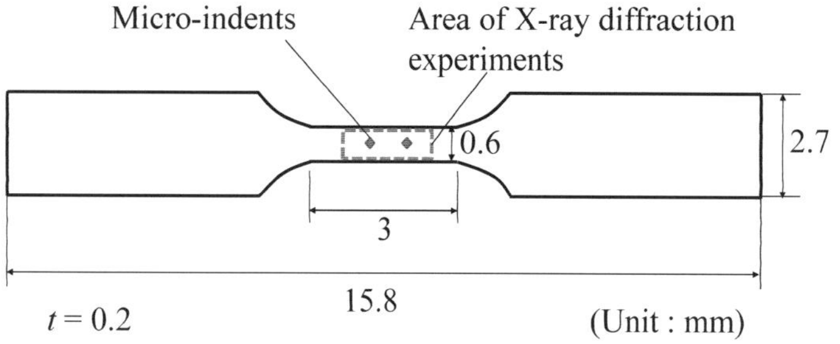

Figure 1. Schematic of tensile sample used for X-ray microbeam diffraction experiments.

For recovery strain measurements, two marks with an initial distance of 1 mm were indented on the gauge part of the sample with microindentation. The recovery strain was calculated by measuring the change in the distance after deformation and recovery heating. Recovery heating was conducted at 350 °C for 10 min.

2.2. White X-ray microbeam diffraction experiments

White X-ray microbeam diffraction experiments were conducted using high-energy synchrotron radiation at the SPring-8 BL28B2 beamline (Hyogo, Japan). Figure 2 shows a schematic diagram of the X-ray microbeam diffraction experiments. The incident beam with a size of 25 × 25 μm was illuminated on the tensile sample and the diffracted beams generated from the tensile sample in the transmission (Laue) geometry were collected on the image detector. X-ray energy spectra for several Laue reflection regions were measured using a Ge solid-state detector. The obtained X-ray energy spectra were used to calculate the lattice spacing of the (*hkl*) planes, d_{hkl} [10]. White X-ray microbeam diffraction experiments were performed during a shape-memory cycle, i.e., before deformation (0 % strain) → tensile deformation (23 % strain, unloaded) → after recovery heating. Details of the experimental setup and experimental procedures are given in reference [15].

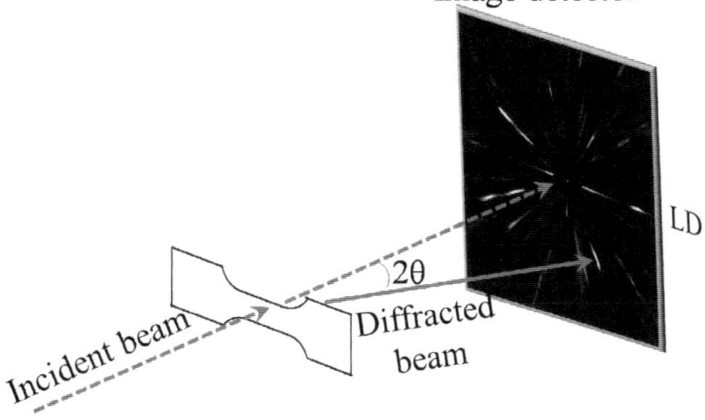

Figure 2. Schematic diagram showing the X-ray microbeam diffraction experiments.

Prior to the X-ray diffraction experiments, the microstructure of the sample was investigated using electron backscatter diffraction (EBSD). Then, the sample with a particular grain orientation was selected to use in the X-ray diffraction experiments. EBSD patterns were obtained using a Nordlys II EBSD detector mounted on a Hitachi SU-6600 scanning electron microscope operating at an acceleration voltage of 20 kV. EBSD data were processed using HKL CHANNEL 5 Flamenco software to obtain the microstructural information. To analyse the microstructural evolution during the shape-memory process, the microstructure of the sample, which was used in the X-ray diffraction experiments, was observed at some strain levels and after recovery heating.

The sample used in the present study had an orientation near <144>, parallel to the loading direction (LD) in tensile tests (<144>//LD), which is among the most preferential orientations for a martensitic transformation. The EBSD orientation map of figure 3a, taken from the undeformed gauge region of the tensile sample, shows that the microstructure consisted of a coarse γ grain with a specific orientation as confirmed by the green colour. Low-angle grain boundaries with misorientations less than 10° were observed, indicating the presence of sub-grains. The image (figure 3b) obtained by X-ray microbeam diffraction from the similar area with the EBSD orientation image shows a somewhat blurred structure because of the presence of sub-grains. The X-ray diffraction experiments were performed on three positions as marked with P1, P2, and P3 in figure 3b.

The EBSD inverse pole figure shown in figure 4a indicates that the grain is oriented near <144>//LD, which has a high Schmid factor (value of 0.5) for a martensitic transformation, as shown in the contour map of figure 4b. We therefore expect a martensitic transformation to be very favourable in that grain. In addition, the transformation strain by martensitic strain is as large as 19 % in this orientation, as shown in figure 4c, meaning that a significant elongation would be obtained by tensile deformation of the sample.

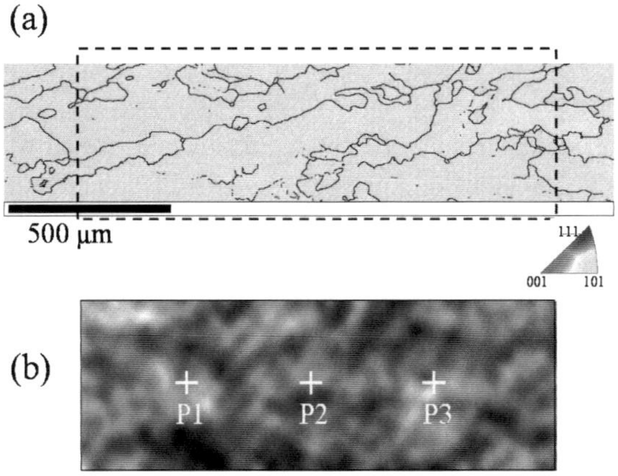

Figure 3. Image showing the microstructure of the tensile sample. a) Orientation image obtained by EBSD. b) Image of the area marked by a rectangle in figure 3a obtained by white X-ray diffraction. Crosses (+) in (b) denote stress measurement points.

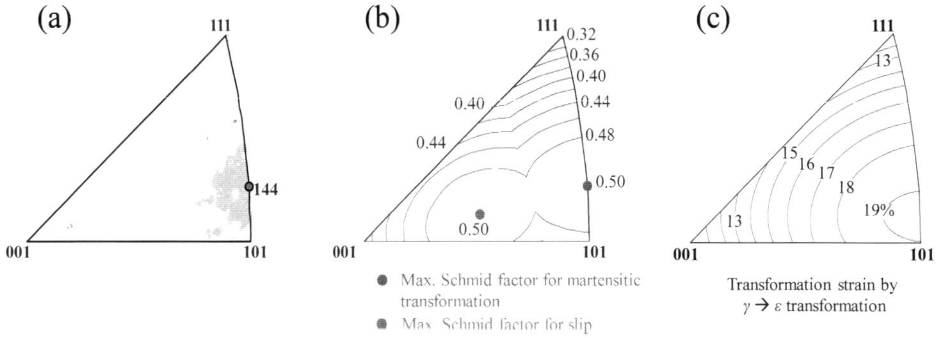

Figure 4. a) EBSD inverse pole figures showing the orientation of the grain in the tensile direction. Stereo-triangles showing b) the Schmid factor, and c) the transformation strain of grain orientation in the tensile direction.

3. Results and discussion

Figure 5 shows the variation of the Laue patterns during a shape-memory cycle obtained at the position marked with P1 in figure 2. Laue spots at 23 % strain show pronounced streaking. Laue spots often display streaked peaks because of the existence of geometrically necessary dislocations and deviation of the crystal orientation [16, 17]. The streaking of Laue spots in this alloy could be mainly attributed to the formation of stress-induced ε martensite. After recovery heating, the Laue spots nearly recovered their initial shape, although the incomplete restoration of their original shape may be due to residual martensite and slip deformation. In shape-memory alloys, the reversible change of the Laue patterns is most likely to be associated with a reversible martensitic transformation during the shape-memory process.

Laue spots were indexed based on their measured lattice spacing values using a stress calculation programme [18]. All of the Laue spots at 0 % strain were indexed to lattice planes of the γ phase, as shown in figure 5a, which is in good accordance with the EBSD phase analysis of the sample. At 23 % strain, Laue spots are mostly indexed to the lattice planes of ε martensite, although some are indexed as γ phase, suggesting that γ phase is almost transformed to ε martensite at this strain. Laue spots indexed as ε martensite were not observed after recovery heating, indicating that the ε martensite on the deformed microstructure reverse-transformed to γ phase after recovery heating. Although some residual martensite can remain in the microstructure after recovery heating, its detection is considered to be difficult due to its low quantity.

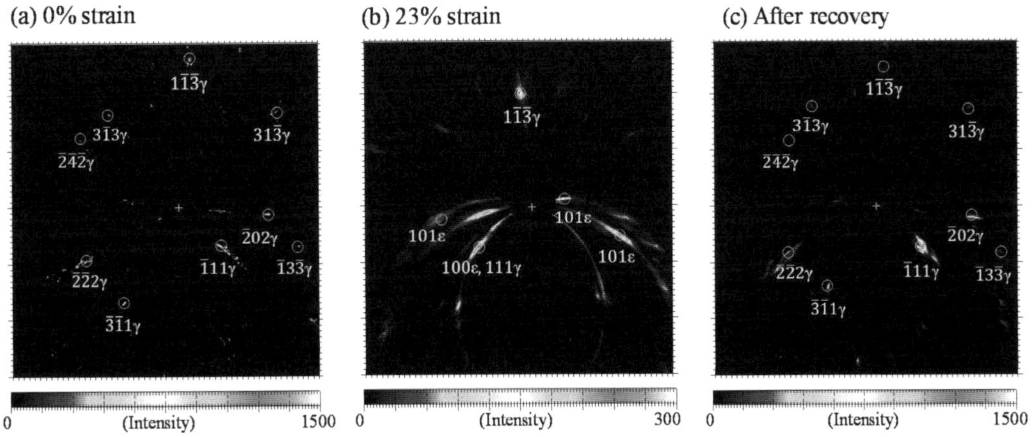

Figure 5. Laue patterns obtained from a point marked by P1 in figure 3b. a) Before deformation; b) at 23 % strain; and c) after recovery heating. Laue spots were indexed based on their lattice spacing values.

The recovery strain of the sample that was subjected to a 23 % strain deformation and subsequent recovery heating is shown in figure 6. For comparison, the recovery strain of a conventional polycrystalline Fe–Mn–Si shape-memory alloy is also presented [2]. It shows that a recovery strain of about 2 % is achievable in a conventional polycrystalline Fe-based shape-memory alloy. Note that a significantly higher recovery strain of 17 % was obtained in this sample. The outstanding shape-recovery properties are attributed to its coarse grain structure with a characteristic orientation of <144>, where a martensitic transformation is favoured and a large elongation is obtained.

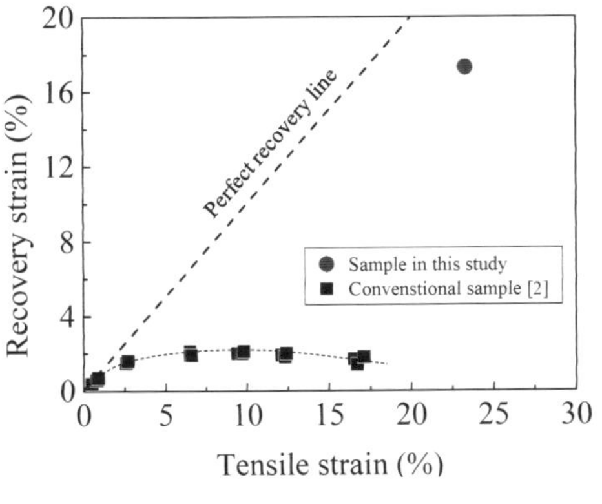

Figure 6. Recovery strain of the tensile sample in the present study and a conventional shape-memory alloy [2].

According to Otsuka *et al.* [19], the amount of martensite is related to the recovery strain as expressed by equation (1):

$$\text{Recovery strain (\%)} = TS \times (V\varepsilon/100)f \qquad (1)$$

where TS is the transformation strain, $V\varepsilon$ is the volume fraction of stress-induced martensite, and f is a factor ($0 < f \leq 1$) that depends on the polycrystalline condition determined by thermal or mechanical treatment. According to eq. (1), the recovery strain is proportional to the volume fraction of

IOP Conf. Series: Materials Science and Engineering **304** (2017) 012008 doi:10.1088/1757-899X/304/1/012008

stress-induced martensite. In particular, the maximum recovery strain can be obtained when the reverse transformation of martensite is complete, i.e., when f is equal to 1. A maximum recovery strain of 19 % can be obtained under the conditions, $f = 1$ and $x = 100$, according to eq. (1). The strain level of 17 % obtained in this sample is fairly large considering a maximum achievable recovery strain of 19 %. This means that the martensite in this sample has a good reverse-transformation ability.

Laue patterns of deformed shape-memory alloys would be composed of many spots that are generated from both the γ phase and stress-induced ε martensite. The presence of ε martensite can be verified by measuring the lattice spacing of the Laue spots. Lattice planes were manually indexed based on the measured lattice spacing and the results of the examples are shown in figure 7. The lattice spacing was calculated from the obtained energy profile of the respective Laue spots according to Bragg's law [14]. Obviously, the lattice spacing profile of figure 7b obtained from the energy profile of figure 7a shows a peak corresponding to the (101) plane of ε martensite. The energy profile sometimes presents two distinct energy peaks, as exemplified in figure 7c. In that case, the measured lattice spacing needs to be compared with the anticipated lattice spacings for γ phase and ε martensite for accurate indexing. Otherwise, accurate measurements of lattice strains/stresses cannot be made due to mis-indexing. As shown in figure 7d, the two peaks are referenced to the (111) plane of γ phase and (100) plane of ε martensite based on the measured lattice spacing, indicating the co-existence of the two phases at 23 % strain. It also means that the overlap of Laue spots occurs during diffraction, making automatic indexing using the calculation programme very difficult. For this reason, clearly distinguishable non-overlapped Laue spots of high intensity were used for strain/stress measurements in this study. The strain/stress was measured only for the γ phase because the calculation programme used is only capable of measurements for γ phase with an fcc structure at present. Development of the calculation programme for strain/stress measurements of hcp martensite is in progress.

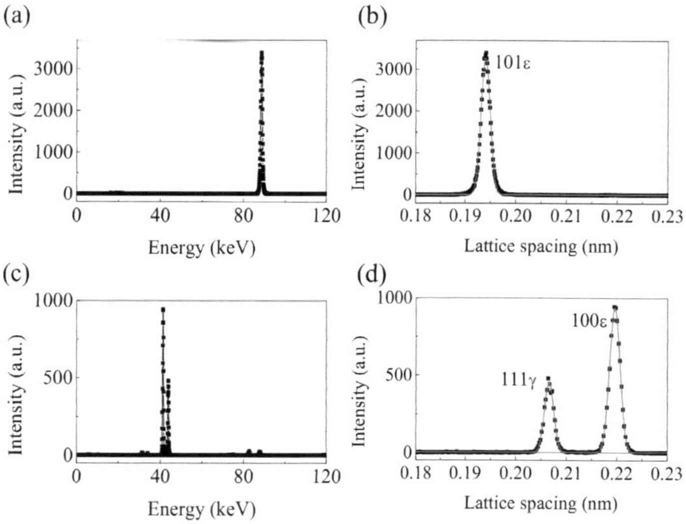

Figure 7. Energy profiles of Laue spots at 23 % strain and the corresponding lattice spacing profiles.

Figure 8 shows the evolution of the principal stresses of the γ phase at three points within a grain during the shape-memory cycle. The red and blue arrows indicate the tensile stress and compressive stress, respectively. The stress data contained a measurement error of about ± 40 MPa, based on an experimental strain resolution of about ± 0.04 % [15]. At 0 % strain (before deformation), there were some small stresses, which might be induced during the production process of the alloy or the preparation of the tensile sample. Stresses in the γ phase were not measured at 23 % strain because there were a limited number of measurable lattice planes for the γ phase, most likely due to the transformation of most of the γ phase to ε martensite. However, large stresses probably exist at deformed states considering the significant Laue spot streaking. In a previous study on microscopic stress analysis in several grains of shape-memory alloys, it was found that large compressive residual stresses were developed in grains with a high Schmid factor for a martensitic transformation at 4 % strain [15]. Furthermore, the magnitudes of the compressive residual stresses in the grains differed due to the different martensitic transformation behaviours in the grains. Considering this fact, large compressive stresses could be formed in the γ phase at 23 % strain. Meanwhile, because martensite newly forms during deformation, there would be a stress redistribution between the parent phase and the martensite. It would also be of interest to study the stress evolution in the martensite phase; this will be explored in our future studies.

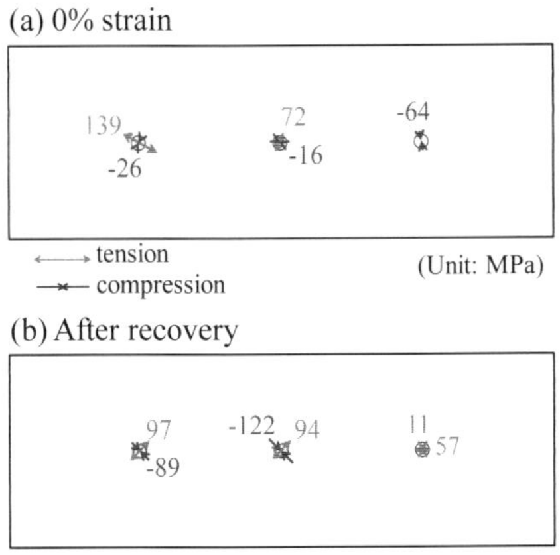

Figure 8. Principal stresses of the γ phase measured at three points at a) 0 % strain, and b) after recovery heating.

Figure 8b shows that the residual stresses after recovery heating in all positions are very small, despite the applied high strain of 23 %. It is considered that stresses developed by deformation are released due to shape recovery. In fact, this sample exhibited a very large recovery strain of 17 % at the applied strain of 23 %. Therefore, one may conclude that the stresses developed by deformation in this sample were mostly released by a pronounced shape recovery, resulting in very small residual stresses, which are comparable with those at 0 % strain.

The reason for the good shape recovery can be revealed by observation of the martensitic transformation behaviour during the shape-memory process. The microstructural evolution during tensile deformation and after recovery heating was observed using EBSD. The results shown in figure 9 reveal that martensite gradually forms with tensile strain and almost disappears after recovery heating due to the reverse transformation of martensite to γ phase. To obtain a shape-memory effect, martensite must be transformed back to the parent phase by a backward movement of the Shockley partial dislocations, these dislocations having operated in the forward transformation process [20]. In the case of the sample with the <144>//LD orientation, only one shear system with the highest Schmid factor of 0.5 operates during both forward and reverse transformations, thereby leading to good transformation reversibility of martensite. The formation of one martensite variant is verified in the EBSD phase map of figure 9 and is indicated by a dashed line. From a crystallographic analysis, the most plausible active shear system among 12 possible shear systems of {111}<112> was found to be $(1\bar{1}2)[1\bar{1}\bar{2}]$, as shown in figure 10c. Therefore, the outstanding shape recovery of this sample is attributed to the good reversibility of the martensitic transformation. Furthermore, the highly reversible martensitic transformation accounts for the observed reversible Laue pattern change and the low residual stresses after recovery heating.

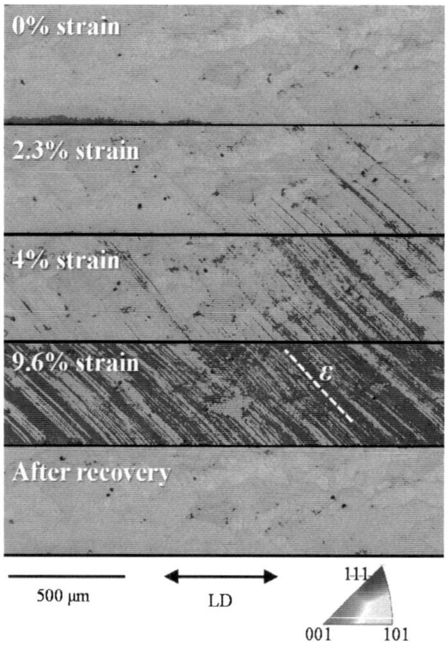

Figure 9. EBSD phase maps showing the evolution of ε martensite with the increase of accumulated tensile strain and after recovery heating. Martensite is coloured in blue.

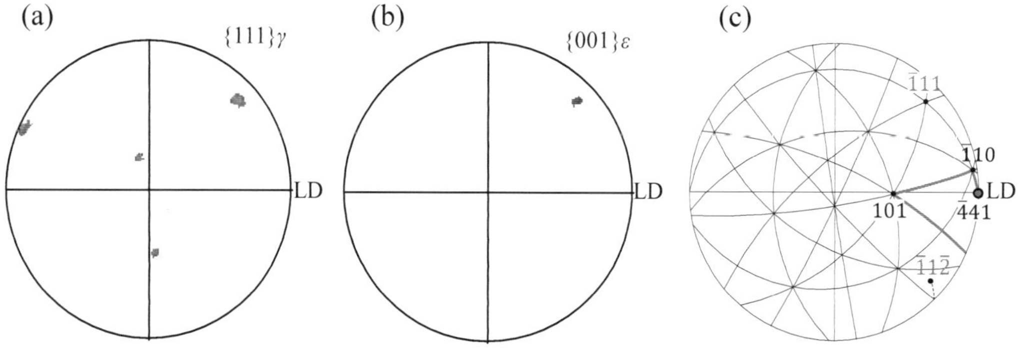

Figure 10. EBSD pole figures of a) γ phase and b) ε martensite at 2.3 % strain. c) Stereographic projection showing configuration of the most plausible active shear system of $(\bar{1}11)[\bar{1}1\bar{2}]$.

4. Conclusions

In conclusion, the residual stress evolution during a shape-memory cycle of a tensile sample in an austenite grain in an Fe–Mn–Si–Cr shape-memory alloy was investigated using white X-ray microbeam diffraction. The X-ray diffraction experiments were conducted on the coarse grain with an orientation close to the <144> direction with a high Schmid factor for a martensitic transformation. The sample, which was subjected to a tensile strain of 23 % and subsequent recovery heating, showed a large recovery strain of 19 %. Microstructural analysis using EBSD revealed that the outstanding shape recovery resulted from a highly reversible martensitic transformation. The residual stresses after recovery heating were found to be very small and comparable with those before deformation despite the applied high level of tensile deformation (23 % strain). The low residual stresses observed in the sample after recovery heating may be attributed to the good reversibility of the martensitic transformation, resulting in the release of the stresses developed due to tensile deformation.

Acknowledgements
This study was financially supported in part by a Grant-in-Aid for Scientific Research Fund from the Japan Society for the Promotion of Science and research projects from the Korea Institute of Industrial Technology. The synchrotron radiation experiments were performed on the BL28B2 beamline at SPring-8 with the approval of the Japan Synchrotron Radiation Research Institute.

References
[1] Sato A, Chishima E, Soma K and Mori T 1982 *Acta Metall.* **30** 1177-1183
[2] Kwon E P, Fujieda S, Shinoda K and Suzuki S 2010 *Mater. Sci. Engng.* A **527** 6524-6532
[3] Nozaki H and Tomota Y 2000 *Mater. Trans.* **41** 727-732
[4] Tomota Y and Maki T 2000 *Mater. Sci. Forum* **327-328** 191-198
[5] Tsuzaki K, Natsume Y, Tomota Y and Maki T 1995 *Scripta Mater.* **33** 1087-1092
[6] Tomota Y, Harjo S, Lukas P, Neov D and Sittner P 2000 *JOM* **52** 32-34
[7] Ice G E, Larson B C, Yang W, Budai J D, Tischler J Z, Pang J W L, Barabash R I and Liu W 2005 *J. Synchrotron Rad.* **12** 155-162
[8] Ice G E, Pang J W L, Barabash R I and Puzyrev Y 2006 *Scripta Mater.* **55** 57-62
[9] Ice G E, Larson B C, Tischler J Z, Liu W and Yang W 2005 *Mater. Sci. Engng.* A **399** 43-48
[10] Kajiwara K, Sato M, Hashimoto T, Hirosawa I, Yamada T, Terachi T, Fukumura T and Arioka K 2009 *Phys. Status Solidi A* **206** 1838-1841
[11] Kwon E P, Sato S, Fujieda S, Shinoda K, Kainuma R, Kajiwara K, Sato M and Suzuki S 2015 *Metals* **5** 1845-1856
[12] Suzuki S, Hotta K, Kwon E P, Fujieda S, Shinoda K, Kumagai M, Kajiwara K, Sato M and Sato S 2015 *ISIJ Int.* **55** 2158-2165
[13] Kubo H, Otsuka H, Farjami S and Maruyama T 2006 *Scripta Mater.* **55** 1059-1062
[14] Pyzalla A 2000 *J. Nondestructive Eval.* **19** 21-31
[15] Kwon E P, Sato S, Fujieda S, Shinoda K, Kajiwara K, Sato M and Suzuki S 2013 *Mater. Sci. Engng.* **570** 43-50
[16] Barabash R I, Ice G E and Walker F J 2003 *J. Appl. Phys.* **93** 1457
[17] Hofmann F, Song X, Abbey B, Jun T S and Korsunsky A M 2012 *J. Synchrotron Rad.* **19** 307-318
[18] Kajiwara K, Sato M, Hashimoto T, Yamada T, Terachi T, Fukumura T and Arioka K 2013 *ISIJ Int.* **53** 165-169
[19] Otsuka H, Yamada H, Maruyama T, Tanahashi H, Matsuda S and Murakami M 1990 *ISIJ Int.* **30** 674-679
[20] Tsuzaki K, Natsume Y and Maki T 1995 *J. Physique* IV **5** 409-414

IOP Conf. Series: Materials Science and Engineering **304** (2017) 012009 doi:10.1088/1757-899X/304/1/012009

Influence of simulation parameters on the speed and accuracy of Monte Carlo calculations using PENEPMA

X Llovet[1] and F Salvat[2]

[1] Scientific and Technological Centres. Universitat de Barcelona. Lluís Solé i Sabarís, 1-3. ES-08028 Barcelona, Spain

[2] Facultat de Física (FQA). Universitat de Barcelona. Diagonal 645, ES-08028 Barcelona, Spain

E-mail: xavier@ccit.ub.edu

Abstract. The accuracy of Monte Carlo simulations of EPMA measurements is primarily determined by that of the adopted interaction models and atomic relaxation data. The code PENEPMA implements the most reliable general models available, and it is known to provide a realistic description of electron transport and X-ray emission. Nonetheless, efficiency (i.e., the simulation speed) of the code is determined by a number of simulation parameters that define the details of the electron tracking algorithm, which may also have an effect on the accuracy of the results. In addition, to reduce the computer time needed to obtain X-ray spectra with a given statistical accuracy, PENEPMA allows the use of several variance-reduction techniques, defined by a set of specific parameters. In this communication we analyse and discuss the effect of using different values of the simulation and variance-reduction parameters on the speed and accuracy of EPMA simulations. We also discuss the effectiveness of using multi-core computers along with a simple practical strategy implemented in PENEPMA.

1. Introduction

The interaction of an electron beam with a sample is the basis of different spectroscopic techniques used for materials analysis, such as electron probe microanalysis (EPMA) [1, 2]. After impinging on the sample, the beam electrons are scattered and slowed down by the atoms of the material until they are stopped or escape from the sample. The interactions of the beam also produce secondary electrons and photons. In EPMA one is interested in the characteristic X-rays emerging from the sample. The elemental composition and stoichiometry of the probed volume of the material can be derived from the measured characteristic-line intensities. To do this, a quantitative understanding of the different processes involved in the electron transport and X-ray generation in the sample is required.

The Monte Carlo (MC) method is now widely recognised as the most accurate method to model the transport of radiation in matter. This is so because it can incorporate realistic physics interaction models and also because of its flexibility to handle complex geometries, making it well suited for the simulation of EPMA measurements, specially for heterogeneous samples such as small particles, inclusions, interfaces or multilayer films. Unfortunately, due to the random nature of the method, the results from MC simulation are affected by statistical uncertainties which, in principle, can only be reduced at the expense of increasing the simulation time. However, significant progress has been made in recent years in the application of the so-called variance-reduction techniques which, combined with the increasing availability of fast computers, has enhanced the attractiveness of MC simulation.

Our group at the University of Barcelona develops and maintains the general-purpose MC subroutine package PENELOPE, which performs simulations of electrons, photons (and positrons) in arbitrary materials, for energies ranging from 50 eV to 1 GeV [3, 4]. PENELOPE has been applied extensively by researchers worldwide over the past two decades to a wide variety of problems in electron microscopy, radiation metrology, radiotherapy, and other fields. PENELOPE uses a combination of numerical and analytical differential cross-sections (DCS) to describe the different interaction mechanisms of electrons and photons with matter. A detailed description of the cross-section models and the electron tracking algorithm implemented in PENELOPE can be found in the code manual [3].

The programme PENEPMA was developed to facilitate the use of PENELOPE in EPMA [5–7]. The distribution package includes a manual that describes the structure and operation of the programme [8]. To set up a simulation, the user has to specify the parameters that characterise the experiment (e.g., electron beam energy, sample composition, detector take-off angle, etc..), a number of simulation parameters, which define the algorithm used for tracking the electrons within the sample, and also a set of parameters that specify the variance reduction techniques that are to be applied. These parameters largely determine the speed of the simulation; tuning these parameters adequately can greatly improve the efficiency of the simulation or, equivalently, reduce the computer time needed to get results with a prescribed accuracy.

In this study we analyse and discuss the effect of using different simulation parameters on the simulation speed for typical cases of interest in EPMA. Our aim is to establish appropriate parameter values with which simulations of X-ray intensities emitted from bulk samples, thin films on substrates and of secondary X-ray fluorescence emitted near phase boundaries may be efficiently modelled with PENEPMA.

2. Simulation parameters

2.1. Simulation algorithm

Electron trajectories and photon histories are simulated from the initial energy down to the electron and photon absorption energies E_{el} and E_{ph}, respectively, to be selected by the user, at which the particles are considered to be effectively absorbed in the medium. The default absorption energies are set to 50 eV. Note that, in the case of samples consisting of several materials, the parameters E_{el}, E_{ph} can be specified for each material. PENELOPE not only transports the primary electrons, but also all secondary electrons and photons produced in interactions of the transported particles (i.e., the complete particle shower originated by each primary electron). Secondary particles (electrons and photons) emitted with initial energy larger than the corresponding absorption energy are simulated (down to their corresponding absorption energies) after completion of each primary electron track.

The absorption energies E_{el} and E_{ph} are generally determined by the characteristics of the experiment. When we are interested only in determining the intensity of a particular X-ray line, to speed up the simulation process we can assign to E_{el} a value slightly lower than the ionisation energy of the active subshell. At the same time E_{ph} can be given a value slightly lower than that of the X-ray energy, in order to avoid spending time transporting photons that are not able to contribute to the line of interest. The X-ray spectrum will then be truncated at an energy equal to E_{ph}.

The simulation of photon tracks follows the detailed procedure, i.e., all the interaction events in a photon history are simulated in chronological order and, hence, the transport scheme is fixed and independent of user-defined parameters. In contrast, electrons can be simulated either detailedly, interaction by interaction, or by using a so-called mixed (class II) simulation algorithm. In the latter schemes, cut-off values of the polar scattering angle (θ_c) and the energy loss (W_c) in single interactions are introduced; interactions with scattering angle $\theta \geq \theta_c$ or energy loss $W \geq W_c$ (the so called hard events), are simulated individually according to the adopted (restricted) single-scattering DCSs. These interactions are responsible for large lateral displacements or large energy-loss fluctuations, which are then described faithfully. The remaining interactions, with $\theta < \theta_c$ and $W < W_c$, are classified as soft interactions; they have a mild effect on the electron trajectories and are simulated by using simple, but still fairly accurate, multiple scattering approximations. With a judicious selection of the cut-off parameters θ_c and W_c, mixed simulation algorithms yield reliable results in moderate computation times, even for electrons with very high energies.

In PENELOPE, the user is spared from having to define the cutoff angle, which is defined internally through a recipe that involves two energy-independent used-defined, parameters, namely the average angular deflection C_1, $C_1 \simeq 1 - \langle\cos\theta\rangle$, produced by multiple elastic scattering along a path length equal to the mean free path between consecutive hard elastic events; and the maximum average fractional energy loss C_2 in a step between hard events. The user also defines the cut-off energy loss for hard inelastic collisions W_{cc}, and the cut-off energy loss for hard bremsstrahlung emission W_{cr}, which are both independent of the energy of the electron.

The parameters C_1 and C_2 determine the simulation of elastic scattering and the tracking algorithm for electrons. Their allowed values are limited to the interval [0, 0.2]. If $C_1 = C_2 = 0$, the simulation of electron elastic scattering is performed in detailed mode, i.e., interaction by interaction. If the values of these parameters are not zero, the mixed simulation algorithm is switched on, and the simulation usually becomes faster. The parameters W_{cc} and W_{cr} have a very weak influence on the accuracy of the results as long as they are both smaller than the width of the bins used to tally energy distributions. When electron energy distributions are not of interest, our recommendation is to set these cut-off energies equal to 1 keV. Note that, for the sake of consistency, W_{cc} should not be larger than E_{el}, otherwise, we would exclude secondary electrons with energies larger than E_{el}. Likewise, W_{cr} should be less than E_{ph}. When $W_{cc} = 0$, the simulation of inelastic collisions is performed in detailed mode. PENELOPE also allows performing detailed simulation of the emission of bremsstrahlung photons with energies higher than 10 eV, disregarding the emission of quanta with lesser energy.

Finally, the user may define a maximum step length s_{max} of electrons in particular material bodies. This parameter has an effect only for very thin materials and when electrons are tracked using mixed simulation; it should be assigned a value of the order of one tenth of the "thickness" of the body where electrons move. This ensures that, on average, there will be more than ~ 10 soft events (hinges) along a typical electron track through that body, which is enough to "wash out" the details of the artificial distributions used to sample these events. For thick bodies, it is not necessary to limit the length of the steps. Note that s_{max} may be given different values for different "bodies" of the same material making up the sample.

2.2. Variance reduction techniques

The direct simulation of EPMA spectra may be very inefficient, even when using mixed simulation algorithms for electron transport. The reason is that radiative processes are much less probable than the dominant interactions of elastic scattering and outer-shell ionisation (or excitation). Consequently, the simulation may require very large computer times, because most electrons will not produce any photons. To cope with this difficulty, PENEPMA takes advantage of several variance-reduction techniques. The ones implemented in PENELOPE are interaction forcing, particle splitting, and Russian roulette. The most effective techniques to improve the efficiency of simulations of EPMA spectra are interaction forcing and particle splitting.

The method of interaction forcing consists of increasing the probability of a given interaction mechanism and, at the same time, assigning appropriate statistical weights to the generated secondary particles so that simulation results remain unbiased. Interaction forcing can be applied to any interaction mechanism. The practical implementation of interaction forcing consists of replacing the mean free path λ_A of the real process by a shorter one, $\lambda_{A,f} = \lambda_A/\mathcal{F}$ where $\mathcal{F} > 1$ is the forcing factor. This implies that between each pair of "real" A interactions we will have, on average, $\mathcal{F} - 1$ "forced" A interactions. We assume that the probability distribution function for the energy loss and the angular deflections (and the directions of emitted secondary particles, if any) in the forced interactions are the same as those for real interactions. When a negative forcing factor is entered, PENELOPE assumes that its absolute value represents the desired average number of forced events in a particle's history, and determines internally the effective forcing factor that yields this number (approximately).

When a particle, with weight w_0, is subjected to splitting, it is replaced by a number $\mathcal{S} > 1$ of identical particles with weights $w = w_0/\mathcal{S}$. Evidently, splitting leaves the simulation unbiased. It should be noted that splitting can be applied in conjunction with interaction forcing and other variance reduction techniques. In PENEPMA particle splitting is employed to increase the number of photons produced in radiative interactions of electrons. All the photons in a radiative event are sampled from the same distribution, and to compensate for the bias in the number of secondary particles, each particle is assigned a specific weight. For instance, if in a bremsstrahlung interaction the electron produces one photon, we can, instead, produce five, each with a weight of 1/5. The benefit is that there are more photons produced per incident electron. This saves CPU time, because the simulation of the electron trajectories is generally slow.

For the practical implementation, we assume that primary particles start moving with unit weight and each secondary particle produced by a primary one is assigned an initial weight equal to that of the primary or determined by the variance-reduction trick at work. To avoid producing particles with very small weights (which would consume CPU time without contributing appreciably to the scores), interaction forcing and splitting are applied only to particles with weight in a window defined by the user. Generally, we set the weight windows so that only primary electrons and photons of the first generations undergo forced interactions.

The effectiveness of the techniques of interaction forcing and particle splitting relies on the adopted values of the corresponding forcing and splitting parameters. In an EPMA spectrum, characteristic X-ray lines arise from radiative transitions of atoms that have been ionised in an inner shell either by electron impact, or by photoelectric absorption or incoherent scattering of characteristic X-rays or bremsstrahlung photons. Therefore we will consider the forcing factors for electron-impact ionisation $\mathcal{F}_{si}$, bremsstrahlung emission $\mathcal{F}_b$, photoelectric absorption $\mathcal{F}_{ph}$ and incoherent scattering $\mathcal{F}_{co}$, as well as the splitting factors for electron-impact ionisation (and the subsequent characteristic X-ray emission) $\mathcal{F}_{ch}$ and for bremsstrahlung emission $\mathcal{S}_b$. Note that in the case of samples consisting of different bodies (of the same or different material), the variance-reduction parameters can be different for each body.

2.3. X-ray detector definition

In addition to the simulation and variance-reduction parameters described above, in sections 2.1 and 2.2, the speed and accuracy of the simulation of an EPMA measurement also depends on the "size" of the defined X-ray detectors. In EPMA measurements, X-ray detectors cover a very small solid angle about a specific direction (θ_d, ϕ_d), a situation that makes the simulation of a real detector very inefficient. In PENEPMA, the user can define up to 25 ideal photon detectors, each one covering a solid angle corresponding to a "rectangle", $(\theta_1, \theta_2) \times (\phi_1, \phi_2)$, on the surface of the unit sphere centred at the origin of the reference frame. Note that $\theta_d = \pi/2 - \chi$, where χ is the conventional take-off angle relative to the sample surface. Of course, the efficiency of the simulation can be improved by considering a detector with a wider angular window. Notice that the different detectors considered in PENEPMA may overlap, a feature that can be used to analyse the effect of replacing the actual detector by angular windows of various sizes in a single simulation.

3. Application to EPMA simulations

In this Section, we analyse the effect of using different simulation and variance-reduction parameters on the speed and accuracy of MC simulations of typical EPMA measurements. These include the simulation of the characteristic X-ray intensity emitted from bulk samples and thin film samples, and of secondary X-ray fluorescence emitted across the boundary of two bulk materials. The simulations were performed on an 8-core personal computer (Intel Core i7 processor having a clock frequency of 2.93 GHz) with 8 GB RAM, running under Windows 7. Unless stated otherwise, all simulations were performed in serial mode.

The different simulation parameters that can be defined in PENEPMA are listed in table 1. To assess the influence of the size of the detector window on the simulation efficiency, the corresponding angular apertures $\Delta\theta$ and $\Delta\varphi$, defined as $\Delta\theta = \theta_2 - \theta_1$ and $\Delta\varphi = \varphi_2 - \varphi_1$, respectively, will be considered as parameters.

Table 1. Summary of simulation parameters.

Parameter	Description
E_{el}	Electron absorption energy (in eV)
E_{ph}	Photon absorption energy (in eV)
C_1	Average angular deflection in a step
C_2	Maximum average fractional energy loss in a step
W_{cc}	Cut-off energy loss for hard inelastic collisions (in eV)
W_{cr}	Cut-off energy loss for hard bremsstrahlung collisions (in eV)
s_{max}	Maximum step length of electrons in material body (in cm)
$\mathcal{F}_{\text{si}}$	Forcing factor for inner-shell ionisation
$\mathcal{F}_{\text{b}}$	Forcing factor for bremsstrahlung emission
$\mathcal{F}_{\text{ph}}$	Forcing factor for photoelectric absorption
$\mathcal{F}_{\text{co}}$	Forcing factor for incoherent scattering
$\mathcal{S}_{\text{ch}}$	Splitting factor for characteristic X-rays
$\mathcal{S}_{\text{b}}$	Splitting factor for bremsstrahlung emission
$\Delta\theta$	Polar aperture (in deg)
$\Delta\varphi$	Azimuthal aperture (in deg)

When variance reduction techniques are employed, the CPU time T required for the simulation of a given number of histories may increase. A practical measure of the effectiveness of a MC simulation run is provided by the efficiency ϵ defined by

$$\epsilon = \left(\frac{I}{\sigma}\right)^2 \frac{1}{T}, \tag{1}$$

where I is the calculated intensity of the considered X-ray line, σ is the standard deviation of I (considering only statistical uncertainties) and T is the CPU time spent in that run, excluding the PENEPMA initialisation time, which is of the order of 5 seconds. Note that PENEPMA reports statistical uncertainties at a confidence level of 99.7 % (3σ level), i.e., the result of the simulation

IOP Conf. Series: Materials Science and Engineering **304** (2017) 012009 doi:10.1088/1757-899X/304/1/012009

is expressed as $I \pm 3\sigma$. It is also worth pointing out that in EPMA the statistical uncertainty of evaluated concentrations is often given at 1σ level, e.g., at the confidence level of 68 %.

To assess the accuracy of simulated intensities, we will also give the relative uncertainty ΔI, defined as

$$\Delta I = \frac{I - I_0}{I_0} \tag{2}$$

where I_0 is the true intensity or a given reference value (see below).

3.1. *Characteristic X-ray intensity from bulk, homogeneous samples*

Table 2 shows the CPU time required to achieve a 1 % statistical uncertainty (at both 1σ and 3σ levels) in the intensity of the Fe K-L$_2$ (Kα_2) line emitted from a bulk Fe sample bombarded with 20 keV electrons at normal incidence, for different combinations of simulation parameters. The simulation efficiencies and relative errors are also listed. In all simulations, we took advantage of the axial symmetry about the beam axis and used an annular detector ($\Delta\varphi = 360°$) that covers an angular aperture $\Delta\theta = 10°$ around a direction of 40° relative to the sample surface. The forcing factors $\mathcal{F}_{\rm ph}$ and $\mathcal{F}_{\rm co}$, not included in table 2, were both set to 5.

Table 2. CPU time (min) required to achieve a 1% statistical uncertainty (at 1σ and 3σ levels) in the simulation of the Fe K-L2 X-ray intensity from a bulk Fe, for different values of the simulation parameters.

Case	$E_{\rm el}$ (keV)	$E_{\rm ph}$ (keV)	C_1	C_2	$W_{\rm cc}$ (keV)	$W_{\rm cr}$ (keV)	$\mathcal{F}_{\rm si}$	$\mathcal{F}_{\rm br}$	$\mathcal{S}_{\rm ch}$	$\mathcal{S}_{\rm br}$	ΔI (%)	$\epsilon(3\sigma)$	$T(3\sigma)$ (min)	$\epsilon(1\sigma)$	$T(1\sigma)$ (min)
1	0.05	0.05	0	0	0	0.01	−10	−5	-	-	+0.00	0.01	16699	0.1	1855
2	1	1	0.1	0.1	1	1	−10	−5	-	-	−0.69	3.8	43.8	34.2	4.87
3	6	6	0.1	0.1	1	1	−10	−5	-	-	+0.16	41.8	3.98	376.3	0.44
4	6	6	0.2	0.2	1	1	−10	−5	-	-	−0.25	60.1	2.77	540.6	0.30
5	6	6	0.2	0.2	6	6	−10	−5	-	-	−0.52	56.5	2.94	508.5	0.32
6	6	6	0.2	0.2	1	1	−10	−5	4	4	−0.50	134.2	1.24	1207.9	0.13
7	6	6	0.2	0.2	1	1	−10	−5	16	16	−0.88	81.7	2.04	735.1	0.22
8	6	6	0.2	0.2	1	1	−100	−5	16	16	+0.06	133.9	1.24	1204.9	0.13

We take as a reference the case of a detailed simulation with default absorption energies of 50 eV (case 1). Notice that for all the considered sets of parameters, the accuracy of the results is practically the same as that of the detailed simulation; in fact, $\Delta I < \pm$ 1 % in all the studied cases. We can see that a precision of 1 % (at 3σ level) can be achieved in about one minute CPU time by setting C_1 and C_2 to their maximum allowed values 0.2, adjusting $E_{\rm el}$ and $E_{\rm ph}$ to be slightly lower than the ionisation energy of the Fe K-shell, setting $W_{\rm cc}$ and $W_{\rm cr}$ to about 1 keV, as well as by using moderately large forcing and splitting factors (case 8). As compared with the case of detailed simulation, the use of these "optimised" parameters represents a gain in speed by a factor of $\sim$ 13,400.

It is worth noting that the efficiency of case 6 ($\mathcal{F}_{\rm si} = -10$, $\mathcal{S}_{\rm in} = 4$, $\mathcal{F}_{\rm br} = 4$) is essentially the same as that of case 8 ($\mathcal{F}_{\rm si} = -100$, $\mathcal{S}_{\rm si} = 16$, $\mathcal{F}_{\rm br} = 16$). This shows that an increase of the forcing and splitting factors does not necessarily lead to an increase in efficiency, as more calculations might be required to generate each shower. Also, the increase in $W_{\rm cc}$ and $W_{\rm cr}$ up to 6 keV (case 5), although not altering significantly the simulation results, does not provide a relevant gain in efficiency, as compared with using the recommended values $W_{\rm cc} = W_{\rm cr} = 1$ keV (case 4).

If the uncertainty is evaluated at 1σ level, a 1 % statistical accuracy can be achieved in only 0.13 min (case 8). This shows that PENEPMA is suitable for being used as the basis of an iterative quantification method in practical EPMA work.

The effect of both forcing and splitting factors on the simulation efficiency is shown in more detail in fig. 1, which displays the efficiency and relative error as functions of $\mathcal{F}_{si}$, for different values of the splitting factor $\mathcal{S}_{ch}$. In these simulations, the following parameters were used: $C_1 = C_2 = 0.2$, $E_{el} = E_{ph} = 6$ keV, and $W_{cc} = W_{cr} = 1$ keV. No other forcing factors or splitting factors were applied. We can see that the efficiency reaches a maximum at around $\mathcal{F}_{si} \sim -75$ and then decreases. It is also evident that there is an important gain in efficiency by applying a splitting factor $\mathcal{S}_{ch}$ of 2. However, increasing the splitting factor from $\mathcal{S}_{ch} = 2$ to $\mathcal{S}_{ch} = 5$ shows no gain. Therefore, before increasing the value of any of the simulation parameters it is advisable to perform short test simulations to verify that with the augmented parameter value not only the results remain essentially unaltered but also that there is indeed a gain in the efficiency of the simulation.

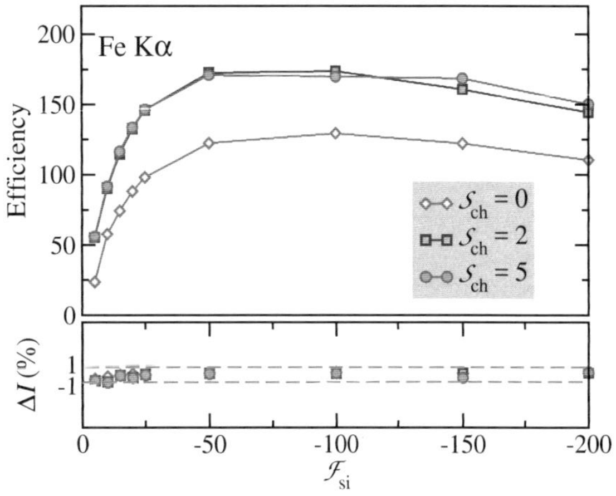

Figure 1. Efficiency and relative error as a function of forcing factor $\mathcal{F}_{si}$ for different splitting factors $\mathcal{S}_{ch}$ in the simulation of the Fe K-L_2 X-ray intensity from a bulk Fe sample.

Figure 2 shows the simulation efficiency as a function of the azimuthal aperture for simulations of the Fe K-L_2 line emitted from bulk Fe and of the Al K-L_3 line emitted from bulk Al, in both cases the samples being bombarded with electron beams of 20 keV at normal incidence. The polar aperture $\Delta\theta$ was set to 10° for all cases. It is seen that the efficiency increases linearly without altering the accuracy, reaching its maximum for $\Delta\phi = 360°$ (annular detector). Note that this kind of configuration can in principle only be used in cases where the flux of emerging photons has axial symmetry about the z-axis (e.g., for plane-surface samples that are homogeneous within the interaction volume, and at normal incidence).

The variation of the efficiency with polar aperture $\Delta\theta$ is also shown in fig. 2 for the same simulation cases. Here, the use of large polar apertures $\Delta\theta$, while increasing the efficiency, do alter significantly the accuracy of results, as shown by the relative error, which rapidly (at $\sim \Delta\theta = 20°$) departs from the 1 % uncertainty. In general, it is recommended to keep $\Delta\theta$ below 20°, especially for materials for which absorption within the target may be important (such as Al in our case).

3.2. X-ray intensity from thin films on substrates

The simulation of the X-ray intensity emitted from a relatively thin film is generally more difficult than for bulk samples because electrons might traverse the film without producing (or producing

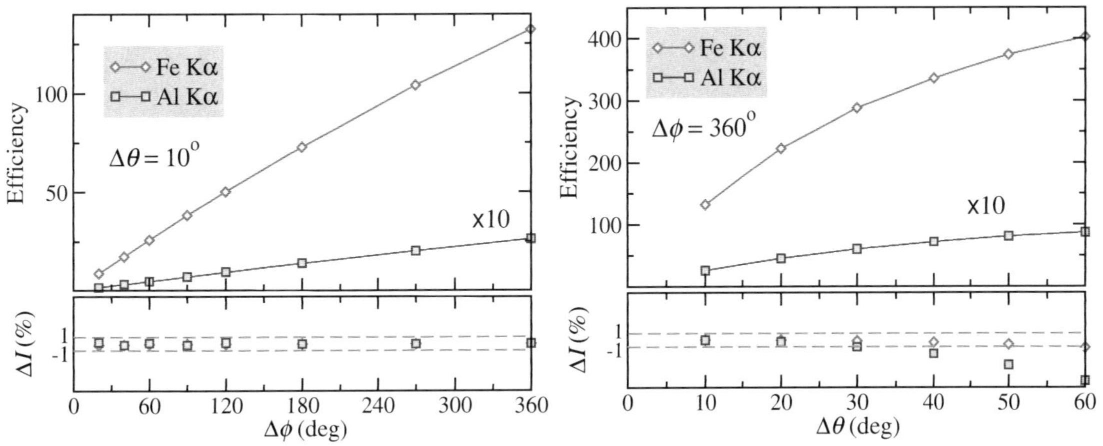

Figure 2. Efficiency and relative error as a function of azimuthal (left) and polar (right) aperture for simulations of the Fe K-L_2 line emitted from bulk Fe and of the Al K-L_3 line emitted from bulk Al.

very few) radiative interactions (see fig. 3a). Thus, the use of variance-reduction techniques is strongly recommended.

In order to determine the effect of simulation parameters on the intensity emitted from thin films, simulations were performed for a 100-nm-thick Al film deposited on Fe bombarded with 20 keV electrons at normal incidence, for different values of the simulation parameters used for both the film and the substrate. The results are tabulated in table 3, which compares the CPU time required to achieve a 1 % statistical uncertainty (at both 1σ and 3σ levels) in the Al K-L3 ($K\alpha_1$) X-ray intensity emitted from the film. In all cases, $s_{\max}$ was set to 10 nm for Al (1/10th of the film thickness), and the values $E_{\rm el}$ and $E_{\rm ph}$ of the electron and photon absorption energies were both set equal to 1 keV (not shown in the Table).

Table 3. CPU time (min) required to achieve an uncertainty of 1 % (at 1σ and 3σ levels) in the Al K-L3 X-ray intensity emitted from a 100-nm-thick Al/Fe film bombarded with an electron beam of 20 keV, for different values of the simulation parameters.

Case	Material	C_1	C_2 (keV)	$W_{\rm cc}$ (keV)	$W_{\rm cr}$	$\mathcal{F}_{\rm si}$	$\mathcal{F}_{\rm br}$	$\mathcal{F}_{\rm ph}$	$\mathcal{F}_{\rm co}$	$\mathcal{S}_{\rm ch}$	ΔI (%)	$\epsilon(3\sigma)$	$T(3\sigma)$ (min)	$\epsilon(1\sigma)$	$T(1\sigma)$ (min)
1	Al	0	0	0	0.01	−10	−5	-	-	-					
	Fe	0.2	0.2	1	1	-	-	-	-	-	+0.00	0.5	311.2	4.8	34.5
2	Al	0	0	0	0.01	−10	−5	10	-	2					
	Fe	0.2	0.2	1	1	-	-	-	-	-	+0.04	1.0	170.6	8.8	18.9
3	Al	0	0	0	0.01	−10	−5	10	-	2					
	Fe	0.2	0.2	1	1	−5	−5	-	-	-	−0.63	0.7	222.1	6.7	24.6
4	Al	0	0	0	0.01	−20	−5	10	-	5					
	Fe	0.2	0.2	1	1	−5	−5	-	-	-	−0.61	2.9	58.1	38.3	4.3
5	Al	0	0	0	0.01	−50	−5	10	10	5					
	Fe	0.2	0.2	1	1	−5	−5	10	10	-	−0.89	5.2	32.1	46.6	3.5
6	Al	0.2	0.2	1	1	−50	−5	10	10	10					
	Fe	0.2	0.2	1	1	−5	−5	10	10	-	−0.62	8.0	20.8	71.9	2.3

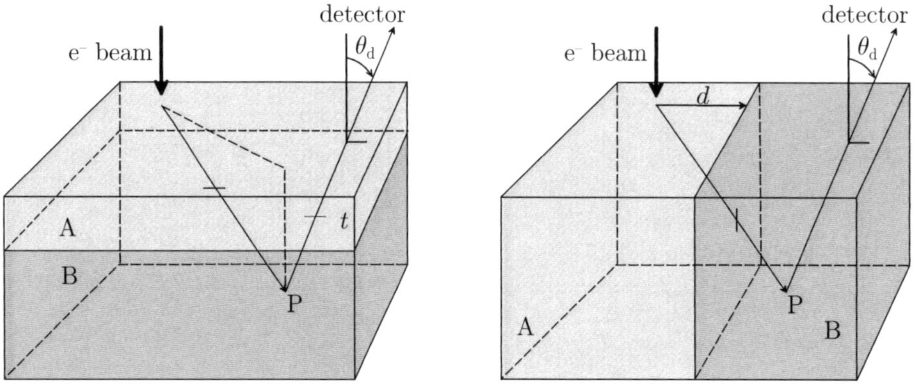

Figure 3. Geometries of the thin film on a substrate (left) and of a material couple (right) used in simulations of SF. Characteristic X-rays and bremsstrahlung photons generated in material A can undergo photoelectric absorption or incoherent scattering and produce X-ray fluorescence emission in material B.

It is apparent that when the different forcing and splitting factors are gradually increased, efficiency improves, with case 6 giving a $\sim$ 15 times speed up as compared to case 1. We see that a considerable reduction in the CPU time can be achieved by using moderately high values of the forcing and splitting factors for both film and substrate materials. The increase in efficiency is significant when going from $\mathcal{F}_{si} = -10$ (case 3) to $\mathcal{F}_{si} = -20$ (case 4), indicating that the forcing factor $\mathcal{F}_{si}$ for the film appears to be the most important parameter to pay attention to. It is also worth mentioning that the use of the maximum allowed values for C_1 and C_2 does not alter practically the accuracy of results, and therefore these values can be safely used even for such relatively thin films.

Case 6 represents the most efficient selection of simulation parameters, which allows the obtention of the X-ray intensity in $\sim$ 2 minutes time (at 1σ level), showing that PENEPMA could be used in practice as the basis of a quantification algorithm.

3.3. Secondary fluorescence from phase boundaries

A more demanding case is the simulation of secondary X-ray fluorescence emitted near the interface of two materials, for which specific simulations have been published [9]. As mentioned earlier, the characteristic lines in an EPMA spectrum arise from radiative transitions of atoms that have been ionised in an inner shell either by electron impact, or by photoelectric absorption or incoherent scattering of characteristic X-rays or bremsstrahlung photons. Characteristic and bremsstrahlung X-rays emitted from interactions of primary electrons can reach distances much larger than the electron range, and can produce X-ray fluorescence through photoabsorption and Compton scattering in a sample region (possibly having a different chemical composition) far away from the electron point of impact. This contribution is usually referred to as secondary fluorescence (SF). Petaccia *et al.* [10] have described the application of particle splitting to simulate X-ray fluorescence in EPMA using PENELOPE.

To study the relevance of simulation parameters for the calculation of SF, we will consider an ideal specimen consisting of two materials A and B separated by a plane interface perpendicular to the surface of the specimen. The electron beam impacts on the left-hand side of the specimen (material A) at a distance d from the interface (fig. 3b). Primary characteristic and bremsstrahlung photons from material A can induce emission of fluorescent photons from materials A and B.

Table 4 shows the CPU time spent to achieve an uncertainty of 1 % (at both 1σ and 3σ levels) in the simulation of the Fe K-L2 X-ray intensity emitted from a Cu/Fe couple bombarded with an electron beam of 20 keV at normal incidence, when the beam impacts on Cu (material A) at 10 µm away from the boundary, for different values of the simulation parameters. Because of the distance of the electron point of impact to the boundary, the Fe K-L2 intensity has a negligible contribution from impact ionisation by primary electrons (the range of electrons of 20-keV energy in Cu is of $\sim$ 1.6 µm). Instead, the Fe K-L2 intensity arises from SF of Cu characteristic X-rays and bremsstrahlung generated at the Cu side of the couple, which are absorbed (and to a lesser extend incoherently scattered) at the Fe side (material B).

Table 4. Comparison of CPU time T required to achieve an uncertainty of 1 % (at 1σ and 3σ levels) in the Fe K-L2 X-ray intensity emitted from a Fe/Cu couple bombarded with an electron beam of 20 keV at 10 µm away from the boundary on the Cu side, for different values of the simulation parameters.

Case	Material	$\mathcal{F}_{\rm si}$	$\mathcal{F}_{\rm br}$	$\mathcal{F}_{\rm ph}$	$\mathcal{F}_{\rm co}$	$\mathcal{S}_{\rm ch}$	$\mathcal{S}_{\rm br}$	ΔI (%)	$\epsilon(3\sigma)$	$T(3\sigma)$ (min)	$\epsilon(1\sigma)$	$T(1\sigma)$ (min)
1	Cu	−10	−5	-	-	-	-					
	Fe	-	-	-	-	-	-	+0.00	0.7	238.2	6.3	26.4
2	Cu	−10	−5	10	10	-	-					
	Fe	-	-	10	10	-	-	+0.01	0.7	237.5	6.3	26.3
3	Cu	−10	−5	10	10	5	5					
	Fe	-	-	10	10	-	-	+0.25	1.8	90.1	16.6	10.0
4	Cu	−20	−5	10	10	10	10					
	Fe	-	-	10	10	-	-	+0.45	1.9	87.6	17.1	9.7
5	Cu	−20	−5	10	10	20	20					
	Fe	-	-	10	10	-	-	−0.51	2.1	79.4	18.8	8.8
6	Cu	−30	−5	10	10	20	20					
	Fe	-	-	10	10	-	-	−0.68	1.7	99.5	15.1	11.0

The forcing factor $\mathcal{F}_{\rm si}$ and the splitting factor $\mathcal{S}_{\rm ch}$ of the material where the electron beam impacts (in this case Cu) appears to play an important role in the optimisation of the simulation, with optimum values of $\mathcal{F}_{\rm si} = -20$ and $\mathcal{S}_{\rm ch} = 20$. Increasing from $\mathcal{F}_{\rm si} = -20$ to $\mathcal{F}_{\rm si} = -30$ results in a slight decrease of the efficiency and CPU time. It is also evident that the simulation efficiency is less sensitive to the forcing factors pertaining to photoelectric absorption and Compton scattering, for both materials B and A.

Simulations of SF are CPU intensive. However, as shown here, they can benefit from the application of variance reduction techniques, to the extent that a typical 10-point simulation profile, with an uncertainty of 1 % (at 1σ level), can be obtained in around 1 h CPU time. Therefore, for such kind of simulations, it is highly recommended not to use the default simulation parameters. Likewise, performance comparisons with other SF calculations should be performed under optimised conditions [11].

4. Parallel execution

The PENEPMA distribution package contains an auxiliary programme called PENEPMA-SUM, which allows the combination of results from independent simulation runs of the same problem. Two simulation runs are effectively independent if the sequences of random numbers used in the two runs are uncorrelated. In PENEPMA this is ensured by initialising each run with seeds of the random number generator (a pair of integer values) taken from a pre-calculated list that describes states of the generator separated by 10^{14} calls in the same long sequence (see e.g., [12]).

That is, the various runs use parts of a long sequence of random numbers that are sufficiently far from each other to avoid overlaps.

This auxiliary programme can be used as a simple parallelisation device in order to improve the efficiency of a simulation problem by running several versions of the same simulation task simultaneously (in parallel) on a multi-core computer. After completing the individual simulations, the programme PENEPMA-SUM combines and averages the partial results producing a unique set of output files. Here we should mention that true parallelised versions of PENELOPE have been coded by other groups (see e.g., the code PENORNL [13], developed for application to SEM analysis).

In order to evidence the effectiveness of the parallel execution of PENEPMA, 2, 4, 6, and 8 versions of the same simulation task (differing only in the seeds of the random number generator) were executed in parallel on a 8-core computer. Specifically, we simulated case 6 from Sec. 3.3 (table 4), that is, the calculations of SF (Fe K-L2 X-rays) emitted from a Fe/Cu couple when a 20-keV-electron beam impacts on the Cu side of the couple at 10 μm from the boundary. The results, shown in table 5, indicate that by running simultaneously 8 versions of the same simulation on a 8-core computer, a gain in efficiency of about a factor of 4.6 is achieved, compared to one single execution (case 1). Thus, by running PENEPMA in parallel on a 8-core computer, it will only take 2.4 minutes to simulate the SF intensity with a precision of 1 % (at 1σ level). Notice that, in contrast with more sophisticated parallelised versions of PENELOPE (e.g., the code PENORNL [13]), no modification of the original PENEPMA code is required.

Table 5. Simulation results as a function of the number of parallel executions.

Case	Number of parallel executions	$\epsilon(3\sigma)$	$T(3\sigma)$ (min)	$\epsilon(1\sigma)$	$T(1\sigma)$ (min)
1	0	1.7	99.5	15.1	11.0
2	2	3.3	49.7	30.1	5.6
3	4	6.1	27.3	54.9	3.0
4	6	7.1	23.3	64.2	2.6
5	8	7.8	21.4	70.0	2.4

5. Conclusion

In this study, we have shown that it is possible to perform MC simulations of typical EPMA experiments using PENEPMA within reasonably short CPU times by adequately choosing the simulation parameters that define the electron transport algorithm, the solid angle covered by the detector, and the adopted variance reduction techniques. The simulation time can further be reduced by taking advantage of multi-core computers by means of a very simple multiple-run strategy.

Acknowlegements

Financial support from the Spanish Ministerio de Economía y Competitividad and ERDF (projects nos. FPA2013-44549-P and FPA2016-77689-C2-2-R) and from the Generalitat de Catalunya (grant 2014 SGR 846) is gratefully acknowledged.

References

[1] Reed S J B 1993 *Electron Microprobe Analysis* (Cambridge: Cambridge University Press)
[2] Rinaldi R and Llovet 2015 *Microsc. Microanal.* **21** 1053

[3] Salvat F 2015 *PENELOPE-2014. A code system for Monte Carlo simulation of electron and photon transport.* (Issy-les-Moulineaux, France: OECD/NEA Data Bank)
[4] PENELOPE is publicly available from the OECD/NEA Data Bank, `http://www.oecd-nea.org`
[5] Llovet X, Fernández–Varea J M, Sempau J and Salvat F 2005 *Surf. Interface Anal.* **37** 1054
[6] Salvat F, Llovet X, Fernández–Varea J M and Sempau J 2006 *Microchim. Acta* **155** 67
[7] Llovet X and Salvat F 2017 *Microsc. Microanal.* **23** 634
[8] Because of its specificity, PENEPMA is not distributed with the PENELOPE package, but it is publicly available from the authors
[9] Llovet X, Pinard P T, Donovan J J and Salvat F 2012 *J. Phys. D: Appl. Phys.* **45** 225301
[10] Petaccia M, Segui S and Castellano G 2015 *Microsc. Microanal.* **21** 753
[11] Ritchie N W M 2017 *Micros. Microanal.* **23** 618
[12] Badal A and Sempau J 2006 *Comp. Phys. Commun.* **175** 440
[13] Bekar K B, Miller T M, Patton B W, Weber C F 2015 *Proc. Joint Int. Conf. on Mathematics and Computation, Supercomputing in Nuclear Applications and the Monte Carlo Method* (LaGrange Park, Illinois: American Nuclear Society)

IOP Conf. Series: Materials Science and Engineering **304** (2017) 012010 doi:10.1088/1757-899X/304/1/012010

Soft X-ray and cathodoluminescence measurement, optimisation and analysis at liquid nitrogen temperatures

C M MacRae[1], N C Wilson[1], A Torpy[1] and C Delle Piane[2]

[1] CSIRO Mineral Resources, Microbeam Laboratory, Private Bag 10, 3169 Clayton South, Victoria, Australia
[2] CSIRO Energy, 26 Dick Perry Avenue, 6152 Kensington, Western Australia, Australia

E-mail: colin.macrae@csiro.au

Abstract. Advances in field emission gun electron microprobes have led to significant gains in the beam power density and when analysis at high resolution is required then low voltages are often selected. The resulting beam power can lead to damage and this can be minimised by cooling the sample down to cryogenic temperatures allowing sub-micrometre imaging using a variety of spectrometers. Recent advances in soft X-ray emission spectrometers (SXES) offer a spectral tool to measure both chemistry and bonding and when combined with spectral cathodoluminescence the complementary techniques enable new knowledge to be gained from both mineral and materials. Magnesium and aluminium metals have been examined at both room and liquid nitrogen temperatures by SXES and the L-emission Fermi-edge has been observed to sharpen at the lower temperatures directly confirming thermal broadening of the X-ray spectra. Gains in emission intensity and resolution have been observed in cathodoluminescence for liquid nitrogen cooled quartz grains compared to ambient temperature quartz. This has enabled subtle growth features at quartz to quartz-cement boundaries to be imaged for the first time.

1. Introduction

Field emission gun equipped electron probe microanalysers (FEG-EPMA) offer high beam power density at a range of accelerating voltages, with excellent beam stability over days of operation. However, with high beam current densities the onset of specimen damage occurs faster [1]. It has been observed that destruction of the specimen and defect transformation is time dependent, and provided spectral measurements can be performed quickly enough, "undamaged" hyperspectral maps can be collected. To slow the onset of beam damage we have integrated a liquid nitrogen (LN) cold stage into a modern FEG-EPMA. The LN cold stage has been optimised for stage mapping, offering an advantage over normal cold stages in that large area maps up to several mm^2 can be collected at cryogenic temperatures using a variety of detectors such as grating optical and X-ray spectrometers for hyperspectral cathodoluminescence (CL) [2-5] and a soft X-ray emission spectrometer (SXES) [6, 7]. Lower accelerating voltages increase the chemical spatial resolution, but also increase the beam power damage, and so reducing the onset of specimen damage through cryogenic cooling allows lower accelerating voltages to be employed.

IOP Conf. Series: Materials Science and Engineering **304** (2017) 012010 doi:10.1088/1757-899X/304/1/012010

2. SXES configuration

The SXES offers the ability to acquire soft X-ray spectra with higher energy resolution than either energy-dispersive (ED) or wavelength-dispersive (WD) X-ray spectrometry. When operated in a mapping mode the SXES can collect a spectra at every pixel of a map, in acquisition times down to 200 ms. The minimum time in this case is governed not by the signal strength but by the time to readout the charge coupled device (CCD). In our system we employ a Princeton CCD which has a 2K × 2K back-thinned detector optimised for the collection of soft X-rays. The CCD is operated at -60 °C using a Peltier cooling device with water cooling to remove the heat and stabilise the temperature of the detector. The CCD is open to the vacuum of the spectrometer, and thus the EPMA, necessitating a clean high vacuum during the operation of the SXES. In our system we collect hyperspectral maps, with the SXES spectra being collected in parallel with other spectral signals acquired from the ED and CL spectrometers, as well as WD spectrometry and electron signals at every pixel. Hyperspectral mapping or analysis has significant advantages over traditional mapping, in that it removes the need to have *a priori* knowledge about what the important X-ray lines are, and when combined with the other signals being acquired from the EPMA, offers a wealth of information.

3. SXES alignment and calibration

For optimum performance the SXES must be correctly aligned through a two-step process. Firstly, the energy range required is set, which involves shifting the CCD up and down using a grub screw. In our case we often select the energy range to allow Li-K (54 eV) X-rays to be observed using the 50XL reflection diffraction grating. Typically the 50XL grating has an operational energy range of 48 - 180 eV [8]. However, in our system we can observe X-rays as low as 44 eV, which allows direct observation of the Mg L-line (49 eV) and up to 195 eV at the high end. For the 200N grating a typical range is 55 - 235 eV, however measurements up to 285 eV are achievable, allowing the first order C K-line to be observed. Thus the SXE spectrometer can have differing energy ranges selected for specific analytical tasks. Once the energy range has been selected the second step is align the SXES to achieve the maximum peak-to-background or resolution. This is achieved by rotating the CCD around its central axis until the best resolution is achieved. We usually perform this task on aluminium metal using the Al-L (77 eV) Fermi edge, which at zero Kelvin would be a vertical X-ray line. At room temperature the Al L-edge has, on our spectrometer, a resolution of 223 meV. These two steps complete the alignment procedure, and next one must calibrate the observed X-ray spectrum which involves converting the 2048 channels of the CCD into energy steps. This calibration takes into account that the crystal diffracts in wavelength and the energy scale conforms to the formula $E = hc/\lambda$, where λ is the wavelength and h is Planck's constant.

For the calibration of the 50XL grating, we use the C-K X-ray line from graphite, the Zr Mζ-line from zirconium metal and the Mo Mζ2-line from molybdenum metal. Given that the SXES has such high energy resolution it can therefore observe shifts in X-ray lines that are due to bonding. Consequently, one must be careful to use single element metals for energy calibration so that the measured X-ray lines from these materials will match X-ray line databases. With the 50XL diffraction grating we can observe the C-K 2nd, 3rd, 4th, 5th and 6th order X-ray reflections. Since the diffraction grating is holographically produced it is highly efficient at reflecting higher order X-ray lines and quite often 8th and 9th order lines are observable. The peak positions of C K reflections together with Zr-Mζ 1st (151 eV) order and the Mo-Mζ_2 1st (193 eV) order are recorded and these positions are then fitted using a polynomial expression to provide a calibration curve for the CCD. As an example, figure 1 which shows the calibration curves for the 50XL and 200N gratings. The non-linear response is a property of the grating and its reflectivity curve. This completes the alignment and calibration of the SXES.

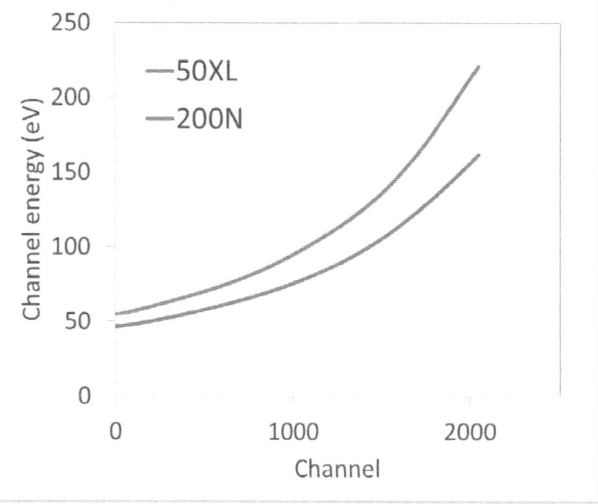

Figure 1. Calibration curves for SXES gratings 50XL and 200N.

4. Soft X-ray emission and transmission though films

To observe soft X-ray emission in poorly electrically conductive materials, such as ionically bonded minerals (e.g., fluorite, CaF_2) or sulphides (e.g., pyrite, FeS_2), one needs to coat with a thin conductive film. To determine whether it is possible to observe soft X-rays (e.g., Li-K) through a carbon coat, we performed a test in which varying thicknesses of carbon were applied to a surface using a carbon arc coater (Cressington 208carbon). An integrated quartz thickness monitor was used to measure the carbon coat thickness and it was assumed the density of the film was 2.2 g/cm^3. The curves in figure 2, are of the Li-K (54 eV) and Al-L (72 eV) peak intensities versus carbon coat thickness, as measured from an aerospace alloy AA2099 at 5 kV accelerating voltage. This alloy contains ~1.8 wt% Li and ~95 wt% Al. At a carbon coat thickness of 80 Å the measured intensity of the Li-K drops by 63 % and the Al-L intensity drops by 75 % compared to the uncoated metal. While the intensity of the Li K-line drops off faster than the Al-L with increasing coat thicknesses, the intensities of these peaks are still sufficiently high to allow mapping and analysis of coated minerals and materials.

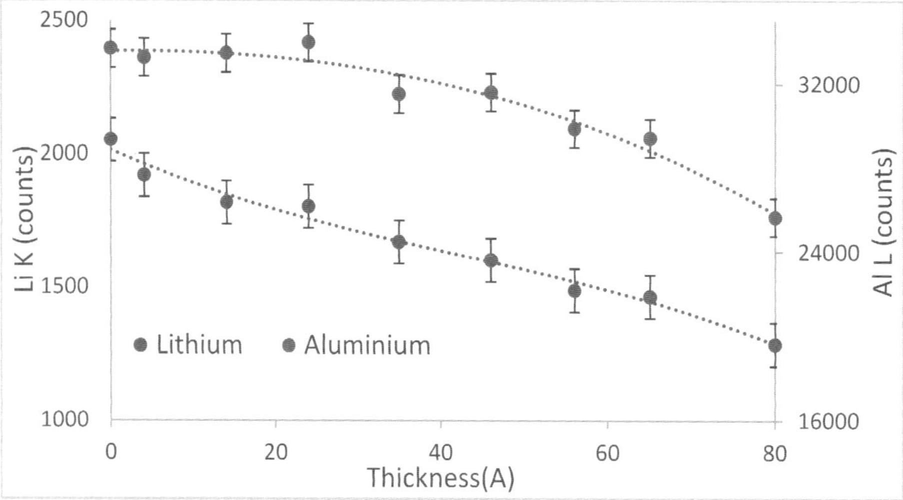

Figure 2. Measurement of Li-K (54 eV) and Al-L (72 eV) emitted intensities through varying film thicknesses of carbon. All measurements were performed at 5 kV with a 10 μm defocussed beam.

X-ray yield curves are not available for many soft X-rays as the mass absorption coefficients (MAC) have either not been either measured or determined. To do this one must measure the X-ray lines of interest (ideally within the material to be examined), from low through to high accelerating voltage. By plotting the emission intensity versus the accelerating voltage one can produce an x-ray yield curve and this can be used to determine the optimum voltage to analyse the X-ray.

Figure 3 shows the effect of changing accelerating voltage on the emitted intensity of Al-L in aluminium and Si-L (91.5 eV) in silicon. A repeat set of measurements were performed from 4.5 - 7 kV on the aluminium to check the reproducibility and both curves show a similar trend with a maximum emission at 6.5 kV. In the case of Si L in silicon the maximum intensity was observed at 9.5 kV. Similar curves have been generated for a wider set of materials and reported in Demers *et al.* [9].

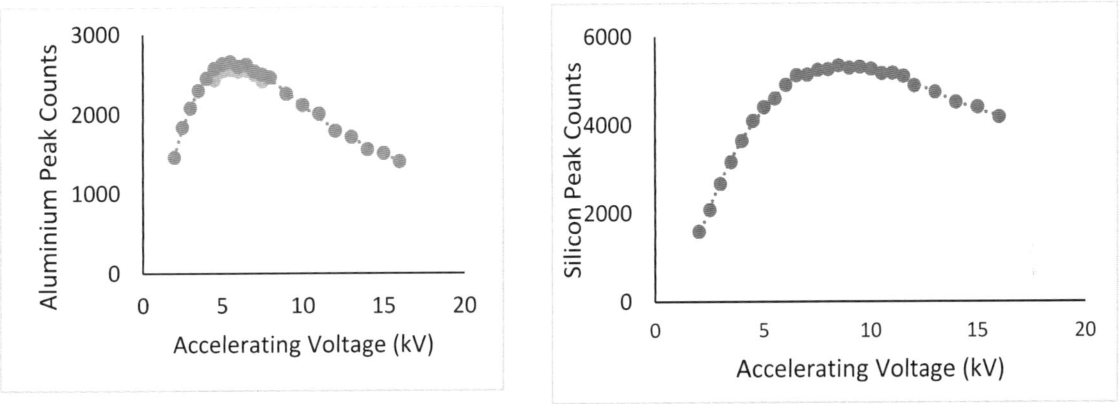

Figure 3. X-ray yield curves for Al-L and Si-L (91.5 eV) measured on aluminium and silicon, respectively. Optimum voltage for Al L-emission is 5 - 7 kV while for Si-L it is 8 - 10 kV. The Al-L measurements, green, were repeated over a smaller range to verify the measurements.

The curvature of the gratings and dispersion gives rise to a change in resolution of the spectrometer with energy. This can be observed by measuring the energy width between the 75 % and 25 % edge height of X-ray peaks that have Fermi edges. Figure 4 shows a plot of the Fermi edge resolution versus X-ray line energy measured using Mg-L (49 eV), Al-L (73 eV) and B-K (183 eV) in all cases the lines were measured on the pure metal at ambient temperature. The Al-L peak position indicated here is the Fermi edge position, 73 eV, which is different from the peak position of 72 eV. The resolution moves from a maximum of 115 meV (Mg-L) through to 1030 meV (B-K).

5. Calibration and alignment of cathodoluminescence spectrometer

We employ a factory aligned, fixed grating spectrometer containing no moving parts. Cathodoluminescence from the specimen is collected by the integrated light optics of the EPMA and transmitted to the optical spectrometer by means of a fibre optic. Our standard optical spectrometer has a wavelength range of 200 - 980 nm and has interchangeable slits with varying sizes ranging from a 25 µm slit giving 1nm resolution and up to 300 nm slit providing 10 nm resolution. Each slit requires a minor adjustment of the alignment optics after replacement, and on the electron microprobe this is usually performed on a non-beam sensitive CL material such as zircon or ruby.

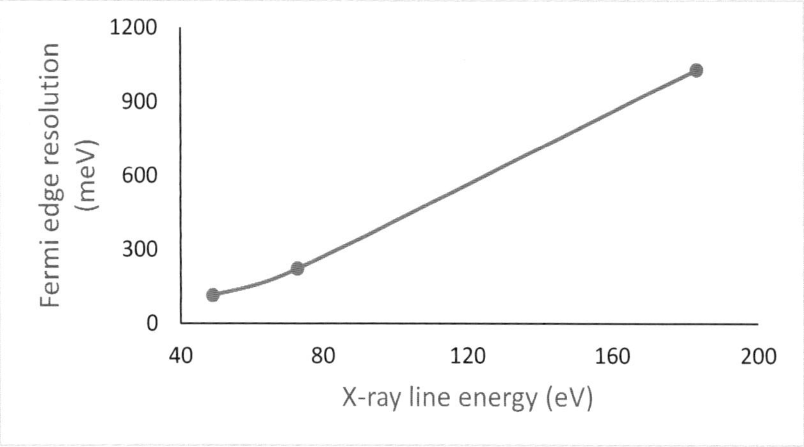

Figure 4. Fermi edge resolution versus X-ray line energy. The curve was generated by measuring the 75 % to 25 % energy difference for Mg-L, Al-L and B-K Fermi edges.

6. Cold stage operation

The cold stage we used in these measurements utilises a GATAN C1001 stage integrated by JEOL (Japan) onto a purpose built carriage and door assemble, figure 5. The design accommodates a 10 mm diameter sample, which is mounted flush within the holder for good thermal conductivity. The sample is pumped down in the airlock to a base pressure of better than 10^{-2} Torr and then inserted into the JEOL 8530II electron microprobe using the supplied high precision stage. The teflon tubing is initially coiled, as in figure 5, and during insertion stretches out but does not interfere with the operation of the stage. Thus the stage movements are uninhibited and either stage or beam mapping can be used. In our case we use the former to cover large areas. To prevent ice build-up on the surface of the sample during use of the cold stage, a liquid nitrogen (LN) cold trap below the pole piece is operated with an auto-filler to allow unattended operation for up to three days. In addition, we have engineered a cascade turbo pumping arrangement so the backing turbo for the main chamber is backed by a smaller turbo pump, with the ultimate pressure of the instrument being below 3×10^{-7} Torr during operation. This minimises surface contamination which is important for detection of low level Li by either windowless SDD or SXES.

Figure 5. JEOL-GATAN cold stage, which uses cryo-cooled nitrogen gas to cool the stage. The white teflon tubing allows the cold stage to be inserted from the airlock into the microscope during operation. Thermocouple wire and heater cabling are visible.

7. Collection of hyperspectral maps

The electron microprobe software and hardware have been modified to allow hyperspectral collection from two Bruker SDDs, a CL spectrometer and the SXES in parallel, together with the wavelength-dispersive and electron signals [10]. This enables single pass mapping with full registration of all signals at the pixel level. The electron beam easily damages many minerals and materials during analysis/mapping and structural retention is as important as elemental stability for CL and SXE spectroscopy, making single pass mapping the key to reducing damage artefacts in the map data. The EPMA has been equipped with a sub-micrometre beam defocus allowing maps with a sub-micrometre step size to be matched with an equivalent beam spot size. This is possible in steps of 100 nm. At the completion of mapping, all data is combined into a single file and offline data processing is performed in the Chimage software [11].

8. SXES mapping

An application of the SXES is in the study of aerospace alloys which have been studied previously by EPMA using a combination of WDS and EDS [12, 13]. The SXES equipped EPMA was used to determine the distribution of alloying elements in an aluminium alloy, AA2099, containing Cu (2.4 %) and Li (1.6 %) as the active alloying components. The focus in this study was on constituent intermetallic (IM) particles (large particles, typically > 1 µm, comprising significant levels of impurities and alloying elements), where corrosion is observed to occur. One gap in the previous studies is that it was not possible to determine whether Li was present in the constituent particles. We have observed a number of IM particles and see copper depletion at high angle grain boundaries but not sub-grain boundaries, figure 6. The map is generated by summing intensities in a region-of-interest centred on Al-L. The map was acquired at 7 kV, 80 nA and 200 nm step size with a 1500 ms dwell per pixel. The Li in these grain boundaries does not appear to vary but fine copper particles are aggregated near or along grain boundaries giving rise to enhanced Al. This illustrates both the high spatial resolution and surface sensitivity of the SXES when coupled to an EPMA.

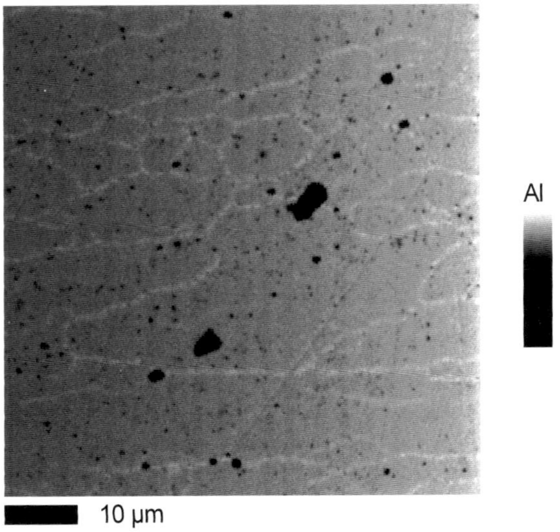

Figure 6. SXES map of an aerospace alloy AA2099 showing Al enrichment at grain boundaries due to copper depletion. The map was generated from a region-of-interest from Al-L.

9. Cold stage analysis using SXES

9.1. Thermal broadening

Measuring spectra at low temperature has the advantage of reducing thermal broadening of peaks and hence increasing the spectral resolution [14]. In the case of SXES we have examined the Al-L Fermi edge (72.5 eV) [15] at both room and LN temperatures (figure 7) and observed that the Al-peak shifts to higher energy by up to 30 meV at LN temperatures, but the base does not move, resulting in a sharper edge. Spectra were collected on the Al specimen at 7 kV, 40 nA, 1.5 s acquisitions averaged over 200 s. Using the energy difference between the 25 % and 75 % of the Al-peak height as a measure of the energy spread, we fit a line to the Fermi edge and at room temperature the energy resolution is measured to be 257 meV, while at LN 246 meV is observed. The broadening seen here is consistent with the Fermi-Dirac distribution at room temperature which follows the 4*k*T rule which for SXES has been described by Terauchi *et al* [6]. Previous work by Skinner has shown changes in Fermi-edge shapes with temperature although with lower spectral resolution [16].

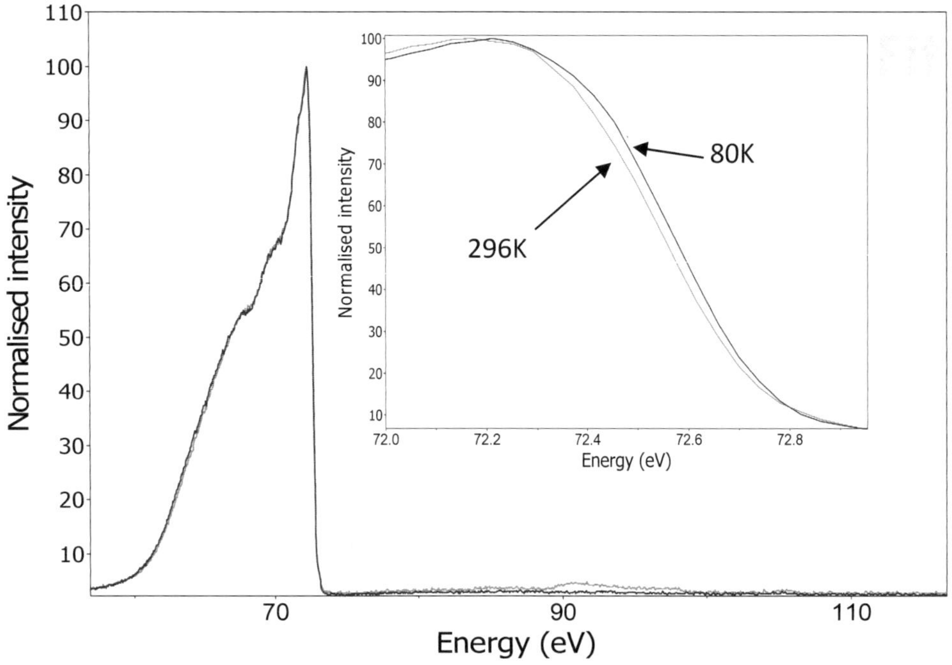

Figure 7. Al-L spectroscopy observed at 296 K and 80 K temperatures showing the Fermi edge at ~72eV. The zoomed in region, insert, shows the Fermi edge at room temperature becoming sharper at LN temperatures, 246 meV compared to 257 meV at room temperature.

9.2. Lithification of LiF

The spectroscopic analysis of ionically bonded materials such as LiF by SXES needs to be performed without inducing beam damage. When a focussed electron beam is employed then we have observed the lithification of the LiF. In figure 8, a Li-K peak at 54 eV is seen on LiF which is associated with the growth of Li metal. In addition we also observe a lower peak which we attribute to a Li oxide peak at 47.8 eV. To observe undamaged spectra from LiF we collected a spectra by defocussing the electron beam and collecting a map. The resulting sum spectra shows a split Li-K peak at 44.5 eV and 61 eV. By comparison the Li-K peak in metal is 54 eV. This measurement is in agreement with Li-K measurements performed using a zone plate installed at BESSEY II, where they reported Li-K in LiF at 46.5 eV and 61 eV [17]. The discrepancy in the lower Li-K peak, 44.5 eV, may be due to an energy calibration issue at the lower end of the SXES.

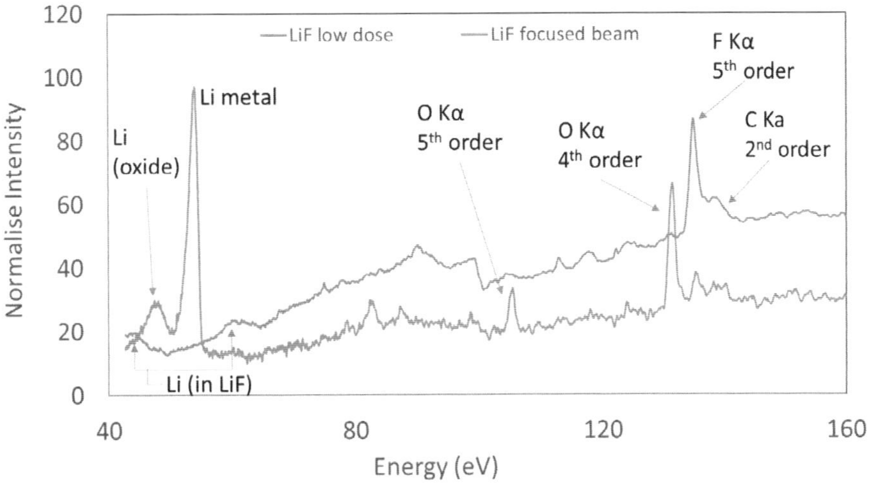

Figure 8. Spectra collected from LiF using a focussed electron beam, orange, which induces lithification resulting in a Li metal peak (54 eV) and Li oxide (47.8 eV). The low dose spectra from LiF reveal two Li-K peaks at 44.5 eV and 61 eV.

10. Cold stage mapping using cathodoluminescence

10.1. Carbonado diamond

Cathodoluminescence analysis performed on a carbonado diamond specimen mapped at room temperature and LN temperatures show marked differences. In figure 9a shows two CL spectra from the carbonado diamond; one collected at room temperature (blue) with the other at LN temperature (red). A slight sharpening of the main peak at 2.8 eV is observed in the LN spectra, while a dramatic sharpening of the nitrogen defects centred at 2.43 eV is observed in the LN spectra compared to the room temperature spectra. The diamond spectral resolution improves at LN temperatures due to the peaks associated with the nitrogen defects becoming narrower [18, 19]. In addition the intensity of the luminescence increases at LN temperature by at least a factor greater than two. Previous work on carbonado diamonds at room temperature showed a wide variety of defect peaks associated with diamond including the nitrogen with a neutral vacancy ($N\text{-}V^0$), the nitrogen pair with a vacancy (N_2V) and the group of three nitrogens with a vacancy (N_3V) [19-21]. The peaks from these defects are complex and are each composed of a zero phonon line and a set of lines from the vibronic system. However, at LN temperatures the finer peak resolution improves the stability of peak fitting. Figure 9b shows a panchromatic map of a carbonado diamond shows the intersection of two distinct crystal planes and enhanced emission is evident at the grain boundaries, possibly due to the incorporation of trace elements. The map was collected at 10 kV, 48 nA with 1 µm step size and the beam was defocussed to 1 µm, with the surface coated with a 3 nm carbon film.

10.2. Quartz-cement sandstone

A Triassic quartz-rich sandstone sampled from a deep well in the Southern Perth Basin has been investigated as a potential target for CO_2 geological storage [22]. Quantitative cathodoluminescence and trace analysis has been used to establish the magnitude and relative timing of diagenetic processes affecting the sandstone and their chemical fingerprint at the microscopic scale. Key to understanding the evolution of the quartz-cement has been the use of cathodoluminescence mapping and X-ray microanalysis to determine the source of the silica in the brine that has precipitated the quartz-cement [23]. The LN cold stage has been used in this study for two reasons; firstly as it offers the ability to map

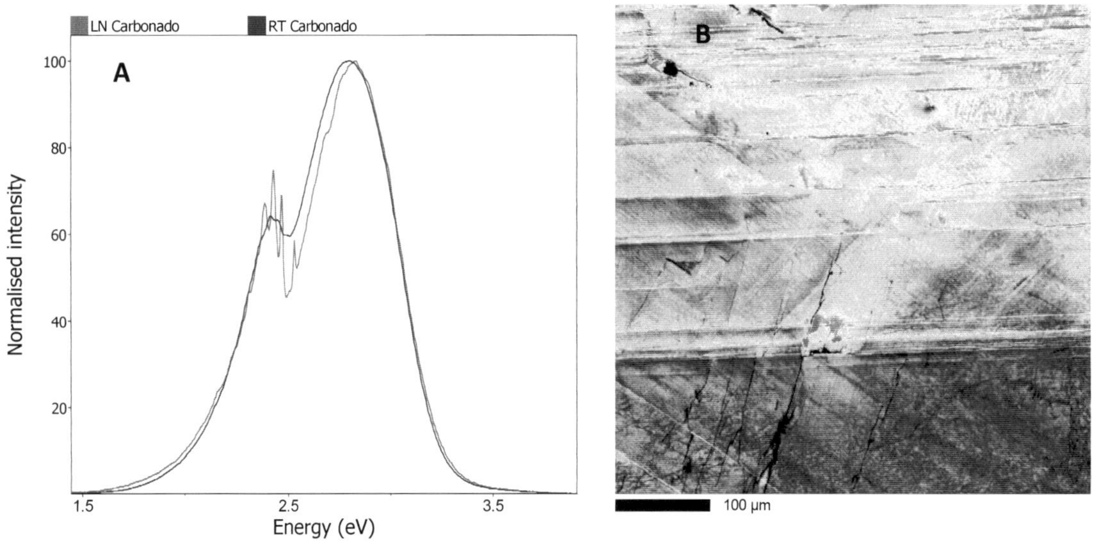

Figure 9. a) Cathodoluminescence spectra at 296 K (RT in blue) and at 80 K (LN in red). The nitrogen vacancies (2.4 eV) become sharper and hence easier to analyse at 80 K. b) Map of carbonado diamond collected at 80 K showing crystal growth planes.

with significantly lower damage compared to ambient conditions. Secondly, at LN temperatures we have observed an increase in CL emission intensity by up to a factor of five in both detrital and diagenetic quartz, figure 10, enabling lower beam currents to be used for mapping. A minor sharpening of the peaks centred at 1.91 eV occurs and the peak centred at 2.7 eV moves down to 2.50 eV.

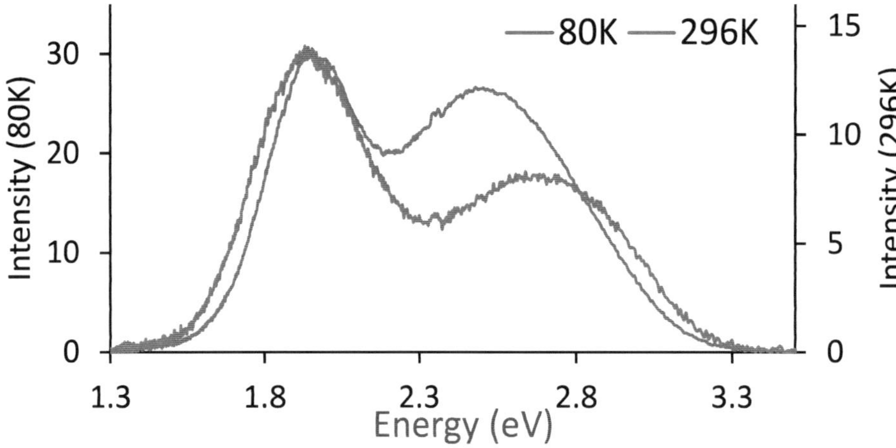

Figure 10. Cathodoluminescence spectra from a quartz-rich sandstone specimen collected at 296 K and 80 K (LN temperature). Both the intensity of the peak and resolution increase at LN temperatures.

The CL associated with the quartz-cement is usually weaker compared to the detrital quartz grains due to its lower formation temperature. To differentiate the quartz-cement from the detrital quartz we use a Si-K versus CL intensity scatter plot to show the different forms of quartz and then select the quartz with low CL intensity cluster, as shown in figure 11. The higher CL intensity clusters are from the two main, detrital quartz grains. The map was acquired at 80 K (cryogenic nitrogen) with beam conditions of 10 kV, 20 nA, 20 ms dwell, 1 µm step and defocus. Typically the CCD in the optical spectrometer is cooled to -15 °C and a long background collection is run immediately prior to collecting the map to determine the CCD response at the regulated temperature. The spectrometer Peltier cooler is regulated to maintain the temperature within a 0.1 °C window but the response of the detector electronics changes slightly with room temperature and so this is tracked pixel-by-pixel and compensated for using the blanked off pixels within the CCD which respond to these subtle temperature fluctuations. In figure 12, all pixels are selected and the growth history within the quartz grains are evident. The highly angular nature of the quartz-cement may indicate it is growing following the grain boundary orientation of the quartz grain. The lower large quartz grain, figure 12, shows some evidence of a crack that has been healed by infilling with quartz-cement (green line through the quartz grain).

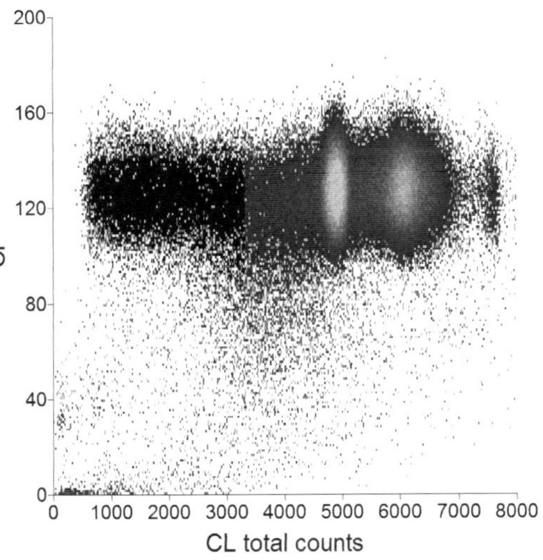

Figure 11. Scatter plot of CL versus Si K (WDS), the various clusters represent different quartz grain types, by choosing pixels from the lower CL intensity cluster we select the quartz-cement. The map shows the selected pixels in blue, which bond the quartz grains together.

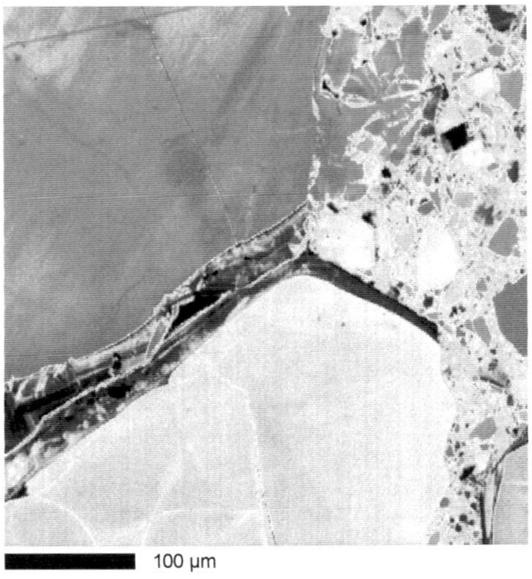

Figure 12. Map of quartz, same as figure 11, with all pixels selected. The infilling between adjacent quartz grains by the quartz-cement is evident.

IOP Conf. Series: Materials Science and Engineering **304** (2017) 012010 doi:10.1088/1757-899X/304/1/012010

To understand the role of trace elements within the quartz, in particular Ti and Al, we have performed quantitative point analysis on a number of quartz grains and the associated quart-cement [24, 25]. The technique we used was firstly to map a large area, 4 × 4mm at 2 μm step size at 20 kV, 30 nA with spot size set to match the step size, 2 μm. Critical for good CL images the surface finish needs to be as defect free as possible and this usually involves finishing the sample with a 20 nm colloidal silica polish. The polish may be further refined using a low energy ion beam to remove the amorphous layer left by the mechanical polishing, which may be desirable for high resolution CL investigations. This procedure is similar to sample preparation for electron backscatter diffraction (EBSD). The ion beam polisher can also clean any residual hydrocarbons or contaminants from the surface prior to coating with 10 - 15 nm of amorphous carbon. The carbon coat provides a conductive path for the electrons and has little absorption of the CL signal. Light absorption in carbon films is approximately linear across a range from 400 – 1,000 nm [1]. Clearly thinner carbon films are more desirable for use in CL due to the lower absorption, however they do not introduce non-linear absorption artefacts.

To quantify trace element concentrations by CL, the specimen is first mapped in the EPMA, with CL, EDS and WDS collected in parallel. Once the map is complete, coordinates for follow-up quantitative analysis are selected from quartz grains that have a range of CL intensities, since higher Ti-levels are generally associated with brighter grains [26, 27]. In addition a series of points on the quartz-cement were also loaded. The points were analysed at 20 kV, 120 nA, 10 μm spot and this gave detection limits on Ti of 20 ppm and Al was 55 ppm. A zero background synthetic quartz was analysed to determine the background offsets and these measurements were then used to correct the trace element levels. The vendor-supplied microanalysis software has been modified to allow negative *k*-ratios to be carried through the correction procedure. This allows a full scatter of trace levels and importantly the "zero" levels are scattered both above and below zero. The effect of allowing negative concentrations through can be seen in figure 13a, were the lowest Ti-levels scatter both above and below zero. To determine trace element correlations with CL peaks one first must determine the peak present, their positions and full width half maximum (FWHM). We use a CL database which has been described previously but now contains reference spectra and these can be used to compare with the collected spectra from the map [28]. The identified peaks in the quartz spectra are shown in Table 1. For fitting we describe the peaks using a Gaussian shape, which for quartz is adequate. Where forbidden transitions are observed in CL, lifetime broadening results in peaks with a Lorentzian or pseudo-Voight profile, which must be fitted appropriately. The described peaks are fitted at each pixel by locking each peaks FWHM but allowing the peak intensity to vary. Previous work has shown that the Ti^{4+} concentration in quartz is associated with the 2.73 - 2.83 eV peak and the Al can be correlated with the non-bridging oxygen hole centre (NBOHC) 1.945 eV peak [29, 30]. Finally to correlate the trace element measurements with the CL spectra, we average the CL spectra from the same 10 μm region in which the analysis was performed. Thus a fitted CL peak versus trace element plot is generated as shown in figures 13a and 13b. The quartz grains show a linear correlation between the Ti^{4+} centre and the Ti trace element analyses of the corresponding points, figure 13a. However, the quartz-cement does not show the same correlation due to the significantly lower formation temperature, resulting in a titanium solubility below measurable levels in the EPMA. The 1.95 eV peak shows a reverse correlation with Al-levels, figure 13b [31]. The detrital quartz and quartz-cement have differing linear gradients indicating that different incorporation mechanisms are likely to be involved, most likely associated with the temperature of formation.

Table 1. Peak centres and associated FWHM for the quartz map shown in figures 11 and 12.

Peak	Attribution	Position (eV)	FWHM (eV)
1	undocumented	1.55	0.79
2	NBOHC & Al	1.95	0.38
3	Defect	2.05	0.49
4	Ti^{4+}	2.73	0.99
5	Dangling NBOHC	2.83	0.85
6	Intrinsic	3.62	0.76

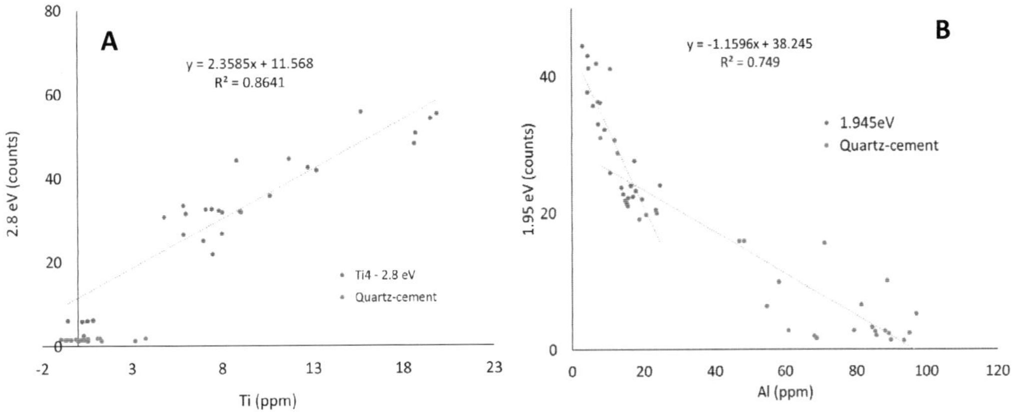

Figure 13. Quantitative elemental microanalyses compared to extracted emission centres. a) Ti^{4+} emission centre at 2.8 eV versus Ti shows a correlation in quartz with R^2 of 0.8641. b) Al emission centre at 1.95 eV shows negative correlations in quartz and to a lesser extent quartz-cement.

11. Summary

The EPMA offers the ability to collect a wealth of both chemical (EDS, WDS, SXES and CL) and structural information (SXES and CL) when used to map minerals and materials. To efficiently gather information from the array of spectrometers and detectors one wants to maximise the collected data but minimise the time to acquire. These conflicting criteria lead to operators employing large beam currents that in the case of the FEG can be quite destructive. To minimise the damage we cool the sample to LN temperatures and always defocus the beam to the step size of the map. Using this strategy we can minimise damage and achieve the desired fast collection and analysis of samples. The use of hyperspectral mapping or analysis, has significant advantages over traditional mapping in that it removes the need to have *a priori* knowledge and together with modern data processing strategies enables previously difficult to analyse elements and lines, i.e., Al-L and Li-K to be detected. This offers new insights into minerals and materials research.

Acknowledgement
The authors acknowledge the support of the Australian Research Council, LE130100087.

References

[1] MacRae C M, Wilson N C and Torpy A 2013 Hyperspectral cathodoluminescence *Mineralogy and Petrology* **107** 429 - 40

[2] MacRae C M, Wilson N C, Johnson S A, Phillips P L and Otsuki M 2005 Hyperspectral mapping - combining cathodoluminescence and X-ray collection in an electron microprobe *Microscopy Research and Technique* **67** 271-7

[3] Götze J 2002 Potential of cathodoluminescence (CL) microscopy and spectroscopy for the analysis of minerals and materials *Analytical Bioanalytical Chemistry* **374** 703-8
[4] Edwards P R, Martin R, O'Donnell K P and Watson I M 2003 Simultaneous composition mapping and hyperspectral cathodoluminescence imaging of InGaN epilayers *Physica Status Solidi C* 2474-7
[5] MacRae C M, Wilson N C and Brugger J 2009 Quantitative Cathodoluminescence Mapping with Application to a Kalgoorlie Scheelite *Microscopy and Microanalysis* **15** 222-30
[6] Terauchi M and Kawana M 2006 Soft-X-ray emission spectroscopy based on TEM—Toward a total electronic structure analysis *Ultramicroscopy* **106** 1069-75
[7] Takahashi H, Murano T, Takakura M, Asahina S, Terauchi M, Koike M, Imazono T, Koeda M and Nagano T 2015 Development of soft X-ray emission spectrometer for EPMA/SEM and its application *IOP Conference Series: Materials Science and Engineering* **109**
[8] Terauchi M, Takahashi H, Handa N, Murano T, Koike M, Kawachi T, Imazono T, Koeda M, Nagano T, Sasai H, Oue Y, Yonezawa Z and Kuramoto S 2012 Ultrasoft-X-ray emission spectroscopy using a newly designed wavelength-dispersive spectrometer attached to a transmission electron microscope *J Electron Microsc* **61** 1-8
[9] Demers H, MacRae C M, Wilson N C, Hovington P, Timoshevskii V, Gauvin R and Zaghib K 2016 Determination of Soft X-ray Emission Spectroscopy Parameters using Experimental Data for Quantitative Microanalysis *Microscopy and microanalysis* **22** 408-9
[10] MacRae C M, Wilson N C, Torpy A, Bergmann J and Takahashi H 2016 Collecting and Analysing -1.6eV -20keV Emission Spectra in an EPMA *Microscopy and microanalysis* **22** 410-1
[11] Harrowfield I R, MacRae C and Wilson N C 1993 Chemical imaging in electron microprobes. In: *Proceedings of the 27th Annual MAS meeting 1993*: Microbeam Analysis Society) pp 547-8
[12] Hughes A E, Glenn A M, Wilson N, Moffatt A, Morton A J and Buchheit R G 2013 A consistent description of intermetallic particle composition: An analysis of ten batches of AA2024-T3 *Surface and Interface Analysis* **45** 1558-63
[13] Hughes A E, MacRae C, Wilson N, Torpy A, Muster T H and Glenn A M 2010 Sheet AA2024-T3: a new investigation of microstructure and composition *Surface and Interface Analysis* **42** 334-8
[14] Callcott T A and Arakawa E T 1977 Temperature Dependence of the K X-Ray Emission Edge of Li *Physical Review Letters* **38** 442-5
[15] Catterall J A and Trotter J 1962 The broadening of soft x-ray emission edges in metals and alloys *Philosophical Magazine* **7** 671-6
[16] Skinner H W B 1940 The Soft X-Ray Spectroscopy of Solids. I. K- and L-Emission Spectra from Elements of the First Two Groups *Philosophical Transactions of the Royal Society A* **239** 95-134
[17] Hafner A, Anklamm L, Firsov A, Firsov A, Lochel H, Sokolov A, Gubzhokov R and Erko A 2015 Reflection zone plate wavelength-dispersive spectrometer for ultra-light elements measurements *Opt Express* **23** 29476-83
[18] Yacobi B G and Holt D B 1990 *Cathodoluminescence microscopy of inorganic solids,* (New York and London: Plenum Press) pp 1-292
[19] Wilson N C, MacRae C M, Torpy A, Davidson C J and Vicenzi E P 2012 Hyperspectral cathodoluminescence examination of defects in a carbonado diamond *Microscopy and Microanalysis* **18** 1303-12
[20] Rondeau B, Sautter V and Barjon J 2008 New columnar texture of carbonado: Cathodoluminescence study *Diamond and Related Materials* **17** 1897-901
[21] Kagi H, Takahashi K, Hidaka H and Masuda A 1994 Chemical properties of Central African carbonado and its genetic implications *Geochimica et Cosmochimica Acta* **58** 2629-1638
[22] Olierook H K H, Delle Piane C, Timms N E, Esteban L, Rezaee R, Mory A J and Hancock L 2014 Facies-based rock properties characterization for CO2 sequestration: GSWA Harvey 1 well, Western Australia *Marine and Petroleum Geology* **50** 83-102

[23] MacRae C M, Torpy A, Delle Piane C and Wilson N C 2017 EPMA characterisation of quartz and quartz-cement from a Triassic Sandstone *Microscopy and microanalysis* **In press**

[24] Leeman W P, MacRae C M, Wilson N C, Torpy A, Lee C A, Student J, Thomas J B and Vicenzi E P 2012 A Study of Cathodoluminescence and Trace Element Compositional Zoning in Natural Quartz from Volcanic Rocks: Mapping Ti Contents in Quartz *Microscopy and Microanalysis*

[25] Vasyukova O V, Goemann K, Kamenetsky V S, MacRae C M and Wilson N C 2012 Cathodoluminescence properties of quartz eyes from porphyry-type deposits: Implications for the origin of quartz *American Mineralogist* **98** 98-109

[26] Götze J, Plötze M and Habermann D 2001 Origin, spectral characteristics and practical applications of the cathodoluminescence (CL) of quartz – a review *Mineralogy and Petrology* **71** 225-50

[27] Wark D A and Watson E B 2004 The TITANiQ: A Titanium-in-Quartz thermometer *Geochimica Et Cosmochimica Acta* **68** A543-A

[28] MacRae C M and Wilson N C 2008 Luminescence database I - Minerals and materials *Microscopy and Microanalysis* **14** 184-204

[29] Stevens-Kalceff M A, Phillips M R, Moon A R and Kalceff W 2000 *Cathodoluminescence in Geosciences,* ed M Pagel, *et al.* (Berlin: Springer) pp 193-224

[30] Stevens-Kalceff M A 2009 Strategies for the Microcharacterization of Structural Defects in Wide Band Gap Materials Using Cathodoluminescence Microanalysis *Microscopy and Microanalysis* **15** 544-5

[31] Vasyukova O V, Goemann K, Kamenetsky V S, MacRae C M and Wilson N C 2013 Cathodoluminescence properties of quartz eyes from porphyry-type deposits: implications for the origin of quartz *American Mineralogist* **98** 98-109

IOP Conf. Series: Materials Science and Engineering **304** (2017) 012011 doi:10.1088/1757-899X/304/1/012011

In-situ TEM study on structural change and light emission of a multiwall carbon nanotube during Joule heating

K Nishikawa, K Asaka, H Nakahara and Y Saito

Nagoya University, Department of Quantum Engineering, Chikusa-ku, JP-464-8603 Nagoya, Japan

E-mail: nishikawa@surf.nagoya-u.ac.jp; ysaito@nagoya-u.jp

Abstract. Structure changes of a multiwall carbon nanotube (MWNT) during Joule heating were studied with simultaneous measurement of light emission spectra. The outer shells of the MWNT peeled off one by one because of excessive heating. All the peeled outer shells finally disappeared and inner shells whose tips were closed emerged, i.e., a new MWNT was formed. Each diameter of the shells comprising the MWNT decreased compared with those before the fracture. Light emission spectra during Joule heating of an MWNT were composed of both the blackbody radiation and characteristic peaks. The peaks in the light emission spectra shifted to higher energies in accordance with shrinkage of the inner shells. The energies of the peaks in the spectra corresponded to energy gaps between van Hove singularities calculated from the diameters of the shells, indicating that the peaks in the spectra are attributed to the interband electron transition in the MWNT.

1. Introduction

Carbon nanotube (CNT) can be regarded as a quasi-one-dimensional nanomaterial and has semiconducting or metallic properties depending on its structure. Electroluminescence [1-5], photoluminescence [6, 7] and thermal emission [8, 9] of CNTs have been reported. For thermal emission of multiwall carbon nanotubes (MWNTs), some broad emission peaks induced by electric current in addition to the blackbody radiation spectrum were observed in a scanning electron microscope (SEM) [10] and a scanning tunnelling microscope (STM) [11]. The emission peaks were interpreted to be derived from electronic transitions between van Hove singularities of the outermost shell of the MWNT. However, the contribution of each inner shell constituting an MWNT to the light emission has not been made clear yet. This is because SEM does not have spatial resolution enough to evaluate detailed structures of an MWNT, and STM cannot observe inner structures of an MWNT. In the present study, the inner structure changes of an MWNT during Joule heating were investigated by transmission electron microscopy (TEM) with simultaneous measurement of the light emission spectra.

2. Experimental

MWNTs were produced by the arc discharge method. MWNTs were attached to a knife edge of a gold (Au) plate by dielectrophoresis of an MWNT suspended in isopropanol. The Au plate was mounted on a movable stage in a specimen holder for TEM (JEM-2010, JEOL). In the microscope, a tip of an individual MWNT protruding from the Au plate (cathode) was brought into contact with an Au-coated tungsten needle which was fixed on a non-movable stage in the specimen holder and used as a counter electrode (anode). Structure changes during Joule heating of the MWNT were observed using a television system. The acceleration voltage of TEM was 120 kV. TEM images, bias voltage, electric current, and light emitted from the MWNT were simultaneously recorded using a home-made *in-situ* observation system. Schematic of the experimental setup of the system is shown in figure 1.

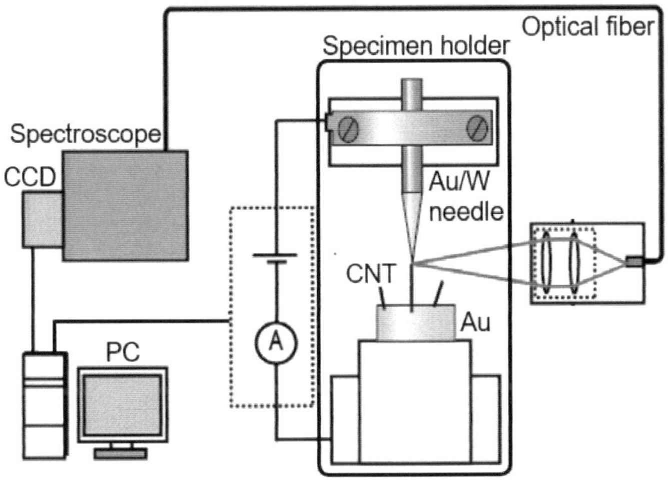

Figure 1. Schematic of our home-made in-situ observation system.

3. Results and discussion

Figure 2 shows TEM images of structure changes in an MWNT during Joule heating. Peeling of the outer shells one by one was observed. Changes of electric conductance are shown in figure 3a. Step-like drops of the conductance were caused by peeling of the outer shells. While peeling the outer shells, conductance did not drop to zero, suggesting that electric current passed through inner shells of the MWNT. The light emission spectra observed simultaneously with the TEM images are shown in figure 3b. Light emission peaks were observed besides blackbody radiation spectra. With the peeling of the outermost shells, the intensity of the spectra and current decreased at the same time, but the peak positions did not change. After the peeling of the outermost shells, the second outer shell peeled off as well. Repeating this process caused the complete fracture of the body of the MWNT. After the fracture of the body of the MWNT, the tips of inner shells maintained closed shell structures. In other words, a new MWNT was formed, as shown in figure 4. This new MWNT is named the second MWNT. After the fracture of an n-th MWNT, the remaining MWNT is named $(n+1)$-th MWNT hereafter.

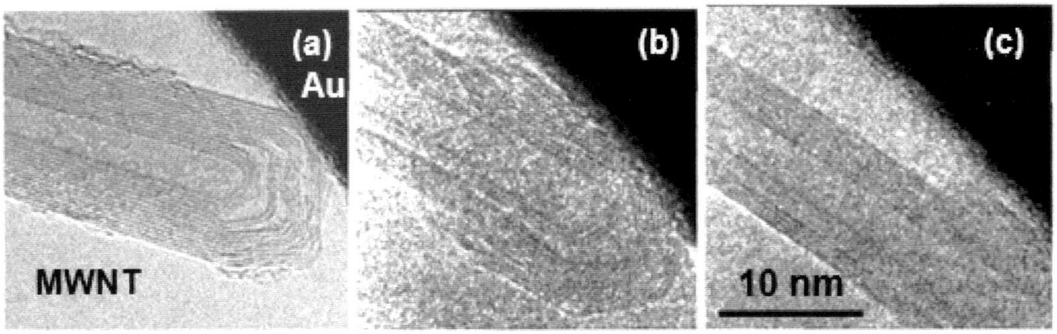

Figure 2. Peeling process of outer shells of an MWNT.

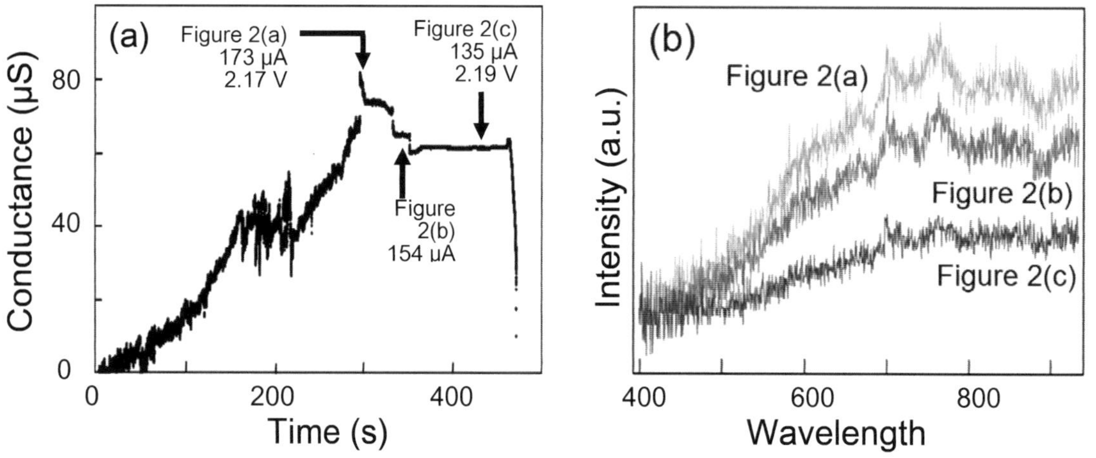

Figure 3. a) Changes in electric conductance of the MWNT shown in figure 2. Step-like drops of the conductance are caused by peeling of the outer shells. b) Spectra of light emission from the MWNT.

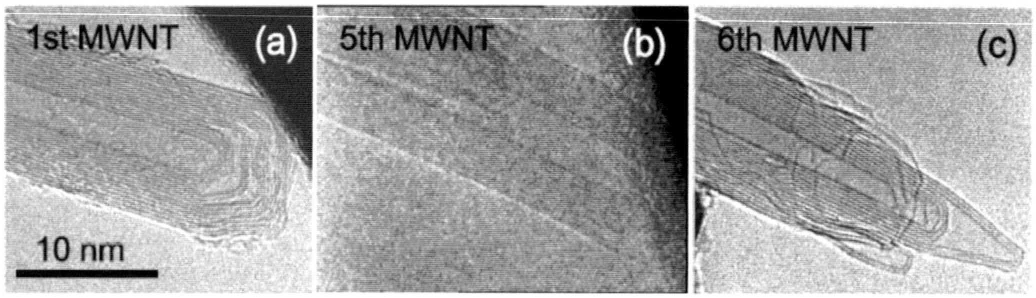

Figure 4. TEM images of 1st, 5th and 6th MWNTs. Shrinkage of the diameter of all shells and decrease of the number of shells composing the MWNT are observed.

TEM images of the 1st, 5th and 6th MWNT are shown in figures 4a, 4b and 4c respectively. The number of the shells in the 1st MWNT was 14. The number of the shells in the 5th MWNT and the 6th MWNT decreased from 14 to 11 and 7, respectively. In addition, the diameters of all shells of the formed MWNT were found to have decreased; the amount of shrinkage for all shells was about 0.4 - 0.7 nm. Figure 5 shows light emission spectra from the 1st, 5th and 6th MWNTs during Joule heating at a constant electric power of 390 μW (2.5 V and 155 μA). In these spectra, some emission peaks were observed. Changes of the peaks with the structure changes of the MWNT are as follows; i) the intensity of the peaks increases, ii) the shape of the peaks becomes sharp, and iii) the peaks slightly shift to higher energy. The peaks in the light emission spectra from the 6th MWNT appear at 550, 590, 677, 729, 771, 790, 815 and 890 nm. In the previous study on light emission from a MWNT using STM, it was suggested that the photon energies corresponded to energy gaps between van Hove singularities in electric density of state of the outermost shell [11]. In the present study, however, photon energies from the 6th MWNT did not completely correspond to estimated energy gaps even if the outermost shell was regarded as either a semiconducting or a metallic shell. Then, the energy gaps were calculated from all the diameters including inner shells, as shown in table 1. The broad peak at around 590 nm in figure 5, for example, can be explained by 575, 585, 600, 606 and 616 nm in this table. Also, the breakdown of

IOP Conf. Series: Materials Science and Engineering **304** (2017) 012011 doi:10.1088/1757-899X/304/1/012011

outer shells was observed during Joule heating. So, the temperature of the MWNT was expected to be higher than 2,200 K [10]. In this situation, not only semiconducting shells of the MWNT but also metallic ones may contribute to the light emission because electrons in the CNT are thermally excited into van Hove singularities above the Fermi level and lead to the recombination with created holes, resulting in the emission of photons.

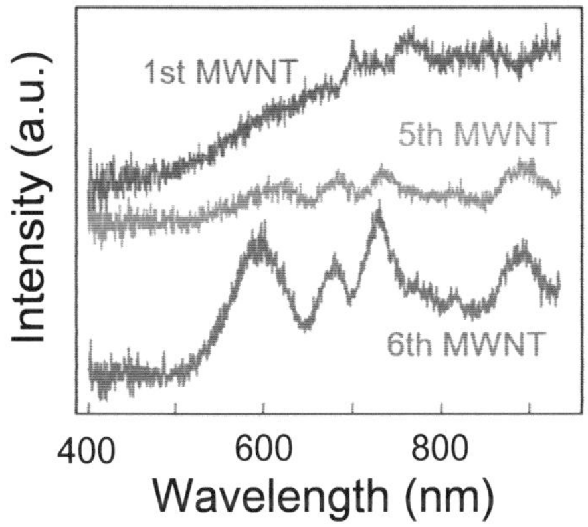

Figure 5. Light emission spectra at a constant electrical power (390 μW). The peak positions shifted to higher energy due to the shrinkage of tube diameters after the fracture.

Table 1. Expected peak wavelengths calculated from diameters of inner shells of the 6th MWNT. a) In case that each shell is semiconducting. b) In case that each shell is metallic.

(a)	SHELL No.						
	1	2	3	4	5	6	7
diameter (nm)	2.17	2.91	3.62	4.43	5.05	5.73	6.38
E_{33}	816	1096	1364	1667	1900	2157	2401
E_{44}	652	877	1091	1334	1520	1726	1921
E_{55}	466	626	779	953	1086	1233	1372
E_{66}	408	548	682	834	950	1078	1201
E_{77}	326	438	545	667	760	863	961
E_{88}		399	496	606	691	784	873
E_{99}			420	513	585	664	739
E_{1010}			390	476	543	616	686
E_{1111}				417	475	539	600
E_{1212}				392	447	508	565
E_{1313}					400	454	506
E_{1414}					380	431	480
E_{1515}						392	437

(b)	SHELL No.						
	1	2	3	4	5	6	7
diameter (nm)	2.17	2.91	3.62	4.43	5.05	5.73	6.38
E_{22}	544	731	909	1111	1267	1438	1601
E_{33}	362	487	606	741	844	959	1067
E_{44}		365	455	556	633	719	800
E_{55}			364	445	507	575	640
E_{66}				370	422	479	534
E_{77}					362	411	457
E_{88}						359	400
E_{99}							356

The aforementioned changes of the emission peaks in the spectra (figure 5) can be explained as follows: reduction of the diameter of the shells brings about the widening of the energy gaps between van Hove singularities, and causes the shift of the peak positions to higher energy. In the new MWNT (after the fracture), the number of the shells which contributes to light emission decreased as compared with the old MWNT (before the fracture). So, the peaks in the spectra related with inner shells after the fracture are expected to be more distinctive than those before the fracture.

Since the inner shells contribute to current flow, as revealed in figure 3a, the inner shells are heated to high temperatures due to Joule heating to excite electrons thermally to high energy levels. The present observation shows that not only outer shells but also inner shells contribute to the light emission from an MWNT through the interband transitions of thermally excited electrons.

4. Conclusion

We investigated the structure changes and light emission mechanism of an MWNT during Joule heating by in-situ TEM. Peeling of the outer shells of the MWNT and fracture of its body were observed. In the light emission spectra, some peaks besides blackbody radiation were observed. After the structure changes of the MWNT, the decrease of the peak intensity and the shifts of the peaks were observed, which are due to the reduction of the diameter of shells contributing the light emission. The wavelengths of the light emission peaks could be explained by energy gaps between van Hove singularities of the MWNT. The present results showed that the light emission from an MWNT during Joule heating contains interband transitions in inner shells of the MWNT as well as outer shells.

References

[1] Freitag M, Martin Y, Misewich J A, Martel R and Avouris Ph 2003 *Nano Lett.* **3** 1067
[2] Hibino N, Suzuki S, Wakahara H, Jobayashi Y, Sato T and Maki H 2011 *ACS Nano* **5** 1215
[3] Janas D, Czechowski N, Krajnik B and Mackowski S 2013 *Appl. Phys. Lett.* **102** 181104
[4] Mueller T, Kinoshita M, Steiner M, Perebeinos V, Bol A A, Farmer D B and Avouris P 2010 *Nat. Nanotechnol.* **5** 27
[5] Zhang Y and Iijima S 1999 *Phys. Rev. Lett.* **82** 3472
[6] Finnie P, Homma Y and Lefebvre J 2005 *Phys. Rev. Lett.* **94** 247401
[7] Lefebvre J, Fraser J M, Finnie P and Homma Y 2004 *Phys. Rev.* B **69** 075403
[8] Liu Z, Bushmaker A, Aykol M and Cronin S B 2011 *ACS Nano* **5** 4634-4640
[9] Mann D, Kato Y K, Kinkhabwala A, Pop E, Cao J, Wang X, Zhang L, Wang Q, Guo J and Dai H 2007 *Nat. Nanotechnol.* **2** 33
[10] Cai X, Akita S and Nakayama Y 2004 *Thin Solid Films* **464-465** 364
[11] Uemura T, Yamaguchi S, Akai-Kasaya M, Saito A, Aono M and Kuwahara Y 2006 *Surface Sci.* **600** L15

IOP Conf. Series: Materials Science and Engineering **304** (2017) 012012 doi:10.1088/1757-899X/304/1/012012

Secondary fluorescence near phase boundaries – Typical cases in steel

M R Rijnders and C J G van Hoek

TATA Steel Research and Development, Postbus 1000, NL-1970 CA IJmuiden, The Netherlands

E-mail: marco.rijnders@tatasteel.com

Abstract. Secondary fluorescence (SF) from across a phase boundary can cause spurious analytical results. Two cases of SF in steel research are presented: Zn in Zn-coated steel across the interface, and Fe in alumina particles embedded in a steel matrix. Wavelength-dispersive X-ray spectrometry (WDS) and energy-dispersive X-ray spectrometry (EDS) measurements were performed both on couples of juxtaposed Fe/Zn as well as equivalent samples that were production-derived (i.e., high-T exposed). Subsequently the SF effect was modelled using CalcZAF to compare the measured profiles with the theoretical ones. For Zn in Fe it is shown that the calculations are in close agreement with the measurements, although some discrepancies need a more detailed research. This work shows the importance of the use of models to support the interpretation of experimental WDS and EDS results on coated steel samples where phase boundaries are involved.

1. Introduction

Secondary fluorescence (SF) near phase boundaries is a well-known but sometimes overlooked phenomenon that causes observed element concentrations to be higher than they actually are. X-rays of energy higher than the absorption edge (E_c) of the measured line, whether they are derived from characteristic lines or from the background continuum, are the cause of such SF. Spurious element concentrations may for example have serious consequences for phase equilibrium models and thermodynamic database development.

We frequently need to measure concentration profiles across or close to a phase boundary, e.g., particle in a steel matrix, coating on a steel substrate or phase boundaries within the bulk. As X-rays can travel tens to hundreds of micrometres through the material, secondary fluorescence will originate from across the boundary, resulting in an apparent concentration that is higher than the actual (as well as resulting in higher than 100 % analytical totals). Due to the fact that any X-ray with energy > E_c may cause fluorescence, bremsstrahlung will contribute to SF. Therefore, near phase boundaries, apparent concentrations of elements heavier than the matrix (e.g., Fe in Al adjacent to Fe or Zn in Fe adjacent to Zn) can be detected.

The effect of SF across boundaries is well-known (in alloys [1], in geochronology [2], in thermobarometry [3, 4], and in metal/metal couples [5-7]), and methods that correct for the effect are available [8-12]. It is especially relevant to the analysis of trace elements if the neighbouring phase also contains a major concentration of the element of interest.

We have analysed fluorescence of iron and zinc near the coating/interface substrate of galvanised material and an undiffused Fe-Zn couple (case 1), and fluorescence of Fe in Al_2O_3 embedded in a steel matrix (case 2).

2. Method

Analyses were performed on a CAMECA SX-100 EPMA (tungsten cathode), or a JEOL 7001F FEG SEM equipped with two silicon drift detectors. Different accelerating voltages (AV) and beam currents (b.c.) were applied depending on the analysis at hand. A focussed beam was used in all measurements. Acquisition times were kept at 60 seconds. The freeware CALCZAF [13], which implements the numerical model described in [10], was subsequently used to model the secondary fluorescence effects and help explain the experimental results.

3. Results and discussion

3.1. Case 1. Fe/Zn: apparent Zn concentration in steel substrate

Steel is hot-dip galvanised for corrosion-protection. The zinc coating thickness usually is a few tens of micrometres (figure 1a). An important characteristic of galvanised iron (GI) material is that the zinc coating consists of almost pure Zn. During analysis of a Zn/steel (DP800) interface in a galvanised sample, we were faced with Zn concentration profiles that extended considerably (tens of μm) into the steel matrix. However when a de-zinced sample from the same material was analysed the effect was not observed (figure 1b). The question then arose to which extent secondary continuum fluorescence could contribute to the measured Zn profile. Because the room temperature solubility of Zn in Fe is low but not zero, the presence of a Zn diffusion profile after galvanising cannot be ruled out. To experimentally test the fluorescence, a non-diffused Fe/Zn couple ("model couple") was produced by juxtaposing flat-polished steel (0.7Mn-0.06Al) and zinc samples in a single mount. A WDS linescan was performed on the model couple under the same instrumental conditions as the GI sample and the results were compared (figure 1c). Significant (a few wt%) concentrations of Zn in Fe and Fe in Zn were measured in the model couple despite it not having reacted. The measured profile from the undiffused couple is for zinc very similar to that of the galvanised sample, with an even higher apparent concentration of zinc (figure 1c). This shows that bremsstrahlung that was excited in the steel many micrometres away from the steel/Zn interface is causing the Zn to fluoresce, and that there was no diffusion of zinc into the steel (galvanised sample). Note that the detection limit [14] for Zn under the current conditions was about 0.034 wt%. At distances over 40 μm away from the interface, the measured Zn-concentration starts to fall below this limit. The Fe concentration profile in Zn was also measured in the model couple. The detection limit for Fe under the current conditions is about 0.019 wt%. At a distance of 50 μm away from the interface the Fe-concentration (0.08 wt%) was still higher than the detection limit.

Both Zn and Fe profiles were consequently modelled with CALCZAF. Figures 1d and 1e show the results of the calculation compared to the analysis results for the undiffused couple. Overall, the agreement between the model and the measured values is satisfactory. It shows that spurious Fe can be detected up to 75 μm into Zn, and spurious Zn up to 37 μm into Fe, respectively. These numbers will depend of course on the analytical conditions.

The apparently lower concentration of Zn in steel in the galvanised samples as compared to the undiffused couple and the calculation (figures 1c and 1d) may be due to the fact that the adjacent phase is only 20 μm thick instead of semi-infinite, so that some X-rays will pass through the coating without causing SF. The semi-infinite couple with a straight boundary is the only geometry available for SF calculation in CALCZAF. For modelling more complex geometries, other software [10, 12] is available.

3.2. Case 2. Fe/Al_2O_3: apparent iron concentration in aluminium oxide inclusions

Alumina (Al_2O_3) particles (d $\leq$ 10 μm) in a steel matrix were studied in relation to clogging. Earlier EDS measurements [15] had shown presence of small amounts of iron that could not be thermodynamically explained, since iron solubility in alumina is negligible. Alumina powder retrieved from the same clogging deposit showed no presence of iron, contradicting the EDS results.

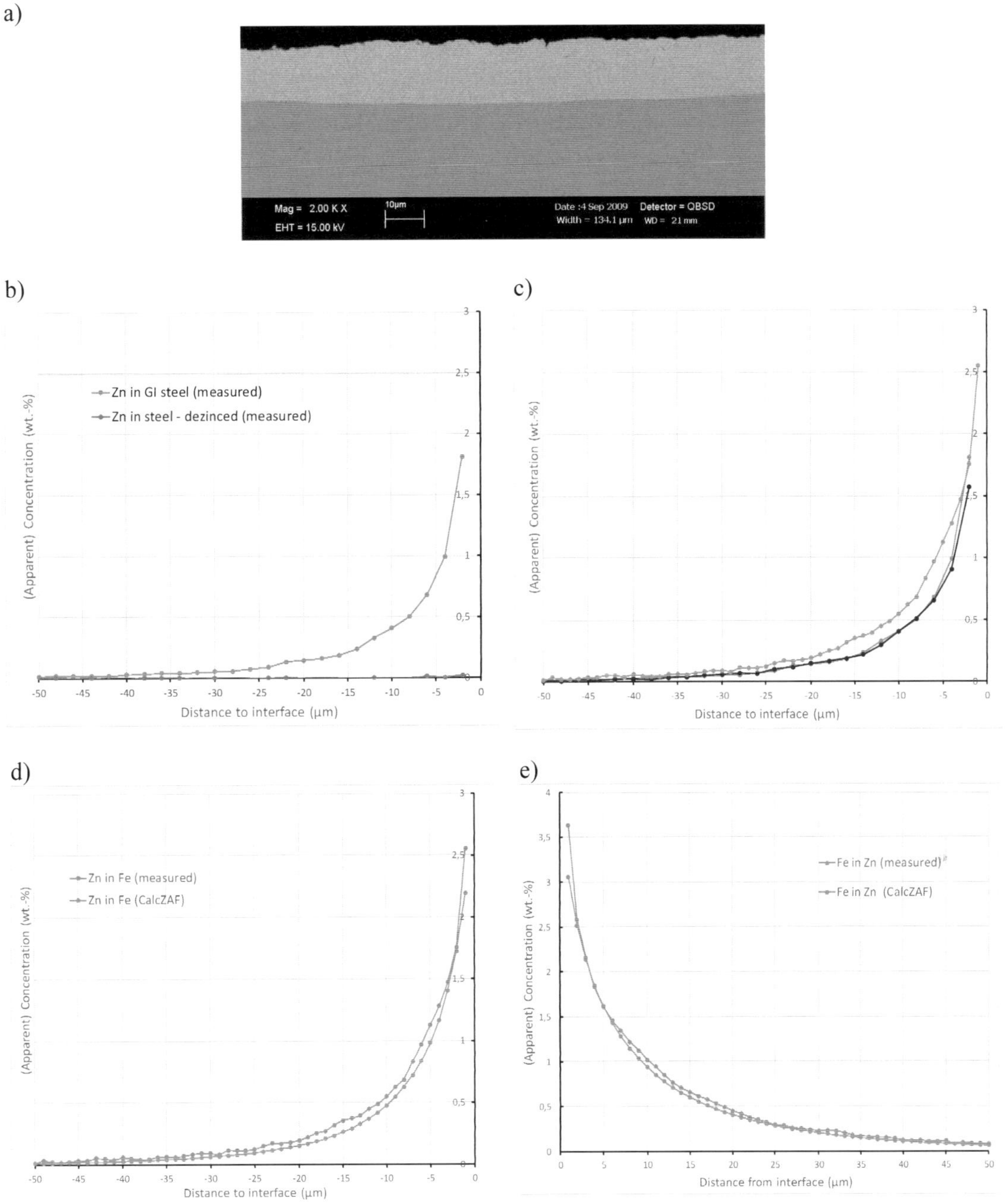

Figure 1. a) SEM BSE image of GI coating, polished cross-section. b) Comparison of zinc concentration profiles of a GI (galvanised, blue) and a de-zinced (red) sample; WDS, 20 kV, 50 nA, peak time 60 s; x = 0 corresponds to the sample surface in the case of the dezinced sample. c) Comparison of measured zinc concentration profiles of a GI sample (blue) and of an undiffused Fe--Zn couple (grey); conditions as in figure 1b. d) Apparent concentration of zinc in iron adjacent to zinc at 20 kV, measurement in undiffused couple (blue) and calculated values (orange). e) Apparent concentration of iron in zinc adjacent to iron at 20 kV, measurement in undiffused couple (green) and calculated values (orange).

EDS linescans were performed at 10 kV and 60 s analysis time per point; both on a bulk steel/alumina couple and on alumina particles embedded in a steel matrix. Results on the Fe/Al_2O_3 interface in a clogging sample (figure 2a) are shown in figure 2b. The same figure also shows the calculated values. Both the Al and the Fe profile depart significantly from the calculated ones. The measured apparent concentration of Fe close to the interface is about 2 wt% and shows a decrease to 0.4 wt% at a position 50 μm away from the interface. However, as may be appreciated from figure 2a, the interface shape is not ideal, and moreover it is not known below the sample surface. The calculation assumes a perpendicular and flat interface between the two materials. The effect of interface orientation is discussed in two recent papers [11, 12].

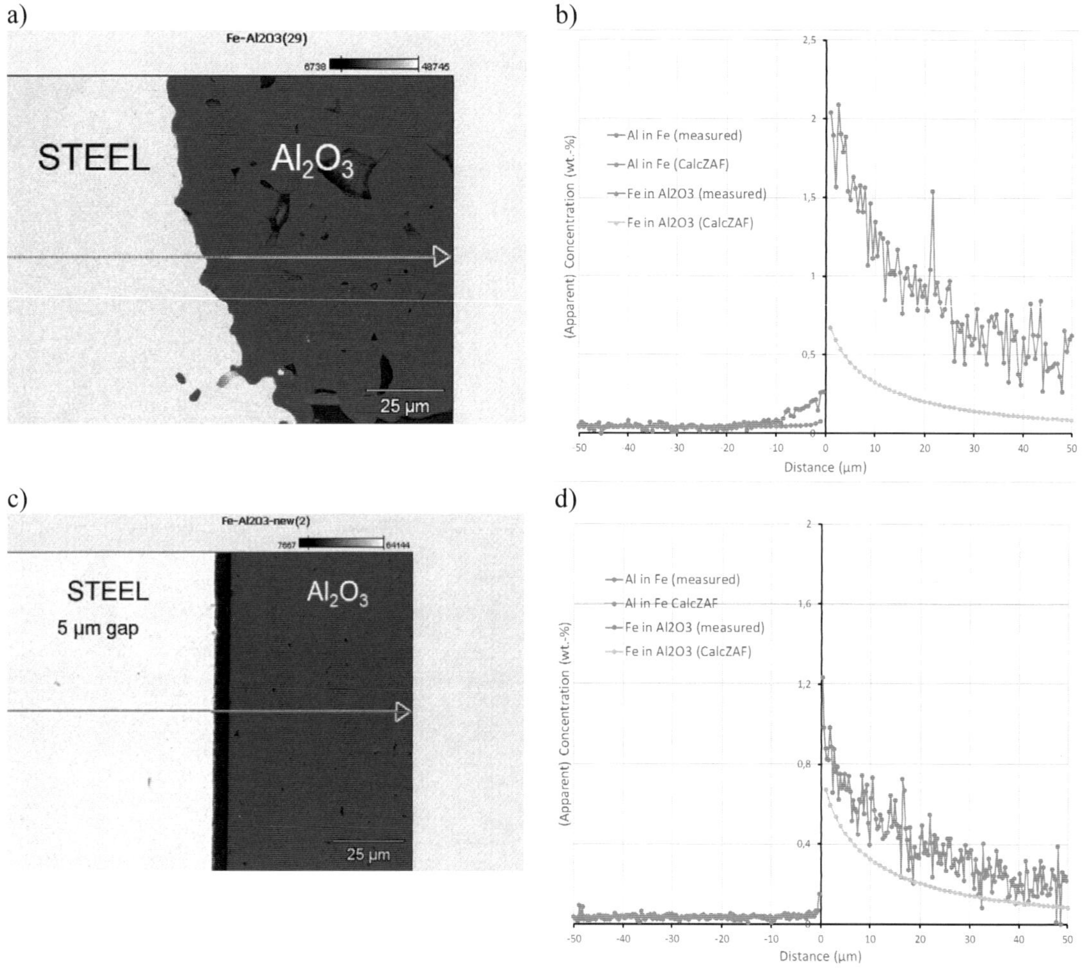

Figure 2. a) Path of the linescan performed on an Fe/Al_2O_3 sample from a steel clog. b) Apparent concentration of Al in steel (blue) and Fe in Al_2O_3 (green) (production sample) and calculated values (orange, yellow); EDS 10 kV, 60 s. c) SE image of undiffused Fe/Al_2O_3 couple; note gap of 5 μm. d) Apparent concentration of Al in steel (blue) and Fe in Al_2O_3 (green) in undiffused couple, and calculated values (orange, yellow).

A model Fe/Al_2O_3 couple (figure 2c) was then constructed and analysed. The results are shown in figure 2d. In this case the apparent Al-profile shows a good fit with the calculation. At distances > 2 μm away from the interface, the secondary Al fluorescence caused by Al_2O_3 is negligible. The apparent Fe-profile still shows a discrepancy with the calculation, albeit less than with the clogging sample. The interfacial measured Fe-concentration lies around 1 wt%, where the model predicts 0.7 wt%, and decreasing to 0.2 wt% at 50 μm away from the interface, where the model predicts 0.08 wt%. Possible causes for discrepancy may be: 1) X-ray intensity reaching the detector is somewhat dependent on the orientation of the interface with respect to the detector [16], 2) inaccuracies in the ZAF correction. Because of the 5 μm gap between the end members (gap filled with epoxy resin), the actual measured interfacial Fe-concentration would be somewhat higher should the gap have been zero.

Note that the spurious Fe concentration is solely due to bremsstrahlung generated in the Al_2O_3.

Linescan profiles across alumina particles (figures 3a-d) show that the apparent Fe concentration is almost constant (~ 2 wt%, comparable to the interfacial concentration measured in the bulk case) throughout the particle, and does only decrease slightly when the beam moves away from the interface, as opposed to the bulk case, where the decrease continues for tens of micrometres. Similar results for spurious Fe in 15 - 20 μm copper particles embedded in Fe-rich slag have been reported by Llovet *et al.* [17]. The size and shape of the particle, which are *de facto* unknown, will certainly influence the amount of SF that is generated. Qualitatively, the particle results can be understood as follows: compared to the case of a flat interface, SF will be higher, since the particle is surrounded by the fluorescing iron matrix. As the beam moves away from one interface, it will approach the other one, i.e., decreasing SF from one side of the particle is balanced by increasing SF from the other side. The subtle difference between the two measured curves might be due to the difference in shape and volume of the particles. It would take more modelling effort to clarify this observation.

4. Conclusions

Steel research often involves the measurement of element concentration profiles across interfaces, where secondary fluorescence effects can affect the calculated element concentrations. Even though the lighter element Kα cannot directly excite Kα of a heavier element, SF can still occur due to the high-energy components of the bremsstrahlung, resulting in overestimation of the element concentration close to the interface. The effect can extend tens of μm away from the interface.

As shown by the analysis of undiffused metal/metal couples, secondary fluorescence across an interface is accurately predicted by CALCZAF for the Fe-Zn couple. The apparent concentration of Fe in Al_2O_3 could not be exactly matched with a CALCZAF calculation.

If accurate analyses near interfaces are needed, the analyst must be wary. Where possible, analytical results should be supported by model results.

Acknowledgement
The authors owe thanks to Bing Doeve and Frank van der Does for sample preparation.

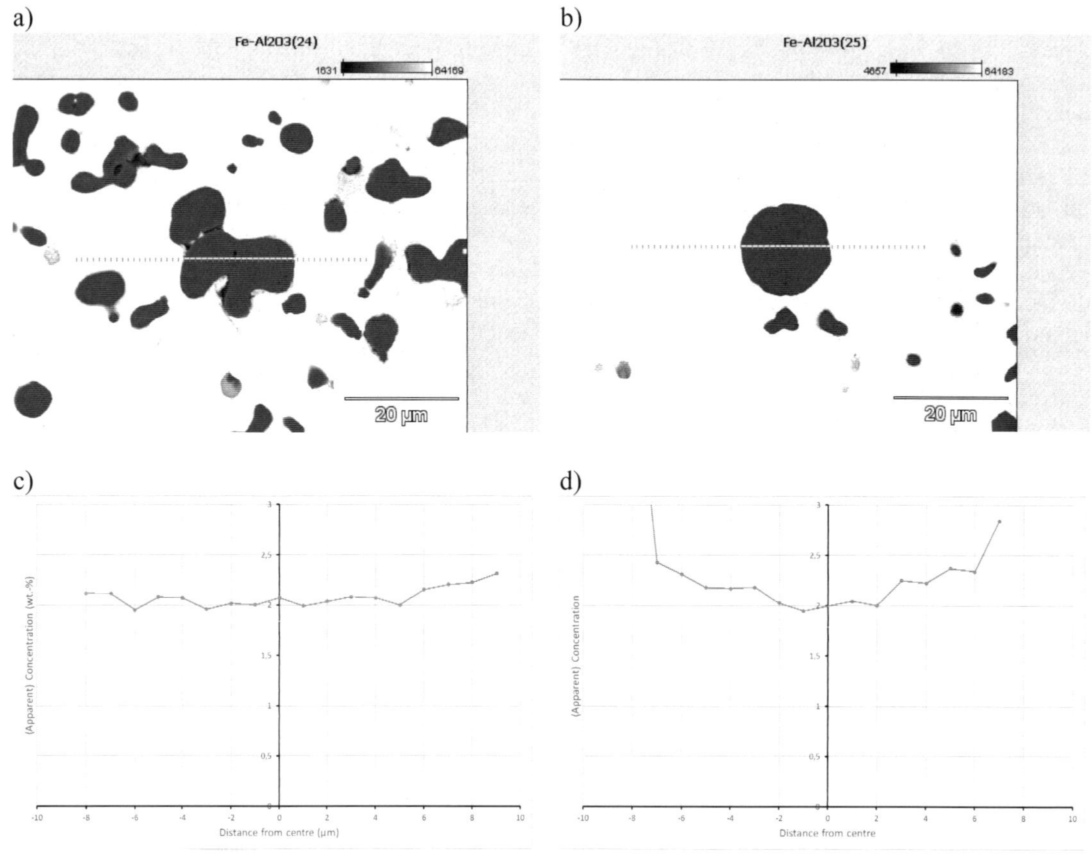

Figure 3. a) Alumina particle 1. b) Alumina particle 2; apparent diameter 16 µm. c) Fe concentration profile of particle 1 shown in figure 3a; essentially a flat profile. d) Fe concentration profile of particle 2 shown in figure 3b; slight decrease towards centre of particle.

References

[1] Fournelle J H, Kim S and Perepezko J H 2005 Monte Carlo simulation of Nb Kα secondary fluorescence in EPMA: comparison of PENELOPE simulations with experimental results. *Surf. Interface Anal.* **37** 1012-1016
[2] Jercinovic M J, Williams M L and Lane E D 2008 In-situ trace element analysis of monazite and other fine-grained accessory minerals by EPMA. *Chem. Geol.* **254** 197-215
[3] Dalton J A and Lane S J 1996 Electron microprobe analysis of Ca in olivine close to grain boundaries: the problem of secondary X-ray fluorescence. *Am. Mineral.* **81** 194-201
[4] Llovet X and Galán G 2003 Correction of secondary X-ray fluorescence near grain boundaries in electron microprobe analysis: Application to thermobarometry of spinel lherzolites. *Am. Mineral.* **88** 121-130
[5] Bastin G F, van Loo F J J, Vosters P J C and Vrolijk J W G A 1984 An iterative procedure for the correction of secondary fluorescence effects in electron-probe microanalysis near phase boundaries. *Spectrochim. Acta B* **39** 1517-1522
[6] Arnould O and Hild F 2002 EPMA measurements of diffusion profiles at the submicrometre scale. *Mikrochim. Acta* **139** 3-10

[7] Zoukel A and Khouchaf L 2014 The secondary X-ray fluorescence and absorption near the interface of multi-material: Case of EDS analysis. *Micron* **67** 81-89
[8] Escuder J A, Salvat F, Llovet X and Donovan J J 2010 Numerical correction for secondary fluorescence across phase boundaries in EPMA. *IOP Conf. Ser. Mater. Sci. Engng.* **7** 1-7
[9] Llovet X, Pinard P T, Donovan J J and Salvat F 2012 Secondary fluorescence in electron probe microanalysis of material couples. *J. Phys. D: Appl. Phys.* **45** 225301
[10] Llovet X and Salvat F 2016 PENEPMA, a Monte Carlo programme for the simulation of X-ray emission in EPMA. *IOP Conf. Ser. Mater. Sci. Engng.* **109** 1-13
[11] Ritchie N W M 2017 Efficient simulation of secondary fluorescence via NIST DTSA-II Monte Carlo. *Microsc. Microanal.* **23** 618-633
[12] Llovet X and Salvat F 2017 PENEPMA: a Monte Carlo program for the simulation of X-ray emission in electron probe microanalysis. *Microsc. Microanal.* **23** 634-646
[13] http://probesoftware.com/Technical.html
[14] Ancey M, Bastenaire F and Tixier R 1977 Statistical control and optimization of X-ray intensity measurements. *J. Phys. D: Appl. Phys.* **10** 817-830
[15] Tata Steel unpublished results
[16] Llovet X 2017 personal communication
[17] Llovet X, Valovirta E and Heikinheimo E 2000 Monte Carlo simulation of secondary fluorescence in small particles at phase boundaries. *Mikrochim. Acta* **132** 205-212

IOP Conf. Series: Materials Science and Engineering **304** (2017) 012013 doi:10.1088/1757-899X/304/1/012013

Exploring the limits of EDS microanalysis: rare earth element analyses

N W M Ritchie[1], D E Newbury[1], H Lowers[2] and M Mengason[1]

[1] National Institute of Standards and Technology (NIST), Gaithersburg, MD, 20899 USA
[2] U.S. Geological Survey, Denver, CO, 80225 USA

E-mail: nicholas.ritchie@nist.gov

Abstract. It is a great time to be a microanalyst. After a few decades of incremental progress in energy-dispersive X-ray spectrometry (EDS), the last decade has seen the accuracy and precision surge forward. Today, the question is not whether EDS is generally useful but to identify the types of problems for which wavelength-dispersive X-ray spectrometry remains the better choice. The full extent of EDS's capabilities has surprised many. Low Z, low energy, and trace element detection have been demonstrated even in the presence of extreme peak interferences. In this paper, we will summarise the state-of-the-art and investigate a challenging problem domain, the analysis of minerals bearing multiple rare-earth elements.

1. Introduction

In 1973, Reed suggested that the then novel energy-dispersive X-ray spectrometer (EDS) could perform quantitative measurements of mass fraction much as the wavelength-dispersive X-ray spectrometer (WDS) had been doing for decades [1]. This was a radical notion. EDS detectors had only been introduced a few years earlier [2] and initial incarnations had resolutions in excess of 500 eV at Mn-Kα. The developer of the filter-fitting technique recalled how much controversy his first paper presenting quantitative results from filter-fit spectra generated [3, 4]. Regardless of the challenges, people have been performing truly quantitative measurements with EDS detectors since the early days. This isn't new. What is new is the accuracy and precision that can be achieved with the EDS detector even when there are severe peak interferences.

Over the decades, there have been a series of incremental and revolutionary improvements in detector and electronics technology. Detector resolutions have improved. As leakage currents were reduced, dead layers trimmed, and analogue electronics and digitizers were improved, resolutions fell from hundreds of eV to approximately 135 eV at Mn-Kα at count rates of around 2,000 counts/s. New types of windows were developed. The initial Be windows were replaced with boron-nitride, diamond, polymer, and more recently silicon nitride windows. These new window technologies permitted lower energy X-rays to reach the detector. In the early days, the Na-K line was the practical limit, but today EDS detectors routinely detect Be-K X-rays and, some even, Li-K X-rays [5]. Digital pulse processors further enhanced throughput and resolution and pushed the best lithium-drifted silicon detectors (Si(Li) detectors) into the 130 eV range at throughputs of up to 5,000 counts/s. Electronic stability improved and spectra became more reproducible.

The latest revolution, the silicon drift detector (SDD), has brought bountiful improvements in throughput, size, resolution, and stability [6–8]. A modern SDD is capable of orders-of-magnitude higher throughput and resolutions that approach 120 eV at Mn-Kα - close to the theoretical limit for Si-based detectors [9].

Each of these improvements in EDS technology has extended the range of measurements that can be performed. By the 1990's, improvements in electronics and windows made measurements of carbides, nitrides, oxides, and fluorides possible. However, it was only with the higher throughput associated with SDD that the EDS detector is truly able to encroach onto the precision and accuracy domain formerly ruled by WDS. For many measurements, the SDD can match or exceed the accuracy and precision of a WDS measurement in a fraction of the instrument and operator time [10]. Element interferences can be handled accurately with high count spectra and careful fitting to measured spectra. Soft X-rays (like carbides, oxides, and nitrides) can be quantified as well as (or even better than) a WDS [11,12]. Even higher level trace measurements (1.0 % mass fraction > C_z > 0.1 % mass fraction, where C_z is the mass fraction of element z) can be performed accurately with an SDD. Regardless, the SDD is unlikely to ever challenge the WDS at the lowest measurable trace levels [13] due to inferior peak-to-background ratios.

Ultimately, limits of detection are constrained by the magnitude of the background (continuum) level relative to the signal (characteristic) level. In the simple case of unobstructed lines, the limits of detection, C_{DL}, can be estimated from the characteristic intensity measured on a standard and the continuum intensity measured on the unknown.

$$C_{DL,z,unk} = \frac{3\sqrt{I_{B,unk}}}{I_{C,std}} \frac{i_{std}t_{std}}{i_{unk}t_{unk}} \frac{[ZAF]_{std}}{[ZAF]_{unk}} C_{z,std} \tag{1}$$

where $I_{B,unk}$ is the bremsstrahlung intensity measured from the unknown, $I_{C,std}$ is the characteristic (background corrected) intensity measured on the standard, i_{std} and i_{unk} are the probe currents measured on the standard and unknown respectively, t_{std} and t_{unk} are the acquisition live times for the standard and unknown respectively and $[ZAF]_{std}$ and $[ZAF]_{unk}$ are the matrix corrections relative to the pure element for the standard and unknown. Since the element of interest is only present at trace levels, the matrix correction for the unknown is largely independent of the mass fraction of the element. The same equation can be used for either EDS or WDS. In the case of EDS, the intensity should be integrated over the same range of channels on the unknown and the standard.

The peak-to-background ratio on a WDS is determined by the convolution of the natural width of the characteristic X-ray line and the dispersive power of the spectrometer. High resolution spectrometers and narrow characteristic line widths tend to produce large peak-to-background ratios, and thus also sensitive limits-of-detection. The peak-to-background ratio on an EDS is determined by the ranges of channels used to estimate the continuum and peak intensity. For modelled background fitting, the range of channels for continuum and peak are distinct. Typically, integrating the peak over the central full-width half-max (FWHM) and an equivalent range of the continuum (above and below) produces close to optimal results. The central FWHM corresponds to approximately 76 % of the total peak intensity. For filtered background fitting, the range of channels is convolved by the fitting filter. In the end, the higher resolution inherent in a WDS permit lower limits-of-detection than an SDD.

2. Metrology

Regardless of the technology, the key to good microanalysis measurements is careful metrology. Historically, practitioners of WDS have been careful metrologists. The importance of sample preparation has long been appreciated by this community. Samples are often mounted in epoxy, and polished to a mirror-like surface finish. Insulating samples are coated with conductive coatings, and care is taken to ensure a well defined conductive path to ground. The orientation of the sample is maintained normal to the incidence electron beam, and the working distance is maintained precisely. Microprobes, the instruments dedicated to WDS measurements, have a number of innovations specifically designed to facilitate careful metrology. An integrated optical microscope assists in maintaining a constant sample-detector geometry. An in-column faraday cup assists in measuring the probe current regularly and reliably. High precision X-Y-Z staging allows for reproducible sample motion.

Most scanning electron microscopes (SEMs) lack some or all of these features. SEM stages often have rotate and tilt axis that hinder reproducible positioning. An accessory Faraday cup and a picoammeter are often required to measure the probe current. Nonetheless, it is possible to make accurate measurements using EDS. Carefully selecting the working distance that maximises X-ray throughput also minimises the sensitivity to small changes in working distance. Maintaining the objective lens current and focusing with Z-stage motion produces more reproducible sample positioning. A retro-reflected laser beam combined with a mirror placed on the sample surface can be used to ensure that the sample is oriented perpendicular to the electron beam axis. A Faraday cup can be used to accurately measure the probe current. While the linearity of a high-quality picoammeter is desirable when a range of probe currents are used, an integrated picoammeter is adequate if the probe current is held constant. All of WDS's requirements for meticulous sample preparation and maintaining consistent measurement conditions remain equally as critical for EDS.

The most common approach to quantification of EDS and WDS data is the k-ratio approach pioneered in the earliest days of microanalysis [14]. In the k-ratio method, the intensity measured from the unknown is compared with the intensity measured from a standard under similar measurement conditions. The ratio, $k = C_{std}\left(I_{unk}/I_{std}\right)$ is the first approximation of the mass fraction of the associated element. The difference between the k-ratio and the actual mass fraction is a correction term called the *matrix correction*. The matrix correction compensates for differences in the way that the electrons and X-rays interact with the sample. There are three predominant mechanisms. The first mechanism ('atomic number' or Z) has to do with how the energetic electrons interact with the material in terms of backscatter losses and energy loss. The second mechanism ('absorption' or A) has to do with how the X-rays, once generated, escape from the sample in the direction of the detector. The third mechanism ('secondary fluorescence' or F) concerns how X-rays can generate addition X-rays through photoabsorption and subsequent X-ray emission. In most modern formulations of matrix corrections the Z and A terms are calculated using an expression based on a $\phi(\rho z)$-curve. A $\phi(\rho z)$-curve is an expression that models the ionisation in an infinitesimally thick layer at depth z relative to the ionisation in an infinitesimally thick film at $z = 0$.

Matrix correction is performed in a similar manner for WDS and EDS measurements *except* that for EDS measurements the signal from multiple lines sum into a single peak that is measured simultaneously. Each of these lines will have slightly different correction factors. This is particularly important when there are multiple ionised shells or when the lines fall on different sides of absorption edges. The absorption can be self-absorption (like Fe L_2-M_4 by Fe L_3) or due to other elements in the material. The appropriate matrix correction for EDS is

$$[ZAF]_{EDS} = \frac{\sum_i W_i\,[ZAF]_i}{\sum_i W_i} \tag{2}$$

where sum over i is over all the X-ray transitions in the integrated interval, W_i represents the nominal relative weight of the ith transition in a pure element and the $[ZAF]_i$ are the traditional single transition matrix correction factors.

Much like WDS, materials used for standards may either be simple (pure elements or simple compounds) or complex (*similar* or "*matrix-matched*"). Simple standards are easier to work with but typically produce less accurate results because the matrix corrections are typically larger and dependent on poorly known quantities like $[\mu/\rho]$, the mass absorption coefficient. Complex standards can be chosen to minimise the dependence of the matrix correction on poorly known quantities. A standard similar in composition to the unknown is likely to have a similar matrix correction so the ratio $[ZAF]_{unk}/[ZAF]_{std}$ is close to unity and careful error propagation demonstrates that the dependences on poorly known physical parameters largely cancels [15]. However, complex standards may not be available or are more difficult to work with because there may be interferences between the characteristic peaks of the elements present in the standard. The strategy used to resolve these interferences in National Institute of Standards and Technology DSTA-II (a pseudo-acronym) [DTSA-II can be downloaded for free `https://goo.gl/MI1Ku`] are additional spectra called *peak-shape references* (or simply *references*), which can be calibrated relative to the standard and fit to the unknown.

Like WDS, it is usually beneficial to collect multiple measurements from each material. It is a good strategy to collect N spectra (with $N \approx 5$ to 10) for a live-time t/N from the standard rather than one for a live-time t. The N spectra can be inter-compared and outliers can be rejected. The remaining spectra can be summed into a single spectrum that is likely to be of high quality. (National Institute of Standards and Technology DSTA-II (a pseudo-acronym) provides a mechanism called the *Standard Bundle* that sums standards and integrates the necessary peak-shape references.) Multiple acquisitions from the unknown permit homogeneity testing and can bring to light sample preparation and other metrology problems.

3. Examples

In previous articles, the ability of EDS to address simple measurements as well as more challenging measurements like major element interferences, light elements, trace elements, and low beam energy measurements has been demonstrated [10–13]. This article will consider the challenging problem of materials containing minor/trace quantities of rare earth elements (REE). Measurements of minerals containing minor and trace quantities ($C_z < 10.0$ % mass fraction) of REE are a challenge for both WDS and EDS. REE rarely come in isolation in nature. When one REE is present, there are likely to be half-a-dozen or more other REE. Furthermore, REE have very complex L-line structures with tens of characteristic lines spread over a range of energies. The complex line structure makes finding suitable positions to make WDS background measurements a challenge [16]. On the other hand, the complex line structure provides significant spectral structure, which is beneficial for EDS measurements.

For the first example of a REE analysis, we considered the four Drake and Weill (DaW) engineered rare earth element containing glasses called REE1, REE2, REE3 and REE4 [17]. These glasses have either 3 or 4 REE at approximately 3 % mass fraction to 4 % mass fraction. The remaining material being a Si-Al-Ca glass. The homogeneity and provenance of these materials are well documented. The nominal mass abundances were confirmed by instrumental neutron activation analysis. The glasses were mounted in epoxy and polished to a fine finish. The surface was thermal evaporatively coated with approximately 10 nm of carbon to ensure conductivity.

The data was collected on a TESCAN MIRA3 equipped with four independent 30 mm^2 PulseTor SDDs [Disclaimer: Any mention of commercial products is for information only; it does not imply recommendation or endorsement by NIST]. The beam energy was 20 keV and the probe current was 0.9 nA. The data was collected under stage automation using the SEMantics [18] instrument control software extension to NIST DTSA-II. The probe current was remeasured between spectra and the spectra from the four independent SDD were combined into a single spectrum for quantification using NIST DTSA-II.

Standards were collected from an SPI Rare Earth Phosphate standard block. Five replicates of each standard were collected for 60.0 s each. The five replicates were combined into a single 300 s standard spectrum. The spectra for each element were combined with the necessary references into a standard bundle to facilitate quantification [19]. Many of the standards also required a reference to resolve interference of the REE M-lines with either the P-K or the O-K lines with GdP_5O_{14} serving as the P K peak-shape reference and MgO serving as the O-K peak-shape reference. Since we were not generally using the REE M-lines in the quantification and since we had a clean P-K reference, this procedure proved adequate, although it did sometimes leave structure in the residual near the P-K peak. In addition, the research glass K411 (NIST Standard Reference Material 470) was used as a standard for Si and Ca, and pure Al for Al. For comparision, the research glass K412 was used as a standard for Si, Al, Ca and O. Oxygen was computed using stoichiometric assumptions.

Nine or ten 600 s spectra were collected from a randomly selected set of points on each unknown. An example of a spectrum from glass REE1 and the residual generated during the quantification is shown in figure 1. The marker lines are shown for the elements O, Si, Al, Ca, Nd, Sm, Yb, and Lu giving an indication of the complexity of the REE L-line structure. The residual under the REE L-lines is seen to be particularly smooth and structureless because of the high quality unknown and standard spectra. The fit under the Ca-K lines has a small but unexplained structure and the fit under the Al-K (1.48 keV), Si-K (1.74 keV), and REE M-lines shows additional structure but remains fairly clean. Overall, this represents an excellent residual and suggests that other elements haven't been overlooked and that we can expect reliable *k*-ratios.

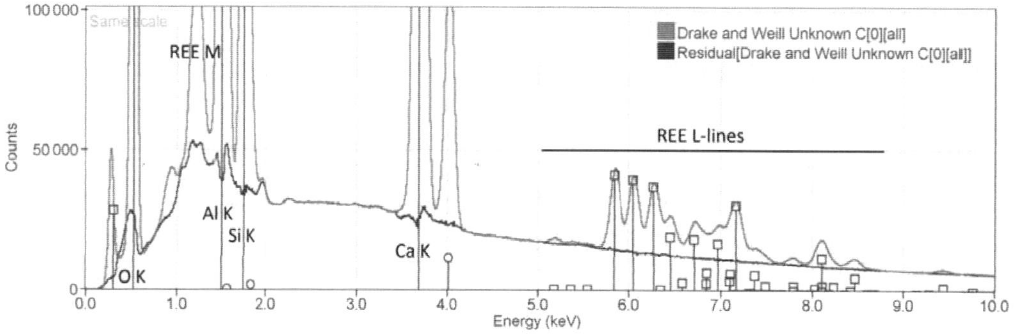

Figure 1. A raw spectrum from Drake and Weill glass REE1 along with the residual generated during quantification. (5.6×10^7 total counts)

The quantitative results from the four REE standards are shown in table 1. The analytic totals were good to excellent ranging from a minimum of 99.61 % mass fraction to a maximum of 102.45 % mass fraction. The element with the largest error was often Al ranging from a relative error of 7.7 % on a nominal value of 16.15 % mass fraction. The relative deviation from expected value (RDEV) will be used for comparison purposes. The RDEV is defined as the $(measured - nominal)/nominal$ and is expressed as a percentage. The largest RDEV for a REE were Nd and Sm at an RDEV of 4.1 % and 4.0 % respectively in REE2. For example in Nd, these corresponded to measured value of 3.8 % mass fraction compared to a nominal value of 3.65 % mass fraction. The point-to-point deviation from expected value (DEV) between spectra of between 0.3 % to 0.4 % speaks to both the homogeneity of the standards and the precision of the measurement.

It is interesting to consider why the analytical totals for two of the four measurements were within 0.5 % mass fraction of unity (REE2 and REE3) and yet two deviated by 1.0 % mass fraction and 2.5 % mass fraction (REE1 and REE4). The majority of the mass is represented by

IOP Conf. Series: Materials Science and Engineering **304** (2017) 012013 doi:10.1088/1757-899X/304/1/012013

Table 1. Quantified results from 9 or 10 spectra collected from the Drake and Weill glasses REE1 to REE4. The nominal values from Drake and Weill are provided along with the absolute deviation from expected value (DEV) and the relative deviation from expected value (RDEV).

Element	O	Al	Si	Ca	Eu	Gd	Tb	Tm	Total
Standards	Stoic	Al	K411	K411	EuP_5O_{14}	GdP_5O_{14}	TbP_5O_{14}	TmP_5O_{14}	
Average (10)	39.04	17.40	12.20	18.38	3.88	3.93	3.83	3.80	102.45
Std. Dev. (10)	0.05	0.03	0.02	0.01	0.01	0.02	0.01	0.02	0.01
Nominal	38.01	16.15	12.60	17.98	3.80	3.87	3.78	3.81	100.00
DEV	0.14%	0.18%	0.15%	0.05%	0.37%	0.46%	0.37%	0.66%	0.01%
RDEV	2.7%	7.7%	-3.2%	2.2%	2.1%	1.6%	1.3%	-0.4%	2.5%
Element	O	Al	Si	Ca	Nd	Sm	Yb	Lu	Total
Standards	Stoic	Al	K411	K411	NdP_5O_{14}	SmP_5O_{14}	YbP_5O_{14}	LuP_5O_{14}	
Average (9)	38.34	16.21	12.70	17.97	3.80	3.82	3.83	3.72	100.39
Std. Dev. (9)	0.05	0.02	0.02	0.01	0.01	0.01	0.01	0.02	0.13
Nominal	38.28	16.21	12.65	18.05	3.65	3.67	3.74	3.75	100.00
DEV	0.13%	0.13%	0.17%	0.07%	0.34%	0.29%	0.26%	0.50%	0.13%
RDEV	0.2%	0.0%	0.4%	-0.4%	4.1%	4.0%	2.5%	-0.8%	0.4%
Element	O	Al	Si	Ca	Y	La	Ce	Pr	Total
Standard	Stoic	Al	K411	K411	YP_5O_{14}	LaP_5O_{14}	CeP_5O_{14}	PrP_5O_{14}	
Average (10)	38.67	16.48	12.15	18.07	3.26	3.72	3.40	3.85	99.61
Std. Dev. (10)	0.06	0.03	0.03	0.02	0.01	0.01	0.01	0.01	0.01
Nominal	38.89	16.25	12.69	18.1	3.21	3.65	3.42	3.79	100.00
DEV	0.16%	0.15%	0.21%	0.10%	0.38%	0.38%	0.36%	0.25%	0.01%
RDEV	-0.6%	1.4%	-4.3%	-0.2%	1.5%	1.9%	-0.5%	1.6%	-0.4%
Element	O	Al	Si	Ca	Dy	Ho	Er		Total
Standards	Stoic	Al	K411	K411	DyP_5O_{14}	HoP_5O_{14}	ErP_5O_{14}		
Average (10)	39.65	17.64	12.86	19.00	3.93	3.99	3.90		100.96
Std. Dev. (10)	0.06	0.03	0.02	0.02	0.02	0.02	0.01		0.01
Nominal	39.43	16.97	13.24	18.90	3.80	3.85	3.81		100.00
DEV	0.15%	0.18%	0.16%	0.08%	0.50%	0.49%	0.27%		0.01%
RDEV	0.6%	3.9%	-2.8%	0.5%	3.3%	3.5%	2.3%		1.0%

the elements Al, Si, Ca, and O with only 11 % mass fraction to 15 % mass fraction in the REE. The same standards for all four samples were used for the bulk of the mass. Yet, in one case (REE2), the mass fraction deviation from expected value was 0.0 % while, on another (REE1), the mass fraction deviation from expected value was 7.7 %. This suggests that the majority of the overage in the analytical total could be a result of inaccuracy in the Al measurement. Using K412 glass as a standard for Al did not change the results substantially further suggesting that the problem may result from a matrix correction challenge. Careful examination of the relevant REE M_5 edges compared with the Al K-L_3 shows that the Al K line is very close to the Tm M_5 edge in REE1 and the Er edge in REE4, the two samples with poor totals. The material mass absorption coefficient (MAC) is known to vary significantly from the weighted sum of the elemental mass absorption coefficients in the range of energies just above the edge energy [20]. This suggests that the large deviations in the Al mass fraction may be a result of poor matrix correction due to large deviations from nominal values for the Al K-L_3 MAC caused by the Tm and Er M_5 edges. In fact, if the total is adjusted by replacing the measured Al mass fraction with the nominal value and adjusting the O mass fraction appropriately, the adjusted analytical totals change from 102.45 % mass fraction to 100.09 % mass fraction and 100.96 % mass fraction to 99.70 % mass fraction. Almost the entire deviation in the analytical total can be attributed to the Al measurement. This amount of error in the Al is surprising given that while the matrix correction for the Al K-L_3 is quite large (≈ 0.56), the fraction of Tm and Er is quite small.

The DaW glasses demonstrate the effectiveness of high count SDD spectra, however, the real interest is in natural materials in which families of REE tend to accumulate in minor and trace quantities. Examples of natural REE samples with established provenance are the

NAM (monazite) and BS1 (xenotime) samples [21]. These materials were measured on the same instrumentation and conditions as the DaW glasses (see figures 2 and 3). The same REE phosphate standards (materials and spectra) were used. For BS1, YP_5O_{14} was used for P and for NAM, CeP_5O_{14} was used for P. NIST SRM-470 K411 glass was used as standard for Si and Ca, and pure As was used for arsenic. Oxygen was calculated from stoichimetry. From eq. 1, we anticipate limits of detection for each of the REE in the 300 ppm to 500 ppm range. These limits are likely to be optimistic as the equation does not account for interferences.

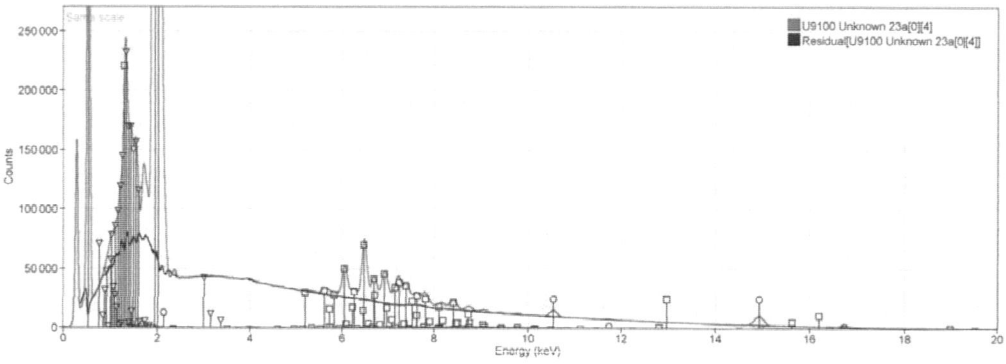

Figure 2. Spectrum BS1[0] (red) with the residual (blue) computed by NIST DTSA-II during the quantification process. The line markers for all the measured elements are displayed without labels. The relatively smooth shape of the residual in both the 1 keV to 2.5 keV range and the 5 keV to 10 keV range suggests that all significant elements have been identified and fit. (8.1×10^7 total counts)

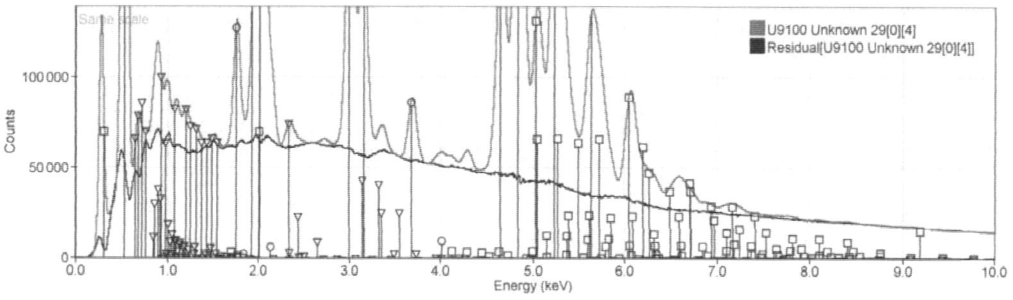

Figure 3. A spectrum representing the sum of the NAM spectra (red) with the associated residual (blue) and line markers. The quality of the residual was not quite as good as in the case of the BS1 material. There was more remaining structure in the 5 keV to 8 keV range. Regardless, this is still quite a clean residual and suggests a reliable quantification. (8.3×10^7 total counts)

The results for the BS1 sample are presented in table 2. BS1 is a xenotime and O, P, and Y represent the bulk of the mass. Si, a bright element in a relatively clean region of the spectrum, was consistently detected at the 600 $\pm$ 100 ppm level. Nd was measured at the 1,190 $\pm$ 100 ppm level compared with a nominal value of 1,700 ppm. Eu was measured at 1,970 $\pm$ 100 ppm when the nominal value was 2,313 ppm. The final two columns list the DEV and RDEV for each element from the nominal values reported by Aleinikoff. The elements La (nominal 5.1 ppm),

Ce (nominal 116 ppm), Pr (nominal 82.2 ppm), and Pb (nominal 101.1 ppm) were not detected and are not reported in the table. The analytic totals for the BS1 sample were consistently a little low at 96.3 % mass fraction to 98.9 % mass fraction. Aleinikoff did not provide nominal values for O, Si, As or Y however when we measure these by conventional microprobe (WDS) we get similar values for Y (34.6 % mass fraction), P (15.1 % mass fraction) and Si (0.06 % mass fraction) but 1.03 % mass fractionb for As by WDS compared with 2.1 % mass fraction by SDD The SDD measured O values compare well with oxygen computed by stoichiometry.

Table 2. The results (in percent mass fraction) from 10 spectra collected from Aleinikoff's BS1 sample quantified using rare earth phosphates. The REE were quantified using the L-lines except for Th and U for which the M-lines were used. Oxygen and phosphorous were measured using YP_5O_{14} as a standard. DEV is deviation from nominal value and RDEV is relative deviation from nominal value. Aleinikoff did not provide nominal values for O, Si, As and Y but the text provides a comparison with the WDS measured values.

	BS1[0]	BS1[1]	BS1[2]	BS1[3]	BS1[4]	BS1[5]	BS1[6]	BS1[7]	BS1[8]	BS1[9]	Average	Std. Dev.	Aleinikoff Nominal	DEV	RDEV
O	31.121	29.972	30.202	31.460	30.842	30.301	30.431	30.306	30.428	30.237	30.530	0.4405			
Si	0.061	0.056	0.052	0.071	0.057	0.056	0.077	0.043	0.047	0.053	0.057	0.0097			
P	14.402	14.236	14.315	14.445	14.404	14.277	14.347	14.385	14.312	14.315	14.344	0.0616			
As	2.182	2.133	2.072	2.203	2.104	2.104	2.045	1.940	2.112	2.032	2.093	0.0723			
Y	33.372	33.015	33.171	33.526	33.244	32.939	33.191	33.240	33.046	33.069	33.181	0.1674			
Nd	0.117	0.133	0.125	0.127	0.111	0.108	0.131	0.101	0.129	0.110	0.119	0.0107	0.17	-0.051	-30%
Sm	0.414	0.386	0.415	0.405	0.446	0.458	0.407	0.385	0.458	0.397	0.417	0.0262	0.5648	-0.148	-26%
Eu	0.206	0.189	0.191	0.206	0.207	0.198	0.200	0.174	0.221	0.179	0.197	0.0134	0.2313	-0.034	-15%
Gd	2.551	2.471	2.522	2.512	2.641	2.677	2.544	2.385	2.648	2.566	2.552	0.0836	2.8646	-0.313	-11%
Tb	0.773	0.728	0.704	0.744	0.733	0.707	0.724	0.683	0.770	0.724	0.729	0.0266	0.7474	-0.019	-2%
Dy	5.777	5.656	5.702	5.721	5.700	5.708	5.711	5.779	5.708	5.642	5.710	0.0414	5.6624	0.048	1%
Ho	1.067	0.964	1.223	1.051	1.140	0.970	1.009	0.905	1.062	1.123	1.051	0.0896	1.1805	-0.129	-11%
Er	3.450	3.432	3.483	3.478	3.386	3.409	3.495	3.570	3.461	3.459	3.462	0.048	3.2739	0.188	6%
Tm	0.450	0.437	0.451	0.479	0.433	0.406	0.452	0.391	0.460	0.434	0.439	0.0246	0.4331	0.006	1%
Yb	2.190	2.220	2.134	2.228	2.151	2.206	2.214	2.232	2.138	2.107	2.182	0.0432	2.2234	-0.041	-2%
Lu	0.158	0.200	0.184	0.184	0.202	0.221	0.219	0.193	0.169	0.181	0.191	0.019	0.2274	-0.036	-16%
Th	0.091	0.064	0.079	0.103	0.067	0.080	0.077	0.081	0.108	0.059	0.081	0.015	0.303	-0.222	-73%
U	0.026	0.004	0.018	0.000	0.000	0.023	0.005	0.005	0.015	0.000	0.010	0.0094	0.0383	-0.029	-75%
Total	98.406	96.294	97.043	98.943	97.868	96.847	97.278	96.798	97.291	96.683	97.345				

The results for the NAM sample are presented in table 3. NAM is a monazite with approximately 20 % mass fraction Ce and approximately 10 % mass fraction La, Nd, and Th. The RDEV for Ce was 3 % and for La, Nd, and Th, 6 %, 6 %, and 14 % respectively. The Th number is not totally surprising as the Th standard was oxidised to an unknown degree. Pr, Sm, and Gd were present at approximately 2 % mass fraction. The RDEV for these elements were 8 %, 9 %, and 2 % respectively. Pr, Sm, and Gd are nominally in the 2 % mass fraction to 3 % mass fraction range with RDEVs of 8 %, 9 %, and 2 % respectively. Dy was nominally around 1 % mass fraction with a RDEV of 2 %. Below 1 % mass fraction, Tb, at nominally 0.251 % mass fraction, was measured 69 % high, U, at nominally 0.127 % mass fraction, was measured 169 % high, and Pb, at nominally 0.239 % mass fraction was measured 26 % low. At nominally 0.239 % mass fraction, Tm should be measurable but was not detected using the L-lines. Furthermore, no structure was observed in the residual near the Tm L-lines but when Tm was quantified using the M-lines, the average result was 0.272 ± 0.017 % mass fraction (a RDEV of 14 %). Based on limit-of-detection (LOD) estimates, Eu and Lu should not be detectable and the fitting affirms this. Ho, at nominally 0.08 % mass fraction, should be detectable but was measured in some spectra but not others but with an RDEV of -56 %.

These are difficult measurements either by SDD or WDS and the results speak to this. At the major / minor levels ($C > 1\%$ mass fraction), the SDD was capable of measuring with RDEVs that were typically better than 10 %. Below these levels, there is no clear story. Most elements in

IOP Conf. Series: Materials Science and Engineering **304** (2017) 012013 doi:10.1088/1757-899X/304/1/012013

Table 3. Ten 600 s spectra collected from Aleinikoff's NAM sample quantified using phosphate standards. The REE were quantified using the L-lines except for Pb, Th, and U for which the M-lines were used. Oxygen was computed from stoichiometry. Composition values are in percent mass fraction.

	NAM[0]	NAM[1]	NAM[2]	NAM[3]	NAM[4]	NAM[5]	NAM[6]	NAM[7]	NAM[8]	NAM[9]	Average	Std. Dev.	Aleinikoff Nominal	DEV	RDEV
O	27.34	27.33	27.29	27.34	27.42	27.34	27.41	27.39	27.35	27.33	27.35	0.04			
Si	0.88	0.88	0.88	0.88	0.88	0.88	0.88	0.89	0.88	0.88	0.88	0.00			
P	11.50	11.50	11.49	11.50	11.54	11.49	11.54	11.53	11.51	11.49	11.51	0.02			
Ca	0.58	0.58	0.59	0.59	0.59	0.59	0.59	0.59	0.58	0.60	0.59	0.01			
Y	1.40	1.40	1.41	1.43	1.43	1.43	1.42	1.42	1.40	1.40	1.41	0.01			
La	8.08	8.02	7.94	8.00	7.96	8.01	7.92	7.99	8.00	8.06	8.00	0.05	7.55	0.45	6%
Ce	20.55	20.55	20.51	20.53	20.55	20.53	20.51	20.53	20.52	20.47	20.53	0.02	19.94	0.59	3%
Pr	2.63	2.64	2.65	2.62	2.69	2.65	2.73	2.68	2.65	2.64	2.66	0.03	2.45	0.20	8%
Nd	9.58	9.57	9.60	9.62	9.62	9.61	9.65	9.65	9.60	9.61	9.61	0.02	9.03	0.58	6%
Sm	3.22	3.24	3.22	3.25	3.30	3.26	3.31	3.24	3.23	3.22	3.25	0.03	2.98	0.27	9%
Eu	-	-	-	-	-	-	-	-	-	-	-	-	0.01	-0.01	-100%
Gd	2.19	2.21	2.19	2.15	2.20	2.20	2.18	2.20	2.21	2.17	2.19	0.02	2.15	0.04	2%
Tb	0.39	0.43	0.44	0.43	0.43	0.43	0.44	0.41	0.42	0.42	0.42	0.01	0.25	0.17	69%
Dy	0.92	0.90	0.87	0.92	0.93	0.92	0.93	0.88	0.92	0.87	0.91	0.02	0.89	0.02	2%
Ho	0.00	0.03	0.03	0.00	0.00	0.00	0.00	0.00	0.22	0.07	0.04	0.07	0.08	-0.04	-56%
Er	0.23	0.24	0.19	0.24	0.25	0.25	0.28	0.21	0.26	0.33	0.25	0.04	0.14	0.11	83%
Tm	0.00	0.00	0.00	0.00	0.00	0.00	0.00	0.00	0.00	0.00	0.00	0.00	0.24	-0.24	-100%
Yb	0.02	0.03	0.02	0.03	0.06	0.05	0.03	0.05	0.02	0.06	0.04	0.02	0.05	-0.01	-25%
Lu	-	-	-	-	-	-	-	-	-	-	-	-	0.00	0.00	-100%
Pb	0.18	0.16	0.18	0.19	0.17	0.18	0.18	0.17	0.18	0.16	0.18	0.01	0.24	-0.06	-26%
Th	11.58	11.56	11.51	11.54	11.59	11.58	11.58	11.60	11.58	11.62	11.57	0.03	10.14	1.43	14%
U	0.34	0.34	0.35	0.37	0.36	0.33	0.35	0.34	0.34	0.32	0.34	0.01	0.13	0.22	169%
Total	101.61	101.58	101.36	101.61	101.97	101.72	101.89	101.77	101.66	101.66					

the 1 % mass fraction to 0.1 % mass fraction range were detected but with large uncertainties. Below 0.1 % mass fraction, detection was hit-or-miss. The most troubling measurement was Tm in NAM. In this case, 0.24 % mass fraction of an element was missed when quantified using the L-line but was measured with an RDEV of 14 % by the M-lines. Independent of the accuracy, the measured values were very consistent (high precision) suggesting both that the sample was homogeneous and that the measurement protocol and quantification procedure were reproducible.

These measurements suggest that an SDD can be used to address some questions about natural REE-bearing minerals. The acquisition times were fairly long (10 points at 10 minutes live-time per point) and the detector configuration (4 separate SDD combined) was unusual. However, there may be economic justification for such a system for discriminating REE at the 0.1 % mass fraction level and above. Further improvements may be possible through better deconvolution, higher throughput or longer count times. However, give current technology and algorithms we seem to be reaching a point of diminishing returns.

4. Conclusion

While the potential to make accurate and precise measurements with an SDD exists, few people are making use of it. The vast majority of EDS measurements remain qualitative or quantitative with large error bars (*"standardless analysis"*). Accurate electron probe measurements have always been a challenge but careful analysts have been doing them for decades with WDS technology and, with similar care, accurate measurements can be performed with an SDD. Because the peak-to-background ratio is higher for WDS than SDD, WDS will always have the lower limits-of-detection. In some challenging problems, the ability to measure the full spectrum can be an advantage. The L-lines in REE are a clear example in which complex line structure enhances our ability to differentiate challenging overlaps. The REE analyses suggest at the utility of the residual spectrum as an important tool for diagnosing quantification problems. Interpreting the residual can be complex and will be the topic of a forthcoming article. Understanding the limitations of the SDD will allow us to build modern microprobes that combine the speed and accuracy of the SDD for major and minor elements with the excellent limits-of-detection of WDS for trace elements. The whole is truly larger than the sum of the parts.

References

[1] Reed S J B and Ware N G 1973 *X-ray Spectrom.* **2** 69–74
[2] Fitzgerald R, Keil K and Heinrich K F J 1968 *Science* **159** 528–530
[3] Schamber F H 2009 *Micros. Microanal.* **15** 491–504
[4] Schamber F H 1977 *A modification of the linear least-squares fitting method which provides continuum suppression.* (Ann Arbor, MI: Ann Arbor Science Publishers) 241–257
[5] Burgess S, Li X and Holland J 2013 *Microscopy and Analysis* **6** S8–S13
[6] Rehak P and Gatti E 1990 *Nucl. Instrum. Methods Phys. Res. A* **289** 410–417
[7] Lechner P, Eckbauer S, Hartmann R, Krisch S, Hauff D, Richter R, Soltau H, Strüder L, Fiorini C, Gatti E, Longoni A and Sampietro M 1996 *Nucl.Instrum. Methods Phys. Res. A* **377** 346–351
[8] Bertuccio G, Sampietro M and Fazzi A 1992 *Nucl. Instrum. Methods Phys. Res. A* **322** 538–542
[9] Niculae A, Bechteler A, Eckhardt R, Hermenau K, Liebel A, Lutz G, Soltau H and Strüder L 2016 *Micros. Microanal.* **22** 40–41
[10] Ritchie N W M, Newbury D E and Davis J M 2012 *Micros. Microanal.* **18** 892–904
[11] Newbury D E and Ritchie N W M 2015 *Micros. Microanal.* **21** 1327–1340
[12] Newbury D E and Ritchie N W M 2016 *Micros. Microanal.* **22** 735–753
[13] Newbury D E and Ritchie N W M 2016 *Micros. Microanal.* **22** 520–535
[14] Castaing R 1951 *Application des sondes électronique à une méthode d'analyse ponctuelle chimique et cristallographique.* Ph.D. thesis (Paris, France: University of Paris) Publication ONERA No. 55 (1952)
[15] Ritchie N W M and Newbury D E 2012 *Anal. Chem.* **84** 9956–9962
[16] Jercinovic M, Williams M, Allaz J and Donovan J 2012 *IOP Conf. Ser.: Mat. Sci. Eng.* **32** 012012
[17] Drake M and Weill D 1972 *Chem. Geol.* **10** 179–181
[18] Ritchie N W M and Filip V 2011 *Micros. Microanal.* **17** 896–897
[19] Ritchie N W M, Mengason M J and Newbury D E 2017 *Micros. Microanal.* **23** 491–500
[20] Chantler C T, Olsen K, Dragoset R A, Chang J, Kishore A R, Kotochigova S A and Zucker D S 2005 *X-Ray form factor, attenuation and scattering tables.* Version 2.1 (Gaithersburg, MD: National Institute of Standards and Technology) downloaded: Sept. 18, 2012 URL `http://physics.nist.gov/ffast`
[21] Aleinikoff J N, Grauch R I, Mazdab F K, Kwak L, Fanning C M and Kamo S L 2012 *Amer. J. Sci.* **312** 723–765

IOP Conf. Series: Materials Science and Engineering **304** (2017) 012014 doi:10.1088/1757-899X/304/1/012014

Monte Carlo simulation and fundamental quantities

F Salvat[1] **and X Llovet**[2]

[1] Facultat de Física (FQA and ICC). Universitat de Barcelona. Diagonal 645, ES-08028 Barcelona, Spain

[2] Scientific and Technological Centres. Universitat de Barcelona. Lluís Solé i Sabarís, 1-3. ES-08028 Barcelona, Spain

E-mail: francesc.salvat@ub.edu

Abstract. Physics interaction models suited for Monte Carlo simulation of coupled electron-photon transport in EPMA and other spectroscopic techniques are briefly described. The considered models are required to be applicable to any material and valid for a wide energy range. Differential cross-sections, the fundamental quantities in Monte Carlo simulation, are presented for the relevant interactions of photons and electrons with energies up to about 500 keV. Approximations and simplifications underlying the interaction models and the particle tracking scheme are discussed.

1. Introduction

It is generally believed that Monte Carlo (MC) simulation provides the most accurate description of the coupled transport of electrons and photons in the energy range of interest in electron-probe microanalysis (EPMA), X-ray fluorescence (XRF), and other X-ray spectroscopic techniques. Although automatic Monte Carlo quantification tools are not widely available, simulation studies have been used to calculate EPMA spectra and characteristic X-ray yields from bulk samples, k-ratios from thin films on substrates, secondary fluorescence near planar interfaces, and the lateral resolution of EPMA measurements [1].

Indeed, MC simulation has distinct advantages over alternative numerical (finite-difference) methods. Firstly, it can describe arbitrary interaction processes, with an accuracy limited only by that of the adopted interaction models, i.e., by the adopted differential cross-sections (DCSs). Secondly, MC simulation is able to track particles through material systems with complex geometries, where deterministic methods would find great difficulties even to define the appropriate boundary conditions. Finally, the stochastic nature of MC methods permits the straightforward evaluation of statistical (type A) uncertainties of simulation results, while numerical methods allow only rough estimations of accumulated errors.

However, MC simulation methods are not free from simplifications and approximations, which limit their applicability and the reliability of the results. A fundamental simplification refers to the structure of the material where radiation propagates, which is assumed to be homogeneous and isotropic, with defined composition and density. The atoms or molecules in the medium are considered to be randomly distributed with uniform density (crystalline ordering is ignored) and molecular aggregation effects are disregarded, i.e., molecules are regarded as sets of individual atoms with *uncorrelated* positions. Only in the case of inelastic collisions of charged particles, molecular binding and aggregation effects are partially accounted for by using empirical values of the mean excitation energies of materials, so as to ensure the correct high-energy collision stopping power. Also implicit in any MC code is the assumption that the various interaction mechanisms are effectively independent; reality is more complex, and the co-existence of multiple time-evolution paths leads to quantum interference effects (such as inelastic absorption and radiative effects in elastic collisions of charged particles), which are frequently disregarded, or treated approximately as corrections to the DCSs of the dominant interactions.

In MC simulations, the trajectories of transported (primary and secondary) particles are generated as sequences of connected free flights, each flight ending by an interaction that may change the energy and direction of the particle. This trajectory picture is justified only for high-energy radiation, whose de Broglie wavelengths are much smaller than the average separation between the atoms in the medium. When the wavelength is comparable to the interatomic distances, interference effects, resulting from the coherent superposition of waves scattered by different atoms, may become important and invalidate the trajectory picture. In addition, the tracking of particles is performed as if they were moving in an infinite, limitless medium having the local composition; this way of operation implies that interfaces are considered to be passive (thus excluding, e.g., surface plasmon excitations).

In the present article we describe physics interaction models suited for MC simulation of coupled electron-photon transport, with the focus on the fundamental quantities and on the underlying assumptions (and simplifications) that determine the reliability of the simulation results. The relevant fundamental quantities are the interaction DCSs, which describe each interaction mechanism, the corresponding total (integrated) cross-sections, and the atomic transition probabilities and energies, which determine the X-rays and Auger electrons released in the relaxation of atoms with vacancies in inner sub-shells. The interaction models described here have been implemented in the general-purpose MC code PENELOPE [2]. More detailed information on theoretical aspects can be found in the article by Salvat and Fernández–Varea [3]. More recently, a review article on electron-impact ionisation and X-ray emission has been published by Llovet *et al.* [4].

Owing to the wide energy range of interest, interactions are described using relativistic quantum mechanics. For electrons (mass $\mathrm{m_e}$, charge $-e$) with kinetic energy E and momentum $\mathbf{p} = \hbar\mathbf{k}$ ($\hbar$ is the reduced Planck constant), we will frequently use the quantities

$$\gamma \equiv \frac{E + \mathrm{m_e}c^2}{\mathrm{m_e}c^2} = \sqrt{\frac{1}{1-\beta^2}} \qquad \text{and} \qquad \beta = \frac{v}{c} = \sqrt{\frac{E(E+2\mathrm{m_e}c^2)}{(E+\mathrm{m_e}c^2)^2}} = \sqrt{\frac{\gamma^2-1}{\gamma^2}}, \tag{1}$$

which are, respectively, the total energy in units of the rest energy $\mathrm{m_e}c^2$ and the velocity in units of the speed of light c. The wave number and wavelength of the electron are

$$k = (c\hbar)^{-1}\sqrt{E(E+2\mathrm{m_e}c^2)} \qquad \text{and} \qquad \lambda = \frac{2\pi}{k} = \frac{12.398\ \text{Å}}{\sqrt{(E/\text{keV})[(E/\text{keV})+1022]}}, \tag{2}$$

respectively. The corresponding quantities for a photon of energy E are

$$k = (c\hbar)^{-1}E \qquad \text{and} \qquad \lambda = \frac{12.398\ \text{Å}}{E/\text{keV}}. \tag{3}$$

2. Interaction processes

A projectile particle (electron or photon) of energy E can interact with matter through a number of different mechanisms. In the energy-range of interest in EPMA and other X-ray spectroscopies (say from 500 eV to 500 keV), photons interact through the following processes:

∘ photelectric absorption (ph),
∘ incoherent Compton scattering (Co), and
∘ coherent Rayleigh scattering (Ra).
The relevant interaction mechanisms of electrons are:
∘ elastic collisions (el),
∘ inelastic collisions (in), and
∘ bremsstrahlung photon emission (br).

Each individual interaction may involve a certain energy transfer W to the material, a change in direction of the projectile, and, occasionally, the prompt emission of a secondary particle. For the sake of concreteness, we assume that before the interaction the projectile moves in the direction of the z axis; the direction of motion after the interaction (as well as the directions of any secondary particles released) are described by the polar and azimuthal angles, θ and ϕ, respectively (see fig. 1). Each interaction process (pr) is described in terms of the so-called differential inverse mean free path (DIMFP), $\mathrm{d}^2\mu_{\mathrm{pr}}(E)/(\mathrm{d}W\,\mathrm{d}\Omega)$, which gives the probability density of having an interaction with energy loss W and angular deflection $\Omega = (\theta, \phi)$ per unit path length of the projectile. Because of the assumed isotropy of the medium, the DIMFP is independent of the azimuthal scattering angle ϕ. The inverse mean free path, defined by

$$\mu_{\mathrm{pr}}(E) = \int \mathrm{d}W \int \mathrm{d}\Omega \, \frac{\mathrm{d}^2\mu_{\mathrm{pr}}(E)}{\mathrm{d}W\,\mathrm{d}\Omega}, \tag{4}$$

gives the interaction probability per unit path length of the projectile. Its reciprocal is the mean free path between interactions of that type. The energy loss and the angular deflection in an interaction are random variables with joint probability density given by

$$P_{\mathrm{pr}}(E; W, \theta, \phi) = \frac{1}{\mu_{\mathrm{pr}}(E)} \frac{\mathrm{d}^2\mu_{\mathrm{pr}}(E)}{\mathrm{d}W\,\mathrm{d}\Omega}. \tag{5}$$

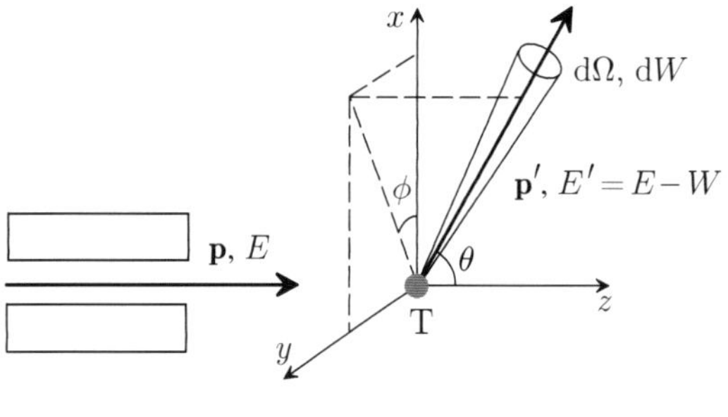

Figure 1. Schematic diagram of an experiment to measure the DCS. Incident particles move in the direction of the z axis; θ and ϕ are the polar and azimuthal scattering angles, respectively.

The details of the DIMFPs for condensed materials depend on the composition and structure (or state of aggregation) of the medium. However, when the de Broglie wavelength of the projectile [for electrons, $\lambda_{\mathrm{B}} = (150\ \mathrm{eV}/E)^{1/2}$ Å] is much less than the inter-atomic distances, the DIMFPs are nearly independent of the molecular or solid structure. Under these circumstances, the DIMFP can be estimated in terms of the atomic DCSs, $\mathrm{d}^2\sigma_{\mathrm{pr}}(Z, E)/(\mathrm{d}W\,\mathrm{d}\Omega)$, of the elements present by means of the additivity approximation. Concretely, the DIMFP for a compound (or mixture) with n_i atoms of the element of atomic number Z_i per molecule is

$$\frac{\mathrm{d}^2\mu_{\mathrm{pr}}(E)}{\mathrm{d}W\,\mathrm{d}\Omega} = \mathcal{N} \sum_i n_i \frac{\mathrm{d}^2\sigma_{\mathrm{pr}}(Z_i, E)}{\mathrm{d}W\,\mathrm{d}\Omega}, \tag{6}$$

where $\mathcal{N}$ is the number of molecules per unit volume. For a material of mass density ϱ and "molecular weight" A_w (g/mol), $\mathcal{N} = N_\mathrm{A}\varrho/A_\mathrm{w}$, where N_A is Avogadro's number. Most general-purpose MC codes utilise the additivity approximation with atomic DCSs defined by numerical databases or given by analytical formulas. The approximation is acceptable for elastic interactions. It is not accurate for inelastic collisions of charged particles, because these are dominated by excitations of conduction and valence electrons whose response depend strongly on the structure of the medium. Compton scattering of photons and photoelectric absorption also present a certain dependence on the state of aggregation. In the following we assume that the additivity approximation, with appropriate corrections, is applicable. That is, we consider interactions of electrons and photons with individual atoms of an element of atomic number Z.

A fundamental quantity, which characterises the response of the material to external electromagnetic fields and the distribution of energy losses in small-angle inelastic collisions of charged particles, is the optical dielectric function $\epsilon(\omega)$, a complex function of the angular frequency ω of the field. It is related to the familiar refraction index, $n(\omega)$, and extinction coefficient, $\kappa(\omega)$, by

$$n(\omega) + \mathrm{i}\kappa(\omega) = \sqrt{\epsilon(\omega)}, \tag{7}$$

where the branch of the square root is the one yielding a non-negative $n(\omega)$. The main source of measured optical dielectric functions is the *Handbook of Optical Constants of Solids* [5–7], which includes tables of $n(\omega)$ and $\kappa(\omega)$ for a number of metals, semiconductors and insulators. These tables contain measured data obtained with different methods, frequently by various groups and with various degrees of accuracy, and cover ranges of excitation energies $W = \hbar\omega$ from about 10^{-3} eV up to an upper energy that depends on the material, typically about 100 eV. For solids for which experimental information is not available, approximate optical functions can be obtained from density-functional theory calculations (see, e.g., [8]).

Because of the scarcity of experimental data, most MC codes use atomic DCSs that are either given by analytical formulas derived from simple models and approximations or defined by numerical databases generated by first-principles calculations. The latter are also affected by a number of approximations. Typically they are based on a relativistic independent-electron model of the atom, in which atomic electrons are assumed to move independently in a common central potential $V(r)$, and atomic wave functions are represented as single Slater determinants built with one-electron central-field orbitals $\psi_i(\mathbf{r})$ that are solutions of the Dirac equation with the eigenvalues E_i. Notice that the one-electron energy spectrum consists of a discrete set of levels ($E_i < 0$) corresponding to bound orbitals, and a continuum ($E > 0$) associated to free orbitals. Many calculations use the self-consistent Dirac-Hartree-Fock-Slater (DHFS) model of the atom (see, e.g., [9]). Compared with the more elaborate Dirac-Fock model [10], DHFS has the advantages of giving one-electron eigenvalues E_i in closer agreement with experimental sub-shell binding energies (see [3]) and of ensuring orthonormality of bound and free one-electron orbitals. The target atom is characterised by its ground-state configuration, i.e., by the set of electron sub-shells Si that are occupied, the number of electrons in each subshell, and the corresponding binding energies $U_i = -E_i$.

2.1. Photon interactions

- **Photoelectric effect**

In the photoelectric effect a photon is absorbed by the target atom and an atomic electron is excited to an empty orbital, bound or free, leaving a vacancy in the active sub-shell. Subsequently, the atom relaxes to its ground state by a sequence of radiative and non-radiative transitions.

Theoretical calculations using the independent-electron approximation (see [9] and references therein) give the partial cross-sections $\sigma_\mathrm{ph}(\mathrm{S}i, E)$ for the electron sub-shells of the atom. The atomic cross-section is the sum of contributions of the various sub-shells,

$$\sigma_{\rm ph}(E) = \sum_i \sigma_{\rm ph}({\rm S}i, E). \tag{8}$$

Most databases give cross-sections for ionisation only, that is, they exclude excitations to bound levels which are very sensitive to aggregation effects. Because photoabsorption is then forbidden for sub-shells with binding energy U_i higher than E, the total cross-section presents abrupt absorption edges. A peculiar feature of the photoeffect is that absorption occurs preferentially in the innermost sub-shell, i.e., in the sub-shell with the highest binding energy U_i that is less than E.

Extensive tables of atomic photoelectric cross-sections are given in the Evaluated Photon Data Library (EPDL) [11], which contain partial sub-shell cross-sections and total atomic cross-sections for all the elements from hydrogen ($Z = 1$) to lawrencium ($Z = 103$) and photon energies E from 10 eV to 100 GeV. For energies from the absorption edge up to 1 MeV, the tabulated cross-sections were computed with the DHFS atomic model using a computer programme of Scofield [12]. For energies higher than 1 MeV, the calculated sub-shell cross-sections were rescaled to match the total cross-sections of Hubbell *et al.* [13]. The XCOM programme [14] gives total atomic cross-sections for photon energies between 1 keV and 100 GeV, which are essentially the same as in the EPDL. Tables of total atomic cross-sections, generated from a compilation of experimental data and theoretical calculations, were published by Henke *et al.* [15] for the elements $Z = 1 - 92$ and for photon energies from 50 eV to 30 keV. Recently, Sabbatucci and Salvat [9] reformulated the theory of the atomic photoeffect to allow the easy calculation of photoelectron angular distributions. They used the DHFS model to generate a complete database, which includes total atomic cross-sections for all elements ($Z = 1$–100) and partial cross-sections for the K-shell and for the L-, M-, and N-sub-shells with binding energies higher than about 50 eV. This database is used in the MC code PENELOPE.

As noted above, atomic calculations do not account for aggregation effects. Photoelectric cross-sections for atoms in molecules and solids differ from the free-atom cross-section mostly because photoelectrons interact with neighbour atoms producing the so-called X-ray absorption fine structure (XAFS), a series of near-edge resonances together with an oscillatory component that decreases with the kinetic energy of electrons and extends up to a few-hundred eV above the edge energy [16]. For photons of small and moderate energies, and for materials with known optical dielectric functions, the photoabsortion cross-section can be obtained from the dielectric function as

$$\sigma_{\rm ph}(E) = \frac{2\omega}{\mathcal{N}c}\,\kappa(\omega), \tag{9}$$

where $\omega = E/\hbar$ is the angular frequency of photons with energy E. For energies E higher than about 100 eV, $n(\omega) \simeq 1$, $\kappa(\omega) \ll 1$, and

$$\sigma_{\rm ph}(E) \simeq \frac{\omega}{\mathcal{N}c}\,{\rm Im}[\epsilon(\omega)]. \tag{10}$$

This relation shows that calculated atomic photoelectric cross-sections can be used to define the dielectric function for frequencies higher than those attainable in measurements.

• Compton scattering

In Compton scattering the incident photon is absorbed by an atomic electron, which is promoted to an empty orbital (bound or free), and a secondary photon is emitted with energy $E' < E$ in a direction forming an angle θ with that of the incident photon. Energy conservation implies that electrons in a sub-shell Si can be excited only if $E - E' - U_i$ is larger than the energy of the lowest empty level.

The DCS for Compton scattering of unpolarised photons by free electrons at rest is given by the Klein-Nishina formula (see, e.g., [17])

$$\frac{\mathrm{d}\sigma^{(\mathrm{KN})}}{\mathrm{d}\Omega} = r_{\mathrm{e}}^2 \frac{1}{2} \frac{E'^2}{E^2} \left(\frac{E}{E'} + \frac{E'}{E} - \sin^2\theta \right). \tag{11}$$

Because of energy-momentum conservation, the energy E' of the scattered photon depends only on the polar scattering angle θ, and is given by

$$E' = \frac{E}{1 + (E/\mathrm{m_e}c^2)(1 - \cos\theta)} \equiv E_{\mathrm{C}}. \tag{12}$$

The Klein–Nishina DCS provides an acceptable description of Compton scattering for those interactions where the energy transfer is much larger than the binding energy of the active electron.

A more realistic description is obtained from the impulse approximation [18], which assumes that the electrons in the active shell Si react essentially as if they were moving with a distribution of velocities represented by the so-called Compton profile, a quantity defined as an integral of the Fourier transform of the atomic orbitals. The motion of the atomic electrons implies that the frequency of the incident photon is Doppler shifted when observed from a Lorentz frame moving with the same velocity as the target electron. The effect of binding is to prevent transitions that would lead the active electron to final bound orbitals. Note that interactions where the final state of the electron is an unoccupied bound orbital are also possible. These interactions correspond to Raman scattering (see, e.g., [19]), a process that involves relatively small energy transfers and is usually disregarded in MC codes. The impulse approximation correctly predicts that photons scattered in directions forming an angle θ with the direction of incidence have a continuous energy distribution, with a maximum near the Compton energy E_{C}, eq. (12). The DCS for Compton scattering by electrons in the active shell Si obtained from the impulse approximation is given by a complicated expression, that depends on the energy transfer $W = E - E'$ and the direction of the scattered photon. Adding the contributions from the various electron sub-shells and integrating over the energy of the scattered photon, the atomic DCS can be expressed as

$$\frac{\mathrm{d}\sigma_{\mathrm{Co}}(Z,E)}{\mathrm{d}\Omega} = \frac{\mathrm{d}\sigma^{(\mathrm{KN})}}{\mathrm{d}\Omega} S_{\mathrm{inc}}(Z;q), \tag{13}$$

where $S_{\mathrm{inc}}(Z;q)$ is the incoherent-scattering function, which depends on the momentum transfer

$$q \sim \frac{1}{c}\sqrt{E^2 + E_{\mathrm{C}}^2 - 2EE_{\mathrm{C}}\cos\theta}. \tag{14}$$

The total atomic cross-section for Compton scattering is

$$\sigma_{\mathrm{Co}}(Z,E) = \int \frac{\mathrm{d}\sigma_{\mathrm{Co}}(Z,E)}{\mathrm{d}\Omega}\,\mathrm{d}\Omega = 2\pi \int_{-1}^{1} \frac{\mathrm{d}\sigma_{\mathrm{Co}}(Z,E)}{\mathrm{d}\Omega}\,\mathrm{d}(\cos\theta). \tag{15}$$

The DCSs for Compton scattering adopted in PENELOPE are based on the impulse approximation, with an analytical approximation to the subshell Compton profiles. To account, at least partially, for aggregation effects the Compton profile of valence and conduction electrons may be replaced by that of an electron gas with the same density (number of electrons per unit volume) as in the material.

- **Rayleigh scattering**

Rayleigh scattering is the process in which the incident photon is absorbed by the target atom and a secondary photon is emitted without excitation of the target atom. Since the recoil energy of the target is very small, the secondary photon has approximately the same energy E as the projectile.

The atomic DCS for Rayleigh scattering of unpolarised photons is given by

$$\frac{\mathrm{d}\sigma_{\mathrm{Ra}}}{\mathrm{d}\Omega} = r_{\mathrm{e}}^2 \frac{1+\cos^2\theta}{2} \left| F(Z,q) + f' + \mathrm{i}f'' \right|^2 , \tag{16}$$

where

$$F(Z,q) = \int \rho(r) \exp(\mathrm{i}\mathbf{q}\cdot\mathbf{r}/\hbar)\,\mathrm{d}\mathbf{r} = \frac{4\pi}{q} \int_0^\infty \rho(r) \sin(qr/\hbar)\, r\,\mathrm{d}r. \tag{17}$$

is the atomic form factor, i.e., the Fourier transform of the atomic electron density $\rho(r)$. Tables of atomic form factors have been published by Hubbell *et al.* [20] and are included in the EPDL [11].

The quantity $f'+\mathrm{i}f''$, the so-called anomalous-scattering factor, accounts for the fast variation of the Rayleigh cross-section for photon energies around absorption edges. The tables of Henke *et al.* [15] include anomalous scattering factors for the elements $Z = 1$–92 and photon energies between 50 eV and 30 keV. An extensive tabulation of anomalous scattering factors, covering all the elements from hydrogen to fermium ($Z = 1$ to 100) and energies from 1 eV to 10 MeV has been prepared by Cullen *et al.* [21].

- **Mass attenuation coefficients**

The primary quantity for simulating photon histories is the (linear) attenuation coefficient, the reciprocal of the mean free path. It is given by

$$\mu = \mathcal{N}\sigma_{\mathrm{T}}, \tag{18}$$

where $\mathcal{N}$ is the number of atoms (or molecules) per unit volume and $\sigma_{\mathrm{T}} = \sigma_{\mathrm{ph}} + \sigma_{\mathrm{Co}} + \sigma_{\mathrm{Ra}}$ is the total interaction cross-section of an atom (or molecule). Assuming that the additivity approximation is applicable, the dependence of μ on the density of the material is practically removed by considering the mass attenuation coefficient,

$$\mu/\varrho = (N_{\mathrm{A}}/A_{\mathrm{w}})\,\sigma_{\mathrm{T}}. \tag{19}$$

To give an indication of the relative importance of the different interaction processes, fig. 2 shows total and partial mass attenuation coefficients of photons in aluminium and iodine as functions of the photon energy.

2.2. Electron interactions

- **Elastic collisions**

By definition, elastic collisions are interactions of the projectile electron that leave the initial state of the target atom unaltered. In an elastic collision the trajectory of the incident electron is deflected, with a very small energy transfer (of the order of a few meV) to the target atom, which recoils after the interaction.

Most MC codes describe elastic collisions by means of DCSs calculated by the relativistic (Dirac) partial-wave expansion method, with a central potential $V(r)$ that represents the electrostatic interaction energy between the projectile and the charge distribution of the target atom, which is assumed to remain "frozen" during the interaction (static-field approximation), plus a generally small correction to account for electron exchange effects. The programme

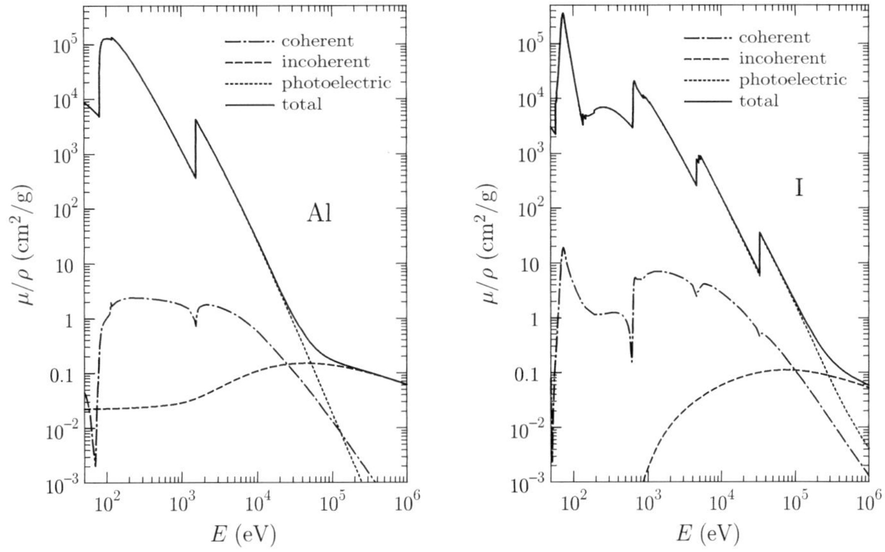

Figure 2. Partial and total mass attenuation coefficients of aluminium metal and iodine as functions of the photon energy. Notice the different low-E behaviour of $\mu_{\rm Co}/\rho$ for insulators (iodine) and conductors (aluminium).

ELSEPA of Salvat *et al.* [22], performs these calculations for electrons and positrons with kinetic energies up to $\sim$ 100 MeV.

The ICRU Report 77 [23] gives a review of theoretical calculation methods and experimental measurements of elastic scattering of electrons and positrons. This report includes an extensive database of DCSs for elastic scattering of projectiles with kinetic energies from 50 eV to 100 MeV by atoms of the elements $Z = 1$ to 103, which was generated by running ELSEPA. It is worth mentioning that DCSs calculated from the partial-wave expansion method are systematically larger than the measured DCSs (see, e.g., [24]) because of inelastic absorption effects (electrons that undergo inelastic interactions are effectively removed from the elastic channel) and also because of radiative corrections [25]. Inelastic absorption reduces the DCS by up to 10 percent in extreme cases. In principle, this correction is accounted for in MC simulations that include a detailed description of inelastic collisions. The static-field approximation fails for low-energy projectiles and small scattering angles, because the electric field of the projectile polarises the charges of the target atom, and the induced dipole acts back on the projectile. This polarisation correction can be described by including an extra term in the interaction potential, but it is usually disregarded in MC simulations on the grounds that small scattering angles do not have much influence on the whole transport process.

In old MC codes it was usual to describe elastic collisions by using the DCS obtained from the Born approximation, which is given by

$$\frac{{\rm d}\sigma_{\rm el}}{{\rm d}\Omega} = \frac{1-\beta^2\sin^2(\theta/2)}{1-\beta^2}\left(\frac{2{\rm m_e}z_0e^2}{q^2}\right)^2 [Z-F(Z,q)]^2 \tag{20}$$

where $F(Z,q)$ is the atomic form factor, eq. (17), and

$$q = 2p\sin(\theta/2) \tag{21}$$

is the momentum transfer.

- **Inelastic collisions**

The theoretical description of inelastic collisions of electrons, and other charged particles, is intrinsically difficult, because of the strong correlations between the energy loss W and the scattering angle θ. The basic theory of the process is based on the first-order plane wave Born approximation (PWBA) in which the effective interaction between the projectile and the target atom is treated as a perturbation to first order, with the initial and final states of the projectile represented as plane waves. The physics of the process is shown more explicitly when the DCS is expressed in terms of the energy loss W and the recoil energy Q defined by (see [26])

$$Q(Q+2\mathrm{m_e}c^2) = (cq)^2 = c^2\left(p^2+p'^2-2pp'\cos\theta\right) \tag{22}$$

where $\mathbf{q} = \mathbf{p}-\mathbf{p}'$ is the momentum transfer ($\mathbf{p}'$ is the momentum of the projectile after the collision). Note that Q equals the relativistic kinetic energy of an electron with momentum equal to the momentum transfer. Evidently, when the energy transfer is much larger than the binding energies U_i of the atomic electrons, these react essentially as if they where free and at rest and, consequently, the most probable excitations are those with $W \simeq Q$. A formal derivation with an independent-electron approximation (see [27]) yields a formula for the DCS which, after neglecting terms that are small for projectiles with energies $\lesssim$ 500 keV, reads

$$\begin{aligned}\frac{\mathrm{d}^2\sigma_{\mathrm{in}}}{\mathrm{d}W\,\mathrm{d}Q} &= \frac{2\pi e^4}{\mathrm{m_e}v^2}\left[\frac{2\mathrm{m_e}c^2}{WQ(Q+2\mathrm{m_e}c^2)}\right.\\ &\left.+\left(\beta^2-\frac{W^2}{Q(Q+2\mathrm{m_e}c^2)}\right)\frac{W\,2\mathrm{m_e}c^2}{[Q(Q+2\mathrm{m_e}c^2)-W^2]^2}\right]\frac{\mathrm{d}f(Q,W)}{\mathrm{d}W},\end{aligned} \tag{23}$$

where v is the velocity of the projectile. The last factor, the so-called generalised oscillator strength (GOS), is defined by an infinite sum of matrix elements of the operator $\exp(\mathrm{i}\,\mathbf{q}\cdot\mathbf{r}/\hbar)$ between orbitals describing the initial and final states of the target electron; the atomic GOS is obtained as the sum of partial GOSs for the various subshells of the target atom. The GOS is independent of the charge and energy of the projectile and embodies a complete description of the dynamics of the interaction (within the PWBA). Bote and Salvat [27] have calculated a complete database of sub-shell GOSs for the elements (Z = 1–99) using DHFS one-electron orbitals.

It is worth mentioning that the PWBA is formally equivalent to a semi-classical formulation in terms of the dielectric function $\epsilon(k,\omega)$, a function of the wave number $k = q/\hbar$ and the angular frequency $\omega = W/\hbar$, which describes the response of the material to external electromagnetic fields. In the limit $k \to 0$, $\epsilon(k,\omega)$ reduces to the optical dielectric function, i.e., $\epsilon(\omega) = \epsilon(k=0,\omega)$. The so-called optical-data models combine empirical optical data with various approximate forms of the dielectric function of the degenerate electron gas to build a model of $\epsilon(k,\omega)$ for finite wave numbers k. The classical theory determines the stopping force on the projectile from the Maxwell equations. The semi-classical interpretation of the quantities $\hbar k$ and $\hbar\omega$ as, respectively, the momentum transfer and the energy loss, allows extracting an expression of the DIMFP. This semi-classical DIMFP is essentially equivalent to the result from the PWBA if one makes the identification

$$\frac{\mathrm{d}f(Q,W)}{\mathrm{d}W} \equiv W\left(1+\frac{Q}{\mathrm{m_e}c^2}\right)\frac{2Z}{\pi\Omega_{\mathrm{p}}^2}\,\mathrm{Im}\left(\frac{-1}{\epsilon(k,\omega)}\right), \tag{24}$$

where

$$\Omega_{\mathrm{p}} = \sqrt{4\pi \mathcal{N} Z \frac{\hbar^2 e^2}{\mathrm{m_e}}} \tag{25}$$

is the plasma resonance energy of an electron gas with the average electron density of the medium. The clear advantage of optical-data models is that, apart from experimental uncertainties of the adopted optical dielectric function, they describe the actual response of the medium, including the effects of aggregation.

The energy-loss DCS is

$$\frac{\mathrm{d}\sigma_{\mathrm{in}}}{\mathrm{d}W} \equiv \int_{Q_-}^{Q_+} \frac{\mathrm{d}^2\sigma_{\mathrm{in}}}{\mathrm{d}W\,\mathrm{d}Q}\,\mathrm{d}Q, \tag{26}$$

where Q_- and Q_+ are the minimum and maximum kinematically allowed recoil energies, which are given by eq. (22) with $\cos\theta = \pm 1$. The total cross-section is the integral of the energy-loss DCS over allowed energy transfers. The collision (or electronic) stopping power S_{col} is defined as the average energy loss per unit path length due to inelastic collisions, and is given by

$$S_{\mathrm{col}} = -\frac{\mathrm{d}E}{\mathrm{d}s} = \mathcal{N} \int_0^{W_{\max}} W \frac{\mathrm{d}\sigma_{\mathrm{in}}}{\mathrm{d}W}\,\mathrm{d}W, \tag{27}$$

where $\mathcal{N}$ is the number of atoms or molecules per unit volume and $W_{\max}$ is the maximum allowed energy transfer. When the energy E of the projectile is much larger than the binding energies of the atomic electrons, the stopping power is given by the celebrated Bethe formula,

$$S_{\mathrm{col}} = \mathcal{N} \frac{2\pi e^4}{\mathrm{m_e} v^2} Z \left\{ \ln\left(\frac{E^2}{I^2} \frac{\gamma+1}{2} \right) + \gamma^{-2} \left[1 - (2\gamma - 1)\ln 2 + \frac{1}{8}(\gamma - 1)^2 \right] \right\}, \tag{28}$$

where I is the mean excitation energy, defined by

$$Z \ln I = \int_0^{\infty} \ln W \frac{\mathrm{d}f(0,W)}{\mathrm{d}W}\,\mathrm{d}W. \tag{29}$$

This quantity has been determined empirically for a large number of materials (see [28] and references therein) from measurements of the stopping power of heavy charged particles and/or from experimental optical dielectric functions. Most MC codes make explicit use of these empirical I values. For materials not included in the ICRU tables [28], I is usually estimated from the additivity approximation. Figure 3 shows stopping powers of electrons in aluminium, silver and gold metals generated by the PENELOPE code. Notice that the Bethe formula (28) is accurate only for energies higher than about 10 keV.

The ionisation of inner atomic shells by electron impact is the main source of X-rays in EPMA. The PWBA yields accurate ionisation cross-sections for electrons with energy higher than about 30 times the binding energy of the active shell. At lower energies, the distortion of the projectile wave functions by the electrostatic field of the target atom, and electron exchange effects, become important. A more elaborate theoretical description of total ionisation cross-sections is obtained from the relativistic distorted-wave Born approximation (DWBA), which consistently accounts for these effects (see e.g., [27] and references therein). DWBA calculations yield total cross-sections in fairly good agreement with measurements for ionisation of K-, L- and M-shells (see [4]). These calculations involve the expansion of free-state wave functions as partial wave series and the subsequent evaluation of multiple radial integrals. Since the convergence of partial-wave series worsens with increasing kinetic energies, DWBA calculations are feasible only for projectiles with relatively small energies, up to about 30 U_i.

The NIST Standard Reference Database 164 [29] contains an extensive numerical tabulation of ionisation cross-sections for K-, L- and M-shells of all the elements from hydrogen ($Z = 1$) to einsteinium ($Z = 99$), for electrons and positrons with kinetic energies from threshold up to 1 GeV, which was calculated by combining the DWBA and the PWBA as described by Bote and Salvat [27]. Also, analytical formulas for the easy calculation of these cross-sections were published by Bote *et al.* [30, 31]. These formulas are coded in a simple Fortran programme that yields cross-sections that agree with those in the numerical database to within about 1 %. The PENELOPE database also includes ionisation cross-sections for N-shells of heavy elements, which were calculated using the same methodology as for the K-, L- and M-shells. Total cross-sections for impact ionisation of the K-, L-, and M-shells of gold are shown in fig. 3.

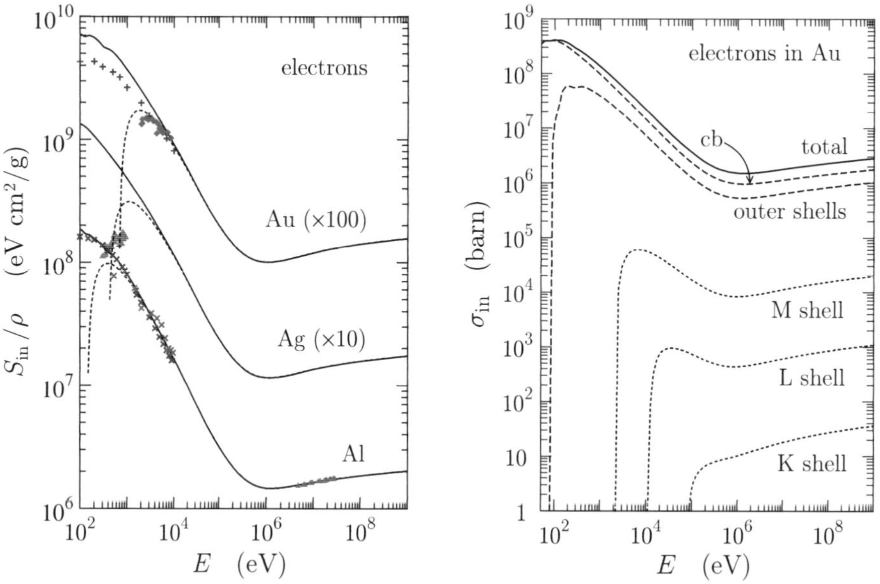

Figure 3. Left: Collision stopping powers of aluminium, silver and gold as functions of the kinetic energy E of the projectile electron. Symbols represent experimental data from different authors. The dotted curves are the stopping powers obtained from the Bethe formula, eq. (28). Right: Sub-shell cross-sections and total cross-section for inelastic collisions of electrons in gold, generated by the PENELOPE code. The dotted curves represent total ionisation cross-sections for the K-, L- and M-shells from the NIST database [29]. Solid curves represent total stopping powers and inelastic cross-sections used in PENELOPE.

• Bremsstrahlung emission

Bremsstrahlung emission is caused by the interaction of the projectile electron with the electrostatic field of the atom or in collisions with atomic electrons. The theory of the process is reviewed in the book of Haug and Nakel [32].

In MC codes the energy of emitted quanta is sampled from an energy-loss DCS, differential in only the energy W of the emitted photon, expressed in the form

$$\frac{d\sigma_{br}}{dW} = \frac{Z^2}{\beta^2} \frac{1}{W} \chi(Z, E, \kappa). \tag{30}$$

where $\chi(Z, E, \kappa)$, the so-called scaled DCS, is a function of the atomic number Z, the energy of the projectile E, and the reduced energy of the photon, $\kappa \equiv W/E$. For a given element, this function is finite and varies smoothly with E and κ. Seltzer and Berger [33, 34] have prepared extensive tables of bremsstrahlung energy spectra, obtained by combining different theoretical models, for electrons with kinetic energies E from 1 keV to 10 GeV incident on neutral atoms of the elements $Z = 1 - 100$.

To complete the simulation of bremsstrahlung emission events, one has to determine the direction of the emitted photon. The angular distribution of emitted photons in electron-nucleus bremsstrahlung (the so-called shape function) has been calculated by Kissel *et al.* [35], using a partial-wave expansion method, for 144 combinations of atomic number Z, electron energy E, and reduced photon energy κ. Acosta *et al.* [36] found that these shape functions can be closely approximated as Lorentz-boosted dipole distributions, from which the emission angle can be sampled analytically.

In the case of compounds, with n_i atoms of the element Z_i per molecule, the molecular DCS is obtained by means of the additivity approximation,

$$\frac{\mathrm{d}\sigma_{\mathrm{br}}}{\mathrm{d}W} = \sum_i n_i \frac{Z_i^2}{\beta^2} \frac{1}{W} \chi(Z_i, E, \kappa). \tag{31}$$

It is worth noticing that the total cross-section for bremsstrahlung emission is infinite, due to the divergence of the DCS at $W = 0$. In MC simulation we consider only the emission of photons with energies higher than a certain cutoff W_{c}, and the associated total cross-section

$$\sigma_{\mathrm{br}}(W > W_{\mathrm{c}}) = \int_{W_{\mathrm{c}}}^{E} \frac{\mathrm{d}\sigma_{\mathrm{br}}}{\mathrm{d}W} \, \mathrm{d}W \tag{32}$$

is finite. The radiative stopping power

$$S_{\mathrm{rad}} = \mathcal{N} \int_0^E W \frac{\mathrm{d}\sigma_{\mathrm{br}}}{\mathrm{d}W} \, \mathrm{d}W = \mathcal{N} \int_0^E W \left(\sum_i n_i \frac{Z_i^2}{\beta^2} \chi(Z_i, E, \kappa) \right) \mathrm{d}W, \tag{33}$$

is finite. Tables of radiative stopping powers, obtained from Seltzer and Berger's scaled DCSs, are given in Ref. [28].

2.3. Atomic relaxation

Interactions of photons and electrons that cause the emission of secondary electrons (photoelectric absorption, Compton scattering of photons, and inelastic collisions of electrons) leave the target atom ionised and in an excited state. The excited ion then relaxes to its ground state through a sequence of radiative and non-radiative transitions, in which photons (fluorescence X-rays) and Auger electrons are emitted, respectively. Atomic relaxation can be simulated by using information from the LLNL Evaluated Atomic Data Library (EADL) of Perkins *et al.* [37], which provides a comprehensive tabulation of transition probabilities and energies for both radiative and non-radiative transitions for all elements. The transition probabilities in this library were calculated theoretically by assuming that the relaxing ion has a single vacancy (an assumption that ceases to hold, e.g., after a non-radiative transition) and transition energies were approximated by means of the DHFS energy eigenvalues for neutral atoms. PENELOPE simulates atomic relaxation using transition probabilities and Auger-electron energies from the EADL. However, the energies of X-rays emitted in radiative transitions are taken from the compilations of experimental values by Deslattes *et al.* [38], for K- and L-lines, and by Bearden [39], for M-lines. In the case of radiative transitions not included in these compilations (e.g., N-lines), the EADL X-ray energies are adopted.

A common practice in studies of X-ray emission is to consider cross sections for emission of characteristic X-rays, $\sigma^{\mathrm{x}}_{\mathrm{S0,S1}}$, defined so that the product $\mathcal{N}\sigma^{\mathrm{x}}_{\mathrm{S0,S1}}$ equals the probability of emission of an S0-S1 X-ray per unit path length of the projectile. These cross-sections can be measured by counting the emitted S0-S1 X-rays when projectile electrons of energy E impinge on a very thin foil of the material. $\sigma^{x}_{S0,S1}$ can be expressed as a weighted sum of electron-impact ionisation cross-sections for the sub-shell S0, and for inner sub-shells, with weights determined by the appropriate partial widths and fluorescence and Coster-Kronig yields. Detailed formulas for the emission cross-sections of K-, L- and M-lines are given, e.g., by Llovet *et al.* [4].

3. Approximations in MC simulation

Within a MC code random particle histories are simulated from the initial energy of primary particles down to a certain cutoff, the absorption energy E_{abs}, at which the particles are considered to be effectively absorbed in the material. Secondary particles that are emitted as a direct result of the interaction, and also in the relaxation of atoms following inner-shell ionisation (by photoelectric absorption and Compton scattering of photons and by electron impact) may be stored in memory and tracked subsequently in the same way as the primary particles.

The simulation of photons can be performed by the usual detailed procedure, where all interaction events in a photon history are simulated in chronological succession. That is, a photon of energy E starts from a certain position, $\mathbf{r}_0$, determined in accordance to the characteristics of the radiation source, moving in a direction defined by the unit vector $\hat{\mathbf{d}}_0$. The distance s to the next interaction is determined by random sampling from the exponential distribution

$$p(s) = \mu(E)\exp[-s\mu(E)], \tag{34}$$

where $\mu(E)$ is the IMFP of the photon. The particle is then moved the distance s along the ray, i.e., to a position $\mathbf{r} = \mathbf{r}_0 + s\hat{\mathbf{d}}_0$, where the next interaction takes place. The type of the next interaction is sampled according to the total cross-sections of the possible interaction mechanisms, and the interaction is simulated from the corresponding DCS. In Rayleigh and Compton scattering, the photon is absorbed by electrons in the medium and a second photon is emitted with energy E' (equal or less than E). When $E' > E_{abs}$, the surviving photon is followed by repeating these steps. Photoabsorption terminates the photon history. Each history is a sequence of a relatively small number of free flights and interactions, which can be simulated rapidly.

This simulation scheme can be readily adapted to real situations, with material structures that may consist of several regions of different compositions separated by mathematical surfaces (interfaces). Photons are simulated in the current material as described above, and when they reach an interface, the tracking is discontinued and resumed with the cross-sections of the new material. That is, to track photons in complex geometries, we only need to determine the distances at which a ray intersects the interfaces.

In principle, electrons can also be tracked using detailed simulation. However, because the average energy loss in each collision is of the order of a few tens of eV, the number of interactions undergone by an electron may be quite large and detailed simulation is inefficient. The code PENELOPE utilises a more efficient tracking strategy that combines detailed simulation of hard events (i.e., interactions with scattering angle or energy loss larger than pre-defined cutoffs) and approximate multiple-scattering methods for describing the accumulated effect of the (usually many) soft interactions that occur between two consecutive hard events. Other codes may have recourse to more drastic approximations, which speed up the simulation at the expense of simplifying the interaction physics. For instance, EPMA codes may use the continuous-slowing-down approximation (CSDA), which consists in assuming that electrons lose energy continuously, at the rate prescribed by the stopping power. Although this approximation yields reasonable results for homogeneous samples, it may introduce considerable distortions for specimens with inclusions at depths slightly larger than the effective range.

Individual interaction events should be simulated by random sampling from the associated DCSs. However, the DCSs for Compton scattering of photons and inelastic collisions of electrons depend explicitly on the energy transfer W and the scattering angle θ (or the recoil energy Q). These variables are strongly correlated and, in addition, the DCSs vary rapidly with the energy of the projectile. Such DCSs are not practicable for MC simulation, not only because of the large memory required to store the numerical DCS (a three-dimensional table) but also because random sampling from multivariate distributions is far more complicated, and slower, than from univariate distributions. A frequent practice is to disregard (or alter artificially) correlations between variables, normally by sampling the most relevant quantity from its marginal probability distribution and using approximate probability distributions for the other variables. Approximations of this kind are employed in virtually all MC codes; they are harmless in the majority of applications, but they can cause visible distortions of the simulation results in cases where particles undergo a small number of interactions (e.g., in transmission through thin foils).

In the MC code PENELOPE inelastic collisions are described by using a simple GOS model consisting of discrete resonances, which reproduces the adopted total cross-sections for ionisation of inner sub-shells and the empirical value of the mean excitation energy I. This model yields very accurate stopping powers at high energies, and the correct number of inner-shell ionisations per unit path length. It also provides a realistic description of energy-loss fluctuation in individual collisions. The description of Compton scattering in PENELOPE is based on approximate analytical forms of the sub-shell Compton profiles, which allow exact random sampling of W and θ.

4. Concluding remarks

The interaction models considered in this review provide a fairly realistic description of coupled electron-photon transport, in spite of the fact that they involve quite drastic approximations. MC simulations based on those and similar models have proved to be practically useful in EPMA and other applications involving X-rays. Still, users of general-purpose MC codes should be well aware of the limited accuracy of the method, which results from uncertainties in the adopted fundamental quantities, from the neglect of possible aggregation effects, and from simplifications in the description of the sample geometry. These limitations give rise to type B uncertainties in the results of MC simulations, which add to the (statistical) type A uncertainties typically quantified by MC codes.

Acknowledgements

Financial support from the Spanish Ministerio de Economía y Competitividad and ERDF (project ref. FPA2016-77689-C2-2-R) and from the Generalitat de Catalunya (grant 2014 SGR 846) is gratefully acknowledged.

References

[1] Llovet X and Salvat F 2017 *Microsc. Microanal.* **23** 634–646
[2] Salvat F 2015 PENELOPE-*2014: a code system for Monte Carlo simulation of electron and photon transport.* (Issy-les-Moulineaux, France: OECD/NEA Data Bank) available from `http://www.nea.fr/lists/penelope.html`
[3] Salvat F and Fernández-Varea J M 2009 *Metrologia* **46** S112–S138
[4] Llovet X, Powell C J, Jablonski A and Salvat F 2014 *J. Phys. Chem. Ref. Data* **43** 013102
[5] Palik E D (Ed.) 1985 *Handbook of optical constants of solids.* (San Diego, CA: Academic Press)
[6] Palik E D (Ed.) 1991 *Handbook of optical constants of solids II.* (San Diego, CA: Academic Press)
[7] Palik E D (Ed.) 1998 *Handbook of optical constants of solids III.* (San Diego, CA: Academic Press)
[8] Werner W S M, Glantschnig K and Ambrosch-Draxl C 2009 *J. Phys. Chem. Ref. Data* **38** 1013–1092
[9] Sabbatucci L and Salvat F 2016 *Radiat. Phys. Chem.* **121** 122–140

[10] Grant I P 1970 *Advances in Physics* **19** 747–811
[11] Cullen D E, Hubbell J H and Kissel L 1997 *EPDL97 the Evaluated Photon Data Library, '97 version. Report UCRL-50400* (Livermore, California: Lawrence Livermore National Laboratory)
[12] Scofield J H 1973 *Theoretical photoionization cross sections from 1 to 1500 keV. Report UCRL-51326* (Livermore, California: Lawrence Livermore Laboratory)
[13] Hubbell J H, Gimm H A and Øverbø I 1980 *J. Phys. Chem. Ref. Data* **9** 1023–1147
[14] Hubbell J H and Seltzer S M 1996 *Tables of X-ray mass attenuation coefficients and mass energy-absorption coefficients from 1 keV to 20 MeV for elements Z = 1 to 92 and 48 additional substances of dosimetric interest. NISTIR 5632 database.* (Gaithersburg, MD: National Institute of Standards and Technology) `https://www.nist.gov/pml/x-ray-mass-attenuation-coefficients`
[15] Henke B L, Gullikson E M and Davis J C 1993 *At. Data Nucl. Data Tables* **54** 181–342
[16] Rehr J J and Albers R C 2000 *Rev. Mod. Phys.* **72** 621–654
[17] Sakurai J J 1967 *Advanced quantum mechanics.* (New York: Addison and Wesley)
[18] Ribberfors R 1975 *Phys. Rev. B* **12** 2067–2074
[19] Baym G 1974 *Lectures in quantum mechanics.* (Boulder, Colorado: Westview Press)
[20] Hubbell J H, Veigele W J, Briggs E A, Brown R T, Cromer D T and Howerton R J 1975 *J. Phys. Chem. Ref. Data* **4** 471–538
[21] Cullen D E 1989 *Program SCATMAN: a code designed to calculate photon coherent scattering anomalous scattering factors and cross sections. Report UCRL-ID-103422.* (Livermore, CA: Lawrence Livermore National Laboratory)
[22] Salvat F, Jablonski A and Powell C J 2005 *Comput. Phys. Commun.* **165** 157–190
[23] ICRU Report 77 2007 *Elastic scattering of electrons and positrons.* (Bethesda, MD: ICRU)
[24] Salvat F 2003 *Phys. Rev. A* **68** 012708
[25] Motz J W, Olsen H A and Koch H W 1964 *Rev. Mod. Phys.* **36** 881–928
[26] Fano U 1963 *Ann. Rev. Nucl. Sci.* **13** 1–66
[27] Bote D and Salvat F 2008 *Phys. Rev. A* **77** 042701
[28] ICRU Report 37 1984 *Stopping powers for electrons and positrons.* (Bethesda, MD: ICRU)
[29] Llovet X, Salvat F, Pujol F S, Jablonski A and Powell C J 2014 *NIST database of cross sections for inner-shell ionization by electron or positron impact.* (Gaithersburg, MD: National Institute of Standards and Technology) available from `http://dx.doi.org/10.6028/NIST.NSRDS.164`
[30] Bote D, Salvat F, Jablonski A and Powell C J 2009 *At. Data and Nucl. Data Tables* **96** 871–909
[31] Bote D, Salvat F, Jablonski A and Powell C J 2011 *At. Data and Nucl. Data Tables* **97** 186
[32] Haug E and Nakel W 2004 *The elementary process of bremsstrahlung.* (Singapore: World Scientific)
[33] Seltzer S M and Berger M J 1985 *Nucl. Instrum. Meth. B* **12** 95–134
[34] Seltzer S M and Berger M J 1986 *At. Data Nucl. Data Tables* **35** 345–418
[35] Kissel L, Quarles C A and Pratt R H 1983 *At. Data Nucl. Data Tables* **28** 381–460
[36] Acosta E, Llovet X and Salvat F 2002 *Appl. Phys. Lett.* **80** 3228–3230
[37] Perkins S T, Cullen D E, Chen M H, Hubbell J H, Rathkopf J and Scofield J 1991 *Tables and graphs of atomic subshell and relaxation data derived from the LLNL Evaluated Atomic Data Library (EADL), Z = 1–100. Report UCRL-ID-50400.* (Livermore, CA: Lawrence Livermore National Laboratory)
[38] Deslattes R D, Kessler E G, Indelicato P, de Billy L, Lindroth E and Anton J 2003 *Rev. Mod. Phys.* **75** 36–99
[39] Bearden J A 1967 *Rev. Mod. Phys.* **39** 78–124

IOP Conf. Series: Materials Science and Engineering **304** (2017) 012015 doi:10.1088/1757-899X/304/1/012015

A tracking algorithm for Monte Carlo simulation of surface roughness in EPMA measurements

D Sánchez-Gonzalo[1], X Llovet[2], R Graciani[1] and F Salvat[1]

[1] Facultat de Física (FQA and ICC). Universitat de Barcelona. Diagonal 647, ES-08028 Barcelona, Spain

[2] Scientific and Technological Centres. Universitat de Barcelona. Lluís Solé i Sabarís, 1-3. ES-08028 Barcelona, Spain

E-mail: francesc.salvat@ub.edu

Abstract.
We present an algorithm for Monte Carlo simulation of particle trajectories through rough interfaces. The microscopic topography of the rough surface is assumed to be described by an altitude map in a dense rectangular grid. The tracking algorithm has been implemented in a Fortran subroutine package, which is coupled to the code system PENELOPE/PENEPMA to perform Monte Carlo simulations of X-ray emission from samples with rough surfaces irradiated by electron beams. To validate the numerical procedure, electron probe microanalysis (EPMA) simulations of an ideal sample with periodic surface roughness have been performed by using 1) the PENGEOM geometry package of the PENELOPE code system and 2) the new tracking algorithm. The results from the two simulations are found to be equivalent, that is, their differences are generally less than the associated statistical uncertainties. Results from simulations of samples with realistic rough surfaces are also presented.

1. Introduction

Conventional quantification procedures for electron probe microanalysis (EPMA) assume flat bulk specimens. However, in some cases specimens must be analysed without any surface preparation either because they cannot be destroyed (e.g., cultural heritage objects) or because the information of interest is in the surface itself (e.g., corrosion studies). The analysis of rough targets is challenging because of the dependence of the measured line intensities on the topography of the sample surface [1,2]. The same type of difficulty is found in X-ray fluorescence (XRF) [3–5], and in the transport of light through scintillating crystals and optical systems with rough interfaces [6].

The effect of surface roughness on radiation transport processes, can be studied by means of Monte Carlo simulation [7,8]. In EPMA and electron microscopy, the area of the sample that is effectively probed has a limited extension, and the relevant portion of the surface can be described numerically, e.g., by using information from stereoscopic or confocal microscopy. Generally, though, the extension of a rough surface is too large to allow a microscopic description, and one must assume that the surface roughness is "homogeneous", in the sense that the statistical roughness parameters do not change appreciably with position. Assuming that curvature is not too large, the rough surface can be described as the combination of a mathematical (smooth) surface and a numerical model representing a finite portion of a nearly planar rough surface with characteristics similar to those of the real surface.

In this study we describe tracking strategies to account for the effect of roughness of the sample surface (and of interfaces between different materials) within a conventional Monte Carlo transport code for material structures defined by mathematical surfaces. In the simulations presented below we use the Monte Carlo code system PENELOPE [9], which describes the material system where radiation propagates by using constructive quadric geometry [10], that is, as a set of homogeneous bodies limited by quadric surfaces.

A rough surface is assumed to be defined within a rectangular cuboid, which will be referred to as "the box", by means of an altitude map on a dense two-dimensional rectangular grid. A continuous representation of the rough surface is obtained by triangularisation of the altitude map. For the simulation of radiation transport through rough surfaces we have adopted a simple scheme in which the neighbourhood of the mathematical surface is replaced by the rough surface box. During simulation, when a particle reaches a mathematical surface that corresponds to a rough interface, an external subprogramme is called to perform the tracking through the associated surface box. Control is returned to the main programme when the incident particle and any secondary particles generated during the process have left the box. This scheme allows keeping the structure of the conventional simulation code practically unaltered.

Simulations of EPMA measurements, that is, of X-ray emission from targets bombarded by electron beams, can be performed with the Fortran main programme PENEPMA [11], which uses PENELOPE as the simulation engine and the quadric geometry package PENGEOM [10]. To allow consideration of targets with a rough surface, we have written a new programme, named EPMAR, which generates the same information as PENEPMA and uses the present tracking algorithm to describe the transport of particles through the rough irradiated surface. To verify the correctness of the proposed tracking algorithm, we consider homogeneous targets with an artificial rough surface consisting of a periodic array of square pyramids. This artificial sample structure can be described by both PENEPMA and EPMAR. The results from the two codes are found to agree within statistical uncertainties. We present also results from EPMAR simulations with a realistic rough surface.

2. Tracking of particles near a rough surface

When the particle being transported approaches a mathematical surface representing a rough interface, the main programme switches to tracking within the corresponding surface box. To simplify the geometrical operations, the motion of particles within the box is referred to a right-handed reference frame with the origin at one vertex of the box and the axes parallel to the edges, so that the box lies in the first octant. The plane $z = 0$ is regarded as the base of the box, and the surface is described by giving its altitude $z(x, y)$ at the points of a regular Cartesian grid with spacings Δ_x, Δ_y. The definition of the surface is provided by the user in an ASCII file with three columns, which specify the coordinates x_i ($i = 1 : N_x$) and y_j ($j = 1 : N_y$) of each grid point, and the surface altitude $z_{i,j}$, respectively. It is assumed that the x coordinate varies first in the file, that is, that the triad $x_i, y_j, z_{i,j}$ is given in the line number $i + (j-1)N_y$ of the file. Since

$$x_i = (i-1)\Delta_x, \quad y_j = (j-1)\Delta_y, \tag{1}$$

the arrays x_i and y_i do not need to be stored in memory. To avoid inconsistencies resulting from numerical roundoff, our subroutines define a box of height $\Delta_z = 1.1S$, where S is the difference in altitudes of the highest and lowest point of the original surface, and the surface is shifted so that the altitudes of the lowest and highest points are $0.05\Delta_z$ and $\Delta_z - 0.05S$, respectively. That is, the shifted values $z_{i,j}$ are all such that $0.05S \le z_{i,j} \le \Delta_z - 0.05S$. A continuous surface is obtained by introducing a triangular mesh. The grid points of each cell $(x_i, x_{i+1}) \times (y_j, y_{j+1})$ define two possible pairs of triangles (see fig. 1). The surface generally looks smoother when using the pair that intersect along the diagonal of the cell that has the largest slope (case b in fig. 1) and, consequently, this is the triangularisation scheme adopted in our algorithm. Thus, each cell defines a column consisting of two sections, separated by the surface triangles. Note that we are assuming that the volumes below and above the surface are filled with two different materials.

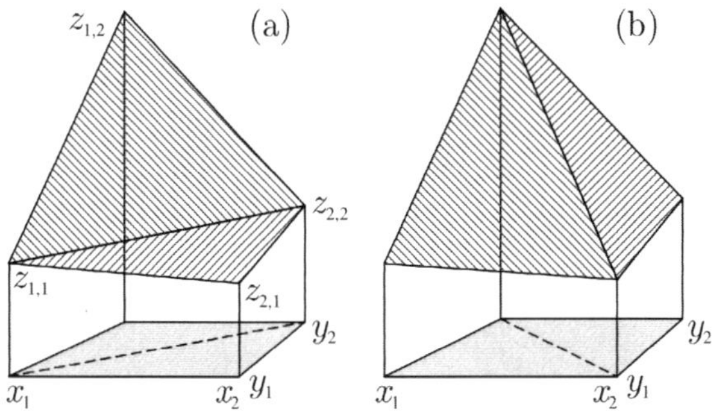

Figure 1. Possible triangularisations of a cell $(x_1, x_2) \times (y_1, y_2)$, characterised by the slopes of the diagonals of the cell.

In Monte Carlo simulations the trajectory of a particle is described as a sequence of connected linear steps, or "jumps", of length Δs determined by the physics subroutines. Each step corresponds to a segment of a ray defined by the equation

$$\mathbf{r}(s) = \mathbf{r}(0) + s\hat{\mathbf{d}}. \tag{2}$$

where the vectors $\mathbf{r}(0) = (x(0), y(0), z(0))$ and $\hat{\mathbf{d}} = (u, v, w)$ are, respectively, the position and the direction of motion at the start of the step. Steps represent free flights of the particle, which end either with an interaction or when the particle arrives at an interface, i.e., a surface separating two different materials.

A conventional Monte Carlo code generates random particle histories sequentially, by sampling the length Δs of the steps, and the energy loss and the deflection angles caused by the interactions. The probability distribution functions of these variables are determined by the adopted interaction model (see, e.g., Ref. [9]). When a particle arrives at a mathematical interface (a quadric, in the case of PENELOPE/PENGEOM), the simulation code halts the particle and resumes simulation in the material behind the interface.

In the case of rough interfaces, the Monte Carlo code can still follow particles up to a distance $\simeq \Delta_z$ from the mathematical surface and then switch the simulation to the surface box. The tracking of particles through the box is performed by using the strategy and the formulas presented in the Appendix.

2.1. Samples with a rough external surface

A situation of interest in EPMA is that of a sample with a rough external surface that is bombarded by the electron beam. Because the probed volume has microscopic lateral dimensions, the relevant portion of the sample surface can be described numerically. We have written a dedicated programme, named EPMAR, which simulates EPMA measurements on targets having the structure sketched in fig. 2, with the rough surface box placed on top of a base sample. The latter is described by means of the PENGEOM subroutines, which allow defining embedded particles, lamellae, etc. Notice that the plane $z = 0$ is assumed to separate the box and the base sample. For simplicity, we also assume that the box and the base sample have the same lateral extensions, i.e., that both are limited by the planes $x = 0$, $x = \Delta_x$, $y = 0$ and $y = \Delta_y$. EPMAR was obtained by modifying the programme PENEPMA [11] to include the option of a rough external surface; the two codes operate similarly and generate the same information.

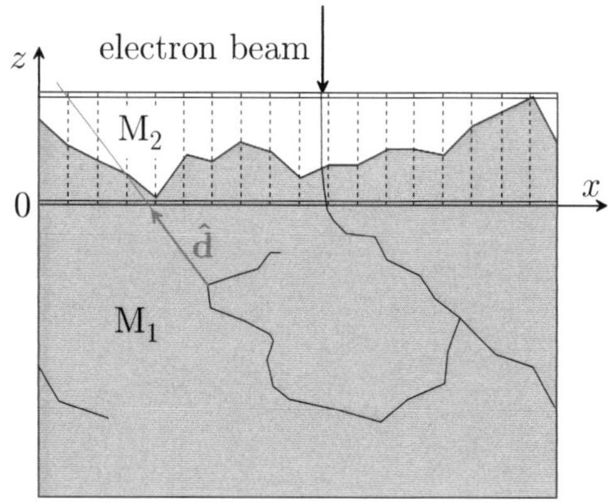

Figure 2. Schematics of the simulation of an EPMA measurement on a sample with a rough surface. The programme EPMAR allows using periodic boundary conditions in the x-y plane.

The simulation of particle histories within the box is performed by calling the PENELOPE subroutines and using the tracking strategy described in the Appendix. When a particle leaves the box through one of the lateral faces, we make it re-enter the opposite face thus imposing periodic boundary conditions in the x and y directions. Each particle is followed until it leaves the box through the lower or the upper plane, or it is absorbed within the material. Because secondary particles may occasionally be produced, they also have to be followed until they leave the box or are absorbed. When the simulation of the shower in the box is completed, the tracking of particles is resumed in the base sample using the PENGEOM subroutines. The process is repeated each time a particle arrives to the rough interface.

In practical EPMA of e.g., geological samples with lamellae, the analysed volume may present complicated structures, which approximately recur with a certain periodicity. The definition of the geometry of these samples with PENGEOM is very laborious. To simplify the work, the programme EPMAR offers the option of applying periodic boundary conditions in the x-y plane also to the base sample. In addition, although PENGEOM uses mathematical surfaces to limit the bodies in the base sample, the method described in Section 2.2 can be used to add roughness to the "internal" surfaces.

2.2. Internal rough interfaces

To obtain a general tracking algorithm we also consider geometries with macroscopic rough interfaces, which are each described by a box representing a finite sample portion of the rough surface. As indicated above, we assume that the main programme tracks particles in the conventional way, i.e., by using only mathematical surfaces, and surface roughness is taken into account only when a particle gets close to the mathematical surface. For the sake of programming simplicity, we let the main programme track a particle until it reaches the mathematical surface, which is then replaced by the surface box. The box is placed "parallel" to the mathematical surface in a position such that the particle crosses the rough surface with randomly selected x and y coordinates (in the box frame). The simulation of the particle is then continued within the box by using the tracking method described in the Appendix.

Let $\mathbf{r}_\mathrm{s} = (x_\mathrm{s}, y_\mathrm{s}, z_\mathrm{s})$ and $\hat{\mathbf{d}}_\mathrm{s} = (u_\mathrm{s}, v_\mathrm{s}, w_\mathrm{s})$ be the position and direction of motion of the particle when it arrives at the mathematical surface. The latter is then replaced by the box oriented so that its z axis is parallel or anti-parallel to the normal $\hat{\mathbf{n}} = (n_x, n_y, n_z)$ to the mathematical surface, making sure that the materials in the box are coincident with those in the PENGEOM geometry (see fig. 3). Since the tracking of particles within the box is performed with reference to the box frame, the position and direction vectors of the particle have to be transformed from the laboratory frame (i.e., from the reference frame adopted in the main programme) to the box frame. Consistently with the assumed "homogeneity" of the rough surface, the particle starts its motion in the box from a random position $\mathbf{r} = (x, y, z)$ where the coordinates x and y are sampled uniformly and z is set equal to the altitude of the triangularised surface at (x, y) plus or minus a small amount, $\epsilon \simeq 10^{-8}$ cm, to ensure that the particle has just passed the surface. The direction vector $\hat{\mathbf{d}} = (u, v, w)$ in the box frame is obtained by rotating the laboratory vector $\hat{\mathbf{d}}_{\rm s}$,

$$\hat{\mathbf{d}} = \mathcal{R}_{\rm B}\, \hat{\mathbf{d}}_{\rm s}, \tag{3}$$

where the rotation $\mathcal{R}_{\rm B}$ is required to transform the unit vector $\hat{\mathbf{z}} = (0, 0, 1)$ into the surface normal vector $\hat{\mathbf{n}}$. We use the following

$$\mathcal{R}_{\rm B} = \mathcal{R}(\phi\hat{\mathbf{z}})\mathcal{R}(\theta\hat{\mathbf{y}}) \tag{4}$$

where θ and ϕ are the polar and azimuthal angles of the normal vector $\hat{\mathbf{n}}$, and

$$\mathcal{R}(\phi\hat{\mathbf{z}}) = \begin{pmatrix} \cos\phi & -\sin\phi & 0 \\ \sin\phi & \cos\phi & 0 \\ 0 & 0 & 1 \end{pmatrix} \quad \text{and} \quad \mathcal{R}(\theta\hat{\mathbf{y}}) = \begin{pmatrix} \cos\theta & 0 & \sin\theta \\ 0 & 1 & 0 \\ -\sin\theta & 0 & \cos\theta \end{pmatrix}, \tag{5}$$

are rotations of angles ϕ and θ around the z and y axes of the laboratory frame, respectively. That is,

$$\mathcal{R}_{\rm B} = \begin{pmatrix} \cos\phi\cos\theta & -\sin\phi & \cos\phi\sin\theta \\ \sin\phi\cos\theta & \cos\phi & \sin\phi\sin\theta \\ -\sin\theta & 0 & \cos\theta \end{pmatrix}, \tag{6}$$

with

$$\cos\theta = n_x, \quad \sin\theta = \sqrt{n_x^2 + n_y^2}, \quad \cos\phi = \frac{n_x}{\sqrt{n_x^2 + n_y^2}}, \quad \sin\phi = \frac{n_y}{\sqrt{n_x^2 + n_y^2}}. \tag{7}$$

When $n_x^2 + n_y^2 \ll 1$, so that $n_z \simeq \pm 1$, we have

$$\mathcal{R}_{\rm B} = \begin{pmatrix} -1 & 0 & 0 \\ 0 & 1 & 0 \\ 0 & 0 & -1 \end{pmatrix} \tag{8}$$

if $n_z \simeq -1$ and $\mathcal{R}_{\rm B} = \mathcal{I}$, the identity matrix, if $n_z \simeq +1$.

The evolution of particle histories within the box is followed by calling the PENELOPE subroutines and using the tracking strategy described in the Appendix. When a particle leaves the box through one of the lateral faces, we make it re-enter the opposite face thus imposing periodic boundary conditions in the x and y directions. Each particle is followed until it leaves the box through the lower or the upper plane, or it is absorbed in the material. The handling of secondary particles is somewhat tricky, because they cannot be sent directly to the secondary stack of PENELOPE (see Ref. [9]), where all particles are referred to the laboratory frame. Secondary particles generated within the box are followed until they leave the box or are absorbed. Only when the simulation of the shower in the box is completed, the particles that have left through the upper and lower faces of the box are transferred to the secondary stack. Their position coordinates and directions of motion in the laboratory frame are obtained by applying the inverse of the rotation, eq. (4), to the vectors $\mathbf{r}$ and $\hat{\mathbf{d}}$ in the box frame, followed by a translation $\mathbf{r}_{\rm s}$ of the position vectors, $\mathbf{r} \rightarrow \mathbf{r} + \mathbf{r}_{\rm s}$.

After completing the simulation of the shower in the box, the main programme resumes normal simulation through the PENGEOM geometry. The process is repeated each time a particle arrives at a rough interface.

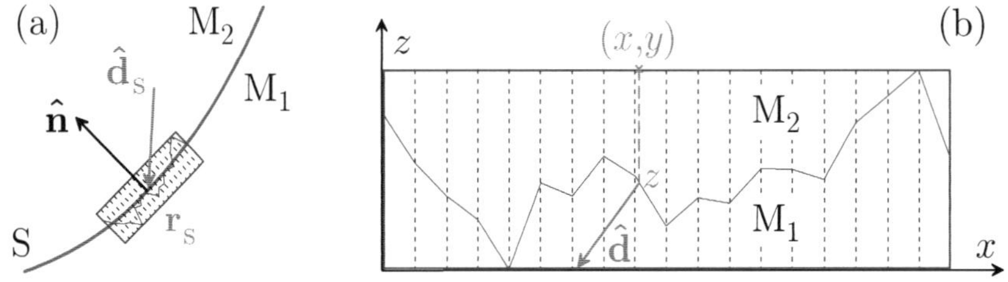

Figure 3. Geometrical transformation from the laboratory frame to the box frame of a rough "internal" surface. The particle reaches the mathematical surface at $\mathbf{r}_S$ moving in the direction $\hat{\mathbf{d}}_S$. The simulation within the rough-surface box starts with the particle at a random position just below the triangularised surface and moving in the direction $\hat{\mathbf{d}} = \mathcal{R}_B\hat{\mathbf{d}}_S$, see eq. (3).

3. Validation and examples

To validate the present simulation scheme, as well as to reveal the effect of surface roughness, we have performed simulations of EPMA measurements on Fe samples with a rough external surface having a simple structure, namely a periodic array of square pyramids with 2 μm base edges and lateral faces forming an angle α with the normal to the surface (fig. 4). This structure can be described by the programme PENEPMA, which uses the PENGEOM geometry subroutines, and by the EPMAR code, which implements the present tracking algorithm for rough surfaces. Notice that the adopted strategy for surface triangularisation (the triangles in each column share the diagonal with the largest inclination of the surface) reproduces the surface geometry exactly.

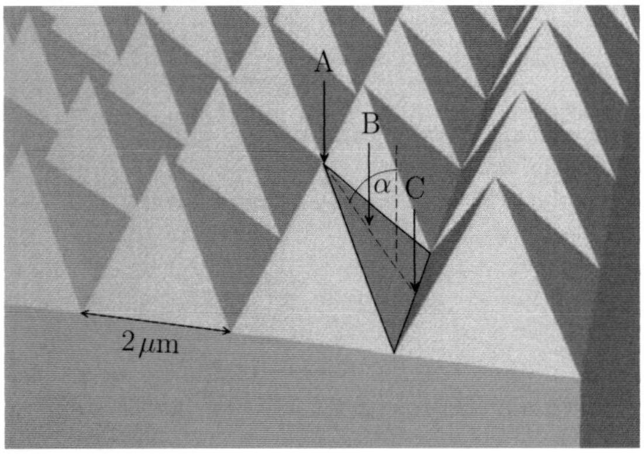

Figure 4. Geometry of the sample with a periodic "rough" surface used to verify the numerical tracking algorithm.

Simulations of identical arrangements were performed with the codes PENEPMA and EPMAR. Line intensities were tallied by using a detector that counts X-rays emitted from the sample in directions with polar and azimuthal angles in the intervals (35°,55°) and (0°,90°), respectively [see Ref. [11] for details]. A pencil beam of 20 keV electrons (with null lateral extent) impinged normally on the surface of the sample, at three different positions of a pyramid near the centre of the structure (to minimise edge effects). The numerical results from the two codes are equivalent, that is, their differences are less than, or of the order of their estimated statistical uncertainties. With the considered sample geometry, EPMAR is slightly more efficient than PENEPMA, and hence the numerical effort devoted to tracking particles through the rough surface box does not slow down the programme significantly.

IOP Conf. Series: Materials Science and Engineering **304** (2017) 012015 doi:10.1088/1757-899X/304/1/012015

Results from simulations for samples with pyramids of two different heights, corresponding to angles α of 30° and 60° are shown in table 1, together with results from the ideal target with a smooth surface ($\alpha = 90°$). The latter case was also simulated with PENEPMA assuming a flat surface of the sample; the results agree with those of EPMAR to within statistical uncertainties. The effect of the surface roughness on the measured intensities is usually manifest.

Table 1. Intensities (number of X-rays per unit solid angle and per incident electron) of the indicated Fe X-ray lines for a 20 keV electron beam at the various impact positions in the pyramid (as indicated in fig. 4). The quoted relative uncertainties correspond to a 3σ confidence interval.

Line	α	A	B	C
K-L2	30	4.14E-5±2.1E-7	4.89E-5±2.4E-7	5.21E-5±2.6E-7
K-L2	60	4.21E-5±2.1E-7	4.81E-5±2.4E-7	5.14E-5±2.6E-7
K-L2	90	4.92E-5±2.4E-7	4.92E-5±2.4E-7	4.92E-5±2.4E-7
K-L3	30	8.13E-5±2.9E-7	9.56E-5±3.4E-7	1.02E-4±4.0E-7
K-L3	60	8.23E-5±3.0E-7	9.42E-5±3.4E-7	1.01E-4±3.6E-7
K-L3	90	9.61E-5±3.4E-7	9.61E-5±3.4E-7	9.61E-5±3.4E-7

Simulations have also been performed for a Fe sample with a realistic rough surface, a small portion of which is displayed in fig. 5. The sample was irradiated with a 20 keV electron beam, impinging normally on the surface box, at the positions indicated in fig. 5. As in the previous example, line intensities were tallied by considering a detector that counts X-rays emitted in directions with polar and azimuthal angles in the intervals (35°,55°) and (0°,90°), respectively. Simulations were performed with the programme EPMAR; each run involved a number of simulated showers such that the statistical relative uncertainty of the Fe K-L2 line intensity was less than 1 %. Figure 5 shows the surface profile along the line scanned by the electron beam, and the calculated intensity (number of X-rays per unit solid angle and per incident electron) of the Fe K-L2 line at the various beam positions.

These examples indicate that line intensities from EPMA measurements in bulk samples with a rough surface tend to be larger at the valleys of the surface and smaller at the peaks.

4. Concluding remarks

The tracking algorithm described in this communication provides a practical solution to account for surface roughness in conventional Monte Carlo transport codes using geometries defined by mathematical surfaces. In our implementation, much care has been exercised to protect the programme against numerical round-off errors. The use of a "movable" surface box allows describing the topography of rough surfaces on the scale of the nanometre.

The programme EPMAR performs Monte Carlo simulations of EPMA measurements on samples with rough surfaces. It not only accounts for the roughness of the irradiated surface of the sample, but also and permits considering "internal" rough surfaces to define material structures (inclusions or lamellae) within the base sample. The programme has been validated by comparison with results from PENEPMA in the case artificial surfaces with periodic roughness.

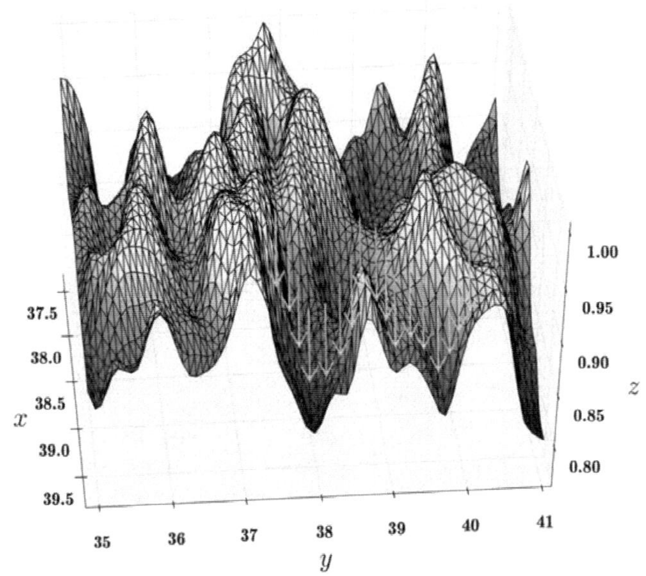

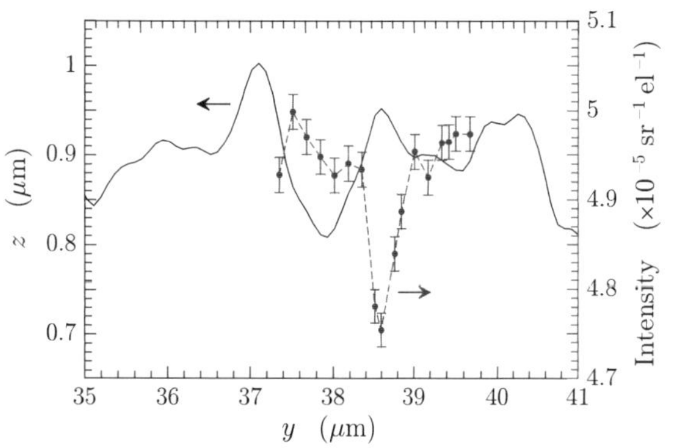

Figure 5. Top: A portion of the rough surface used in the EPMAR simulations, with all lengths in micrometres; the arrows indicate the positions of the electron beam. Bottom: Line scan of the surface (solid black line, left axis) and values of the simulated Fe K-L2 line intensities (circles with error bars, right axis) at the various positions of the beam. The quoted relative uncertainties correspond to a 3σ confidence interval.

Acknowledgements

Financial support from the Spanish Ministerio de Economía y Competitividad and ERDF (projects nos. FPA2013-44549-P, FPA2014-57896-C4-2-R, and FPA2016-77689-C2-2-R) and from the Generalitat de Catalunya (grants 2014 SGR 846 and 2014 SGR 769) is gratefully acknowledged. DS-G wishes to thank the Industrial Doctorate Plan of the Generalitat de Catalunya for mobility support (Ref. 2016DI014).

References

[1] Yamada A, Fons P, Matsubara K, Iwata K, Sakurai K and Niki S 2003 *Thin Solid Films* **431–432** 277–283
[2] Newbury D E 2004 *Scanning* **26** 103–114
[3] Kuhn W K and Andermann G 2001 *Appl. Surf. Sci.* **185** 84–91
[4] Brunetti A and Golosio B 2014 *Spectrochim. Acta B* **94–95** 58–62
[5] Naito M, Hasebe N, Kusano H, Nagaoka H, Kuwako M, Oyama Y, Shibamura E, Amano Y, Ohta T, Kim K J and Lopes J A M 2015 *Nucl. Instrum. Meth. A* **788** 182–187
[6] Kilimchuk I V, Tarasov V A and Vlasova I D 2010 *Radiat. Meas.* **45** 383–385
[7] Gauvin R and Lifshin 2000 *Michochim. Acta* **132** 201–204.
[8] Sorbier L, Rosenberg E and Merlet C 2001 *Advanced Monte Carlo for radiation physics, particle transport simulation and applications.* (Kling A, Barao F, Nakagawa M, Tavora L and Vaz P; Eds.) (Berlin: Springer-Verlag) 389–394

[9] Salvat F 2015 *PENELOPE-2014: a code system for Monte Carlo simulation of electron and photon transport.* (Issy-les-Moulineaux, France: OECD/NEA Data Bank)
[10] Almansa J, Salvat-Pujol F, Díaz-Londoño G, Carnicer A, Lallena A M and Salvat F 2016 *Comput. Phys. Commun.* **199** 102–113
[11] Llovet X and Salvat F 2017 *Microsc. Microanal.* **23** 634–646

Appendix. Tracking particles in the surface box

As indicated above, to describe the motion of particles within a surface box we use a reference frame with the origin at the box corner and axes parallel to the box sides. The tracking is simplified by considering the box as the set of elementary volumes defined by the "vertical" planes $x = x_i$, $y = y_j$ that pass through the grid points, eq. (1), and the triangularised surface. Thus, each cell $(x_i, x_{i+1}) \times (y_j, y_{j+1})$ determines a column volume with two sections of materials M_1 and M_2 separated by two surface triangles.

Given the initial position $\mathbf{r}(0)$ and direction of motion $\hat{\mathbf{d}}$ of a particle, the indices of the column that contains the particle are

$$i = [1 + x(0)/\Delta_x] \quad \text{and} \quad j = [1 + y(0)/\Delta_x], \tag{A.1}$$

where $[x]$ denotes the integer part of x. The altitudes of the surface at the four vertices of the cell determine the two surface triangles (which share the diagonal of the cell corresponding to the largest slope). The material where the particle moves or, equivalently, the position of the particle relative to the surface (below or above), is determined by the altitude of the triangularised surface at the point $x(0), y(0)$, which can be readily calculated by simple linear interpolation of the three vertices of the corresponding triangle.

When the particle tries to jump a distance Δs, decided by the physics subroutines, we must determine whether the flight ends within the current material, the particle changes material by crossing the surface, or it leaves the box. To determine the actual move we consider the intersections of the ray $\mathbf{r}(0) + s\hat{\mathbf{d}}$ with the column planes, the two triangles, and the upper and lower sides of the box; evidently, only the intersections that occur ahead of the particle (i.e., at positive s) are relevant. When the first crossing is with one of the vertical planes, we move the particle to the corresponding column and repeat the process. If the first intersection is with one of the triangles, the particle is halted at the surface and the simulation is resumed in the new material. If the particle leaves the box, control is returned to the steering main program to continue tracking any secondary particles generated within the box, or to proceed with the simulation of particles outside the box. Evidently, the operation that is more laborious is the calculation of intersections with triangles, which is performed as follows.

A triangle is defined by the coordinates of its three vertexes, $\mathbf{V}^{(0)}$, $\mathbf{V}^{(1)}$, and $\mathbf{V}^{(2)}$. The points $\mathbf{R}$ of the plane of the triangle are conveniently represented by using the so-called barycentric coordinates (η, μ) defined by

$$\mathbf{R} = \mathbf{V}^{(0)} + \eta\left(\mathbf{V}^{(1)} - \mathbf{V}^{(0)}\right) + \mu\left(\mathbf{V}^{(2)} - \mathbf{V}^{(0)}\right). \tag{A.2}$$

Notice that the point $\mathbf{R}$ is inside the triangle only when $\eta \geq 0$, $\mu \geq 0$, and $\eta + \mu \leq 1$. The intersection of the ray and the plane of the triangle is determined by the vector equation

$$\mathbf{r}(0) + s\hat{\mathbf{d}} = \mathbf{V}^{(0)} + \eta\left(\mathbf{V}^{(1)} - \mathbf{V}^{(0)}\right) + \mu\left(\mathbf{V}^{(2)} - \mathbf{V}^{(0)}\right). \tag{A.3}$$

Rearranging terms we have

$$-s\hat{\mathbf{d}} + \eta\left(\mathbf{V}^{(1)} - \mathbf{V}^{(0)}\right) + \mu\left(\mathbf{V}^{(2)} - \mathbf{V}^{(0)}\right) = \mathbf{r}(0) - \mathbf{V}^{(0)} \tag{A.4}$$

or, in matrix form,

$$\begin{pmatrix} -u & V_x^{(1)} - V_x^{(0)} & V_x^{(2)} - V_x^{(0)} \\ -v & V_y^{(1)} - V_y^{(0)} & V_y^{(2)} - V_y^{(0)} \\ -w & V_z^{(1)} - V_z^{(0)} & V_z^{(2)} - V_z^{(0)} \end{pmatrix} \begin{pmatrix} s \\ \eta \\ \mu \end{pmatrix} = \begin{pmatrix} x(0) - V_x^{(0)} \\ y(0) - V_y^{(0)} \\ z(0) - V_z^{(0)} \end{pmatrix}. \tag{A.5}$$

The solution can be obtained by using Cramer's rule. It is worth noting that the determinant of the system,

$$\begin{aligned} D &\equiv \begin{vmatrix} -u & V_x^{(1)} - V_x^{(0)} & V_x^{(2)} - V_x^{(0)} \\ -v & V_y^{(1)} - V_y^{(0)} & V_y^{(2)} - V_y^{(0)} \\ -w & V_z^{(1)} - V_z^{(0)} & V_z^{(2)} - V_z^{(0)} \end{vmatrix} \\ &= -\hat{\mathbf{d}}\cdot\left[\left(\mathbf{V}^{(1)} - \mathbf{V}^{(0)}\right)\times\left(\mathbf{V}^{(2)} - \mathbf{V}^{(0)}\right)\right], \end{aligned} \tag{A.6}$$

represents the volume of the parallelepiped formed by the sides of the triangle and the direction of motion of the particle. The vector

$$\mathbf{N} = \left(\mathbf{V}^{(1)} - \mathbf{V}^{(0)}\right)\times\left(\mathbf{V}^{(2)} - \mathbf{V}^{(0)}\right)$$

is perpendicular to the triangle. Evidently, the system of equations eq. (A.5) has a solution only when $D \neq 0$ or, equivalently, when the direction $\hat{\mathbf{d}}$ of the ray is not parallel to the triangle ($\hat{\mathbf{d}}\cdot\mathbf{N} \neq \mathbf{0}$).

To avoid unnecessary calculations, it is convenient to calculate first the distance s to the plane of the triangle, which is given by

$$s = \frac{1}{D}\begin{vmatrix} x(0) - V_x^{(0)} & V_x^{(1)} - V_x^{(0)} & V_x^{(2)} - V_x^{(0)} \\ y(0) - V_y^{(0)} & V_y^{(1)} - V_y^{(0)} & V_y^{(2)} - V_y^{(0)} \\ z(0) - V_z^{(0)} & V_z^{(1)} - V_z^{(0)} & V_z^{(2)} - V_z^{(0)} \end{vmatrix} = \frac{1}{D}\left(\mathbf{r}(0) - \mathbf{V}^{(0)}\right)\cdot\mathbf{N}. \tag{A.7}$$

The particle may cross the triangle only when s is positive and less than the step length Δs.

IOP Conf. Series: Materials Science and Engineering **304** (2017) 012016 doi:10.1088/1757-899X/304/1/012016

Calculation of catalyst crust thickness from full elemental laser-induced breakdown spectroscopy images

L Sorbier[1], F Trichard[2], S Moncayo[3], C P Lienemann[1] and V Motto-Ros[3]

[1] IFP Energies nouvelles, Rond-point de l'échangeur de Solaize, P.O. Box 3, 69360 Solaize, France
[2] Ablatom SAS, Bâtiment Kastler, Domaine Scientifique de la Doua, 10 rue Ada Byron, 69622 Villeurbanne Cedex, France
[3] Institut Lumière Matière UMR 5306, Université Lyon 1 - CNRS, Université de Lyon, 69622 Villeurbanne Cedex, France

E-mail: loic.sorbier@ifpen.fr

Abstract. We propose a methodology to compute the crust thickness of an element in an egg-shell catalyst from a two-dimensional elemental map. The methodology handles two important catalyst shapes: infinite extrudates of arbitrary section and spheres. The methodology is validated with synthetic analytical profiles on simple shapes (cylinder and sphere). Its relative accuracy is shown close to few percent with a decrease inversely proportional to the square root of the number of sampled pixels. The crust thickness obtained by this method from quantitative Pd maps acquired by laser-induced breakdown spectroscopy are comparable with values obtained from electron-probe microanalysis profiles. Some discrepancies are found and are explained by the heterogeneity of the crust thickness within a grain. As a full map is more representative than a single profile, fast mapping and the methodology exposed in this paper are expected to become valuable tools for the development of new generations of egg-shell deposited catalysts.

1. Introduction

Selective hydrogenation is an important process in petrochemistry. The process is employed for the stabilisation of pyrolysis gasoline and vapocracking products. For either processes, the unsaturated carbon-carbon bonds need to be partially hydrogenated. Conjugated double bonds or triple bonds should be removed while keeping intact non-conjugated double bonds. To obtain a good activity and selectivity, heterogeneous catalysts are employed. These catalysts have the shape of cylinders or multilobed extrudates or are nearly spherical pellets. Typical sizes of grains are around a few millimetres. The catalyst support is often composed of a chemically inert oxide such as alumina, silica or titanium oxide. Active phases are noble metals (Pd, Ag, Au) deposited on the support at concentration ranging from 0.1 to 0.5 wt%. As hydrogenation of carbon bonds has a very fast kinetics, these catalysts are working in internal diffusion limited regime: chemical kinetics occur quicker than diffusion of products in the grain.

For catalysts under strong internal diffusion regime, the distribution of active element along the pellets diameter is correlated with the catalyst performance 0. Therefore, for selective hydrogenation catalysts, noble metals are deposited as an egg-shell distribution on the support, otherwise said on a crust close to grain's surface. This crust enhances the activity and selectivity 0 and limits the deactivation rate 0 compared to flat repartition of the active element. It is therefore important to have adequate characterisation techniques to check the deposit of metals at the grain scale. For palladium selective hydrogenation catalyst, a descriptor, the Pd crust thickness, has been defined and correlated with hydrogenation activity 0. This crust thickness was calculated from elemental profiles obtained by electron probe microanalysis (EPMA) and trivial pellet shapes (slab, cylinder, sphere) 0. It is defined as the thickness of the crust containing a fixed proportion (80 % or 98 %) of the total Pd content in the grain. As alumina supports are white and deposited Pd appears darker, a crust thickness may also be evaluated by optical microscopy and image analysis 0. This method, however, is impeded to the optical visibility of the crust and the definition of this crust thickness is not based on a quantitative Pd content criterion.

The methodology exposed in reference 0 does not handle multilobed shapes used for some industrial processes. Using a proper strategy based on the distance transform, long extrudates of arbitrary simply connected section should however be handled 0. Besides, optical micrographs of catalyst cross-sections clearly show that the Pd deposits are often irregular 0. A single profile along the grain's cross-section is then less representative of the catalyst batch than a full two-dimensional (2D) map of the section of the grain. The interest of using 2D mapping is increased by the development of fast laser-induced breakdown spectroscopy (LIBS) mapping. LIBS allows to acquire elemental maps of materials with a lateral resolution of a few micrometres at a sampling rate of 1 to 1,000 points per second. The reviews of Kaiser *et al.* for biological applications 0 and Piñon *et al.* for material science 0 give a good sight of the possibilities and drawbacks of the technique. One of the drawbacks of LIBS is that it needs a proper calibration strategy. There is no universal correction procedure like for EPMA. We have recently proposed to calibrate the intensities measured by LIBS with concentrations obtained by EPMA on the same profiles acquired with the same lateral resolution 0. It leads to quantitative mapping with a lower limit of detection (about 4 times) and higher speed (about 240 times) compared to EPMA. This opens the possibility to compute crust thicknesses from 2D elemental maps.

Nevertheless, computing the crust thickness on 2D map of a 3D object may not be straightforward as for example, for a 2D map of a spherical grain. The aim of this paper is then to propose a convenient strategy for computing crust thickness from 2D elemental map of a catalyst grain for two important catalyst pellet shapes of industrial importance: long extrudates and spherical grains. The paper is organised as follows. In a first section, we summarise the experimental methods: synthesis of the alumina supported Pd catalyst and analysis methods by EPMA and LIBS. The calculation of the quantity of an element contained at a given distance to the grain's border is discussed for two important shapes: infinitely long extrudates and spherical grains. The algorithm to compute the crust thickness from 2D maps is then proposed. The section ends up with the calculation of crust thickness from empirical analytical concentration fields on simple shapes (slabs, cylinder, sphere). The second section is devoted to the presentation of results: accuracy of the methodology on analytical profiles, 2D Pd LIBS maps, EPMA profiles and comparison of crust thickness measured either from profiles or from full maps. We close with a discussion and a conclusion.

2. Materials and methods

2.1. Pd hydrogenation catalyst

Synthesis of Pd catalyst was done by incipient wetness impregnation of an aqueous solution of palladium nitrate to target a 0.274 wt% Pd content. Details of the catalyst preparation can be found in reference 0. Catalyst beads were embedded in conductive resin (Struers Polyfast), polished with SiC paper under water solvent up to their diameter and finished with a 1 μm diamond suspension.

IOP Conf. Series: Materials Science and Engineering **304** (2017) 012016 doi:10.1088/1757-899X/304/1/012016

2.2. Analysis by EPMA

A 20-nm thick carbon layer was coated on the sample before EPMA analysis. EPMA analysis was performed on a JEOL JXA 8100 fitted with five wavelength-dispersive X-ray spectrometers. The electron beam (20 kV, 240 nA) was defocussed to obtain a 15 μm diameter probe. The measurement times on peak and background on each point was set to 20 s. Oxygen was dosed using a stoichiometric method assuming metallic Pd and Al_2O_3 oxide. Pure Pd (Goodfellow, wire 99.95 % purity) and pure alumina (Goodfellow, sapphire pellet diameter 3 mm, 99.9 % purity) were used as standards. Profiles along the beads diameters were acquired with a 15 μm step. The first and last points of the profiles were in the surrounding resin to ensure a full diameter measurement. In these conditions, a profile lasted about 70 minutes with a detection limit at 84 ppm for Pd. Three profiles on three different catalyst grains of the same batch were acquired. From each profile, the crust thickness is calculated using the method described in reference 0 assuming a spherical shape.

2.3. Analysis by LIBS

Before LIBS analysis, the sample was slightly polished and finished with 1 μm diamond suspension in order to remove the carbon coating. LIBS analysis was performed on a system dedicated to LIBS imaging that has been described in detail elsewhere 0. The laser source, a Nd:YAG was used at the fundamental wavelength (1,064 nm) at a 10 Hz repetition rate. The typical pulse energy after shaping and attenuation was set to 600 μJ.

The plasma radiation was collected with a lens-fibre system positioned at a take-off angle of 35°. The collected light was analysed using a Czerny-Turner spectrometer (Shamrock 303, Andor Technology) fitted with an intensified charge-coupled device camera (iStar ICCD, Andor Technology). The intense Pd(I) 340.45 nm and less intense Pd(I) 343.34 nm lines were used for quantitative imaging. Two calibration curves for these lines were constructed from the same analysis points acquired on the same sample using intensities of LIBS and local concentration of EPMA. The calibration curve of Pd (I) 340.45 nm line was used for low local Pd contents ($c \leq 0.1$ wt%) and the one of Pd(I) 343.34 nm for high local Pd content ($c > 0.1$ wt%). Maps of 250 × 210 pixels with a 15 μm resolution were acquired on the same three catalyst grains as EPMA was conducted on, each mapping lasting about 90 minutes with a detection limit of 18 ppm for Pd.

2.4. From concentration map to crust thickness

2.4.1. General equation. Let $c(x, y, z)$ be the concentration field of an element on the grain G. Let M be the mask of the grain. The mask M is computed by binarisation of the concentration map of an element contained in catalyst but not in surrounding resin (Pd). The distance transform of the mask M^{DT} is computed 0. It contains in each pixel, the distance (in pixels) from the border of the mask. For pixels outside the mask, the distance transform is equal to zero. Let K be the maximum distance to the border of the mask, $K = max\, M^{\mathrm{DT}}$.

The quantity $q(a, b)$ of the element contained at a distance from the border of the grain comprised between a and b is equal to:

$$q(a,b) = \iiint_G \chi_{a,b}(x,y,z)c(x,y,z)\rho(x,y,z)\mathrm{d}x\mathrm{d}y\mathrm{d}z \tag{1}$$

where $\chi_{a,b}(x, y, z)$ is an indicator function equal to one if the point of coordinates (x, y, z) lies at a distance comprised between a and b from the border of grain and ρ is the volumic mass.

2.4.2. Infinitely long extrudate. For an infinite long extrudate having its extrusion axis on axis z, neither the concentration field nor the indicator function depends on z. Besides, the concentration field is only sampled on a finite number of pixels such that eq. (1) is evaluated using rectangle method. Hence, if we suppose that $\rho(x, y, z)$ is constant, the quantity of element by unit length along z axis is equal to:

$$q_{\mathrm{E}}(np,(n+1)p) \approx p^2\rho\sum_{(x_i,y_i)}\chi_{np,(n+1)p}(x_i,y_i)c(x_i,y_i) \tag{2}$$

where n is an integer and p the pixel size. Yet, the indicator function $\chi_{np,(n+1)p}(x,y)$ is related with M^{DT} the distance transform of the mask of the grain, namely we have $\chi_{np,(n+1)p}(x,y) = \delta\big(M^{\mathrm{DT}}(x,y)-np\big)$ with δ the Dirac function. Hence:

$$q_{\mathrm{E}}(np,(n+1)p) \approx p^2\rho\sum_{(x_i,y_i)}\delta(M^{DT}(x_i,y_i)-np)c(x_i,y_i) \tag{3}$$

2.4.3. Sphere. For a sphere of radius R where the concentration field is sampled on a plane containing the centre of the sphere, and a radial symmetric concentration field, the quantity of an element at a distance between a and b from the border of the grain is:

$$q_S(a,b) = \rho\int_0^R \chi_{a,b}(r)c(r)4\pi r^2\mathrm{d}r \tag{4}$$

Let $x = r\cos\theta$ and $y = r\sin\theta$, since $r = \sqrt{x^2+y^2}$ and $\mathrm{d}x\mathrm{d}y = r\mathrm{d}r\mathrm{d}\theta$, we have:

$$\begin{aligned} q_{\mathrm{S}}(a,b) &= \rho\int_0^R\int_0^{2\pi}\chi_{a,b}(r)c(r)2r\mathrm{d}\theta r\mathrm{d}r \\ &= 2\rho\iint_{x^2+y^2\leq R^2}\chi_{a,b}(x,y)c(x,y)\sqrt{x^2+y^2}\mathrm{d}x\mathrm{d}y \end{aligned} \tag{5}$$

The concentration field is only sampled in a finite number of pixels such that rectangle method gives:

$$q_S(np,(n+1)p) \approx 2p^2\rho\sum_{(x_i,y_i)}\chi_{np,(n+1)p}(x_i,y_i)c(x_i,y_i)\sqrt{{x_i}^2+{y_i}^2} \tag{6}$$

As $\chi_{np,(n+1)p}(x,y) = \delta\big(M^{\mathrm{DT}}(x,y)-np\big)$ and at a given distance np to the border of the grain, $\sqrt{{x_i}^2+{y_i}^2}$ is constant and approximately equal to $\frac{p}{2\pi}\sum_{(x_i,y_i)}\delta\big(M^{\mathrm{DT}}(x_i,y_i)-np\big)$, we obtain:

$$q_S(np,(n+1)p) \approx \frac{p^3\rho}{\pi}\Big(\sum_{(x_i,y_i)}\delta\big(M^{\mathrm{DT}}(x_i,y_i)-np\big)\Big)\Big(\sum_{(x_i,y_i)}\delta\big(M^{\mathrm{DT}}(x_i,y_i)-np\big)c(x_i,y_i)\Big) \tag{7}$$

2.5. Calculation of crust thickness

Let set $\Delta Q_{\mathrm{F}}(k) = q_{\mathrm{F}}\big((k-1)p,kp\big)$, where $\mathrm{F}=\mathrm{E}$ for an extrudate or $\mathrm{F}=\mathrm{S}$ for a sphere, be the quantity of the element at a distance between $(k-1)p$ and kp from the border of the grain. The crust thickness t_X, defining the distance to the border of the grain containing the proportion X of an element is computed in five steps:

1. From the concentration map, compute the mask M of the grain and its distance transform M^{DT};
2. Compute the quantity of the element at a distance between $(k-1)$ and k pixels from the border of the grain $\Delta Q_{\mathrm{F}}(k)$ for each distance $k \in \{0,\cdots,K\}$ using either eq. (3) for a cylinder or eq. (7) for a sphere;
3. Compute the relative proportion of element present at a distance lower than k pixels from the border of the grain $Q_{\mathrm{F}}(k) = \frac{\sum_{i=0}^{k}\Delta Q_{\mathrm{F}}(i)}{\sum_{i=0}^{K}\Delta Q_{\mathrm{F}}(i)}$ for each distance $k \in \{0,\cdots,K\}$;
4. Search for the distances k_0 and k_0+1 in pixels which are bracketing the relative quantity X: look for k_0 in $\{0,\cdots,K-1\}$ such that $Q_{\mathrm{F}}(k_0) \leq X < Q_{\mathrm{F}}(k_0+1)$;
5. Compute the crust thickness in pixels with sub-pixel accuracy by linear interpolation and convert it in physical unit length by the use of pixel size: $t_X = p\left(k_0 + \frac{X-Q_{\mathrm{F}}(k_0)}{Q_{\mathrm{F}}(k_0+1)-Q_{\mathrm{F}}(k_0)}\right)$.

2.6. Accuracy estimation from analytical concentration fields

The calculation of crust thickness from a synthetic analytical concentration field on simple shapes (infinite cylinder, sphere) may help to validate the approach and estimate the accuracy of the method.

Suppose that the concentration field does only depends on the distance to the border of the grain, consequently to the distance to the centre of the grain r. We also suppose that the concentration field $c_m(r)$ obeys the following law:

$$c_m(r) = \left(\frac{r}{R}\right)^m \tag{8}$$

where R is the radius of the grain and $m \geq 1$ is a real. Such concentration field leads to hollow profiles for $m \geq 1$. The relative quantity of element $Q_m^n(y)$ contained in a ball of radius y in $\mathbb{R}^n$ ($n = 2$ infinite cylinder, $n = 3$ sphere) is equal to:

$$Q_m^n(y) = \frac{\int_0^y c_m(r)\mathrm{d}V^n(r)}{\int_0^R c_m(r)\mathrm{d}V^n(r)} \tag{9}$$

where $\mathrm{d}V^n(r)$ is the infinitesimal volume in $\mathbb{R}^n$. Replacing the expression of $c_m(r)$ of eq. (8) and as $\mathrm{d}V^n(r) = \pi^{\frac{n}{2}}\Gamma\left(\frac{n}{2}+1\right)^{-1} nr^{n-1}\mathrm{d}r$ 0 we obtain:

$$Q_m^n(y) = \frac{y^{m+n}}{R^{m+n}} \tag{10}$$

The crust thickness $T_m^n(X)$ is defined as the thickness of the corona of external radius R and internal radius $R - T_m^n(X)$ which contains the relative fraction X of the element. The ball of radius $R - T_m^n(X)$ should then contain the relative quantity $1 - X$. Therefore, using eq. (10), $T_m^n(X)$ is the solution of:

$$1 - X = \frac{[R - T_m^n(X)]^{m+n}}{R^{m+n}} \tag{11}$$

and we obtain:

$$\frac{T_m^n(X)}{R} = 1 - (1 - X)^{\frac{1}{m+n}} \tag{12}$$

To evaluate the accuracy of the algorithm proposed in section 2.5, we simulate a concentration map following the eq. (8) on a ball of radius R with a pixel size p. On the synthetic concentration map we calculate t_X using the algorithm exposed in section 2.5. We define the relative error ε_t by:

$$\varepsilon_t = \frac{|t_X - T_m^n(X)|}{T_m^n(X)} \tag{13}$$

3. Results

3.1. Accuracy on analytical profiles

To evaluate the accuracy of the algorithm proposed in section 2.5, we have calculated the crust thickness for different values of the parameter m for $R = 1.5$ mm and $p = 15$ µm. The results of the algorithm are compared to the theoretical value given by eq. (12) in figure 1. The relative error given by equation (13) is then computed and presented in figure 2.

The influence of the number of pixels n_p employed to sample the concentration field is shown in figure 3 for a value of $m = 10$. This value of m gives a relative crust thickness ($t_{98\%}/R$) close to the one measured in section 3.2. Relative errors for a cylinder and sphere shapes are both decreasing inversely proportional to the square root of the number of pixels n_p.

3.2. Crust thickness from LIBS Pd maps and EPMA profiles

Figure 4 shows the quantitative LIBS Pd maps for the three sampled catalyst grains and figure 5 the EPMA profiles on these same grains. We notice a clear egg-shell repartition of Pd. LIBS maps show a rather heterogeneous distribution of Pd in the crust among grains and even within the same grain.

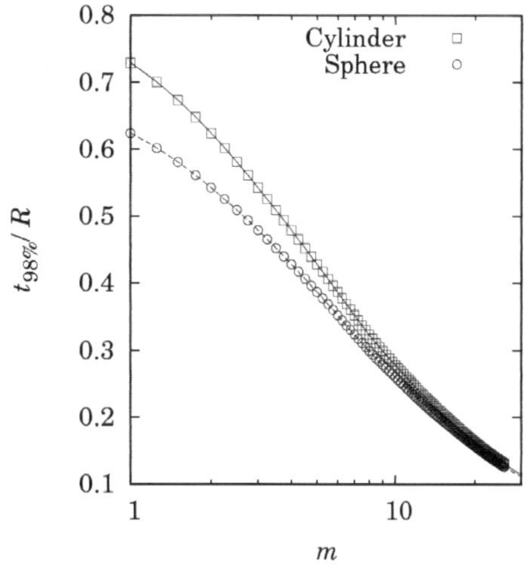

Figure 1. Crust thickness computed by algorithm of section 2.5 (symbols) compared to theoretical value of eq. (12) (lines).

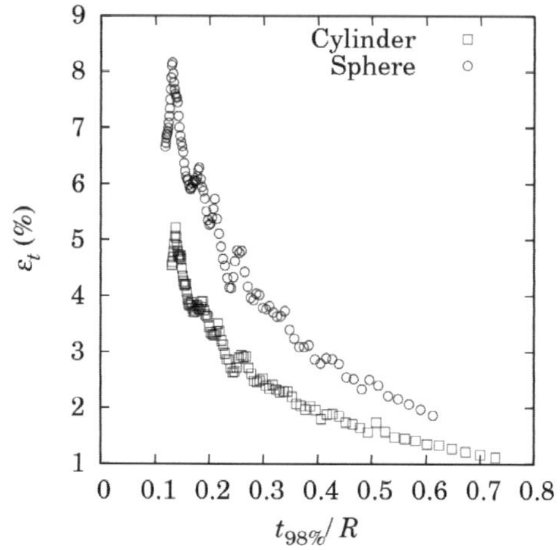

Figure 2. Relative error in function of crust thickness for $R = 1.5$ mm, $p = 15$ µm and a concentration field following eq. (8).

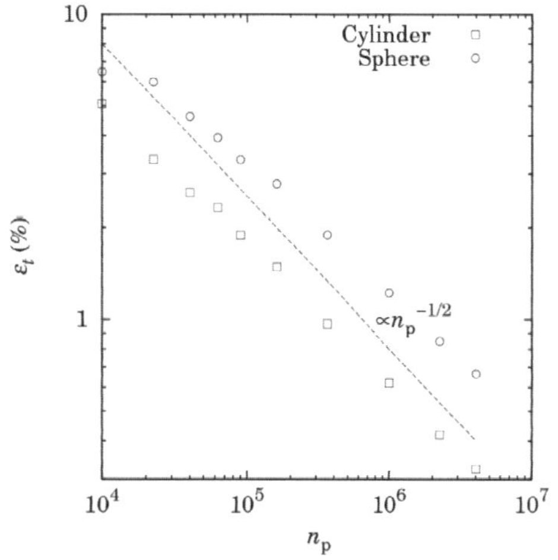

Figure 3. Relative error of $t_{98\%}$ for $m = 10$. The dotted line is decreasing as $n_p^{-\frac{1}{2}}$.

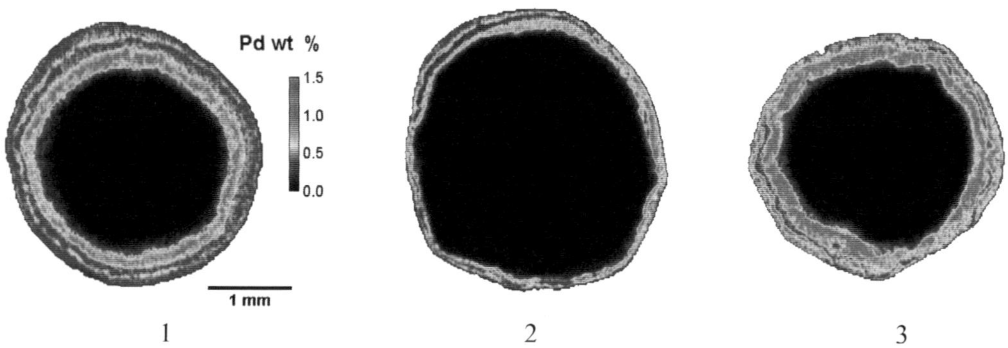

Figure 4. LIBS Pd maps for the 3 sampled grains.

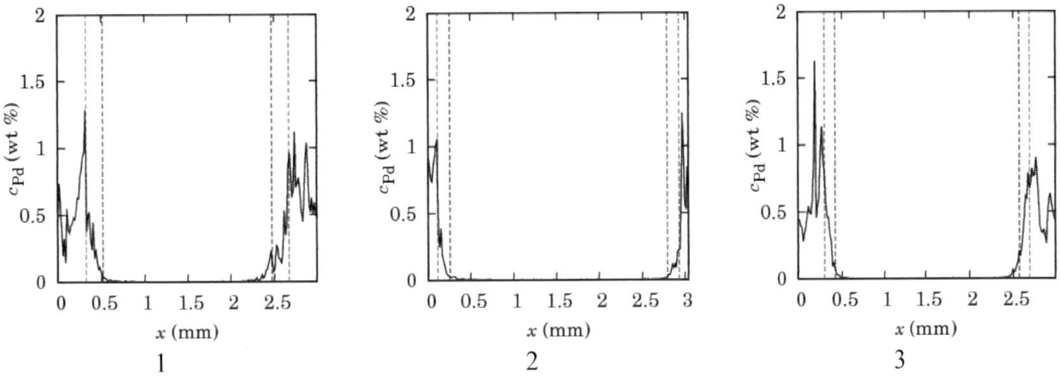

Figure 5. EPMA Pd profiles along the diameter of the 3 sampled grains. Dotted lines show crust thicknesses $t_{80\%}$ (red) and $t_{98\%}$ (blue).

Table 1 reports the crust thickness measured either from profiles (EPMA) or from 2D maps (LIBS) on the three sampled grains. Comparison between the two methods are shown in figure 6.

Table 1. Crust thickness measured on profiles by EPMA and on maps by LIBS for the three catalyst grains.

	$t_{80\%}$		$t_{98\%}$	
Grain	EPMA	LIBS	EPMA	LIBS
1	320	337	516	531
2	113	166	257	406
3	300	262	419	423

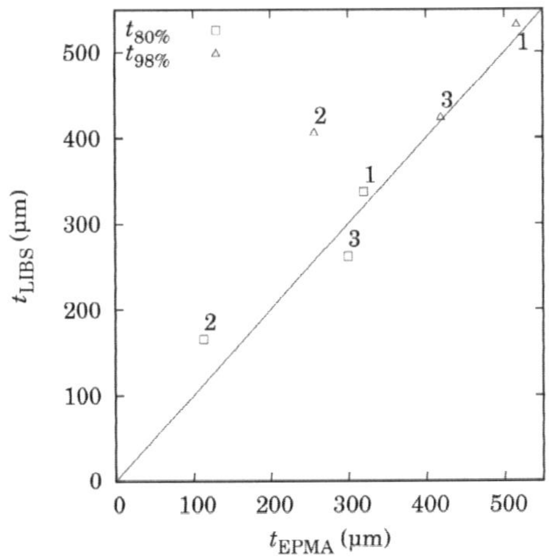

Figure 6. Thickness measured from profiles (EPMA) and from 2D maps (LIBS). Numbers identify grain number. The straight line is the first bisector.

4. Discussion

Figure 1 shows that the algorithm proposed is a good estimator of the crust thickness, from slightly hollow profiles (small values of parameter m, thick crust) to very hollow profiles (high values of

IOP Conf. Series: Materials Science and Engineering **304** (2017) 012016 doi:10.1088/1757-899X/304/1/012016

parameter m, thin crust). Figure 2 clearly shows that the relative accuracy increases when the crust thickness decreases (high value of the parameter m). This is explained by the poor sampling of the concentration field where the concentration is appreciable (when m is large). Figure 3 confirms this explanation. For a fixed profile, increasing the number of points on the concentration map decreases the relative error. The decrease is inversely proportional to the square root of the number of points, which is expected for the two-dimensional rectangle method (see Appendix A). Besides, figure 3 shows that the proposed methodology converges to the expected value, otherwise said that we have proposed an unbiased estimator of the crust thickness. The accuracy predicted for geometry and crust thickness close to the experimental one of figure 4 is in the order of a few percent. This is quite acceptable as it is the same order of magnitude than the accuracy of the measure of apparent activity of a catalyst.

The comparison of crust thickness obtained from 2D LIBS maps with the reference values obtained from EPMA profiles shows an overall good agreement. Discrepancies between the two methods are observed for the grain 2. This grain has the thinnest and less regular crust (see figure 4). It is worth noticing that the crust thickness calculation from a 2D map is more representative that the one calculated from a single profile. The reference values obtained from EPMA profiles are then less reliable than the one calculated from LIBS maps. Interaction volumes are slightly different between EPMA and LIBS. For an alumina with 60 % porosity at 20 kV, the analysis volume is in the order of 5 μm large and 5 μm deep. Since a defocussed probe of 15 μm diameter was used, the analysis volume is about 1,500 μm^3. With the laser conditions employed in this work, LIBS leads to crater sizes of about 10 μm large by 2.5 μm deep thus to an analysed volume of about 200 μm^3. LIBS probes a smaller volume (due to the electron probe defocus) but both lateral resolutions are comparable and well below the measured crust thicknesses.

We now want to compare the advantages and limitations of the three methodologies available to compute the crust thickness of a catalyst: from elemental 2D maps as proposed in this paper, from profiles, and from optical microscopy. It is clear that the use of 2D information (elemental or optical) is more representative than a single profile. The main drawback of optical microscopy is that it is restricted to a visible crust, in other words to elements colouring an uncoloured catalyst support. The definition of crust thickness from optical microscopy is also questionable for low active phase concentrations. The advantage of optical microscopy is the speed of the process (about 10 minutes per sample). The main drawback of the calculation of crust thickness from a 2D map is the long acquisition time of the map. This methodology should therefore be reserved to fast and sensitive analytical techniques. Fortunately, fast LIBS mapping at 10 Hz already leads to a map acquisition time (90 minutes) close to one EPMA profile acquisition time (70 minutes) and with a better limit of detection. Some LIBS mapping systems with a 100 Hz sampling rate already exists 0 and sampling rates of 1 kHz are conceivable in the future. This will enhance the interest of the methodology we have proposed in this paper. This methodology and such fast mapping techniques should reduce the development times of new generations of egg-shell catalysts.

5. Conclusion

We have proposed a methodology to compute the crust thickness of an egg-shell catalyst from 2D elemental maps. This method handles important industrial catalyst shapes: long extrudates of arbitrary section and spheres. This methodology has been validated by analytical synthetic profiles for simple shapes (cylinder and sphere). We have shown that the proposed estimator is unbiased and converge to the true crust thickness value with an error inversely proportional to the square root of the number of sampled pixels.

We have demonstrated the applicability on real data obtained from quantitative LIBS mapping. The calibration of the LIBS intensities has been obtained from EPMA performed with the same lateral resolution on the same profiles of a sample. Comparisons of crust thickness calculated from LIBS 2D maps with the ones calculated from EPMA profiles show a satisfactory agreement except for irregular crust where profiles lack representativeness. The methodology presented in this paper is going to be a valuable tool for the development of new generations of egg-shell catalysts thanks to the development of fast LIBS mapping.

Appendix A. Error bound for rectangle rule

Let $f : \mathbb{R}^2 \to \mathbb{R}, (x,y) \mapsto f(x,y)$, a function that we suppose continuous, locally summable and 1-differentiable. We want to integrate f on the domain $D = [x_1; x_n] \times [y_1; y_n]$. The integration error $\varepsilon_{i,j}$ for the rectangle method on the pixel (x_i, y_j) is:

$$\varepsilon_{i,j} = \left| \int_{x_i}^{x_{i+1}} \int_{y_j}^{y_{j+1}} \left(f(x,y) - f(x_i, y_j) \right) \mathrm{d}x\mathrm{d}y \right| \tag{A1}$$

Obviously:

$$\varepsilon_{i,j} \leq \int_{x_i}^{x_{i+1}} \int_{y_j}^{y_{j+1}} \left| f(x,y) - f(x_i, y_j) \right| \mathrm{d}x\mathrm{d}y \tag{A2}$$

Now, right hand side of (A2) lets appear the first term of the Taylor expansion of f. It is possible to have the Lagrange remainder of this Taylor expansion, more precisely $\forall(x,y) \in [x_i; x_{i+1}] \times [y_j; y_{j+1}]\ \exists(x_\xi, y_\xi) \in [x_i; x_{i+1}] \times [y_j; y_{j+1}]$ so that 0:

$$f(x,y) = f(x_i, y_j) + \frac{\partial f}{\partial x}(x_\xi, y_\xi)(x - x_i) + \frac{\partial f}{\partial y}(x_\xi, y_\xi)(y - y_j) \tag{A3}$$

Injecting (A3) into (A2) and considering that $|a+b| \leq |a| + |b|$ we obtain:

$$\varepsilon_{i,j} \leq \int_{x_i}^{x_{i+1}} \int_{y_j}^{y_{j+1}} \left(\left| \frac{\partial f}{\partial x}(x_\xi, y_\xi)(x - x_i) \right| + \left| \frac{\partial f}{\partial y}(x_\xi, y_\xi)(y - y_j) \right| \right) \mathrm{d}x\mathrm{d}y \tag{A4}$$

Now, we suppose that f is sufficiently regular to have its first derivative bounded in D so that $\forall(x,y) \in D, \exists K \in \mathbb{R}^+, \left|\frac{\partial f}{\partial x}(x,y)\right| \leq K$ and $\left|\frac{\partial f}{\partial y}(x,y)\right| \leq K$. Taking into account that $(x - x_i) \geq 0$ and $(y - y_j) \geq 0$ in (A4) we have:

$$\varepsilon_{i,j} \leq K \int_{x_i}^{x_{i+1}} \int_{y_j}^{y_{j+1}} \left[(x - x_i) + (y - y_j) \right] \mathrm{d}x\mathrm{d}y \tag{A5}$$

The integral in right hand side of (A5) may be explicitly calculated. Setting $x_{i+1} = x_i + p$ and $y_{i+1} = y_i + p$, p being the pixel size, we obtain:

$$\varepsilon_{i,j} \leq \frac{K}{2} p^3 \tag{A6}$$

The error of integration in D, ε writes:

$$\varepsilon \leq \sum_{i=1}^{n} \sum_{j=1}^{n} \varepsilon_{i,j} \tag{A7}$$

Injecting (A6) into (A7) and using the relations $np = x_n - x_1 = y_n - y_1$ and $n_p = n^2$, we obtain:

$$\varepsilon \leq \frac{K}{2} \frac{(x_n - x_1)^3}{{n_p}^{1/2}} \tag{A8}$$

with n_p the number of pixels. The observed behaviour in figure 3 of the paper is recovered.

References

[1] Shadman-Yazdi F and Petersen E 1972 *Chem. Engng. Sci.* **27** 227-237
[2] Berenblyum A, Mund S, Karelskii V, Goranskaya T, Zolotukhin V and Lakhman L 1986 *Kinet. Catal.* **27** 184-187
[3] Lin T B and Chou T C 1994 *Appl. Catal. A* **108** 7-19
[4] Sorbier L, Gay A-S, Fécant A, Moreaud M and Brodusch N 2012 *IOP Conf. Ser. Mater. Sci. Engng.* **32** 012023

[5] Sorbier L, Gay A-S, Fécant A, Moreaud M and Brodusch N 2013 *Microsc. Microanal.* **19** 293–299
[6] Sorbier L, Bazer-Bachi F, Blouët Y, Moreaud M and Moizan-Basle V 2016 *Microsc. Microanal.* **22** 422-431
[7] Kaiser J, Novotný K, Martin M, Hrdlička A, Malina R, Hartl M and Kizek R 2012 *Surf. Sci. Rep.* **67** 233-243
[8] Piñon V, Mateo M and Nicolas G 2013 *Appl. Spectrosc. Rev.* **48** 357-383
[9] Trichard F, Sorbier L, Moncayo S, Blouët Y, Lienemann C P and Motto-Ros V 2017 *Spectrochim. Acta B* **133** 45-51
[10] Motto-Ros V, Negre E, Pelascini F, Panczer G and Yu J 2014 *Spectrochim. Acta B* **92** 60-69
[11] Rosenfeld A and Pfaltz J 1966 *J. ACM* **13** 471-494
[12] Sommerville D 1929 *An introduction to the geometry of n dimensions.* (London: Methuen and Co.) 136
[13] Cáceres J, Pelascini F, Motto-Ros V, Moncayo S, Trichard F, Panczer G, Marín-Roldán A, Cruz J, Coronado I and Martín-Chivelet J 2017 *Sci. Rep.* **7** 5080
[14] Coleman R 2012 *Calculus on normed vector spaces.* (New York: Springer) 111

A convenient method for X-ray analysis in TEM that measures mass thickness and composition

P Statham[1], J Sagar[1], J Holland[1], P Pinard[1] and S. Lozano-Perez[2]

[1] Oxford Instruments NanoAnalysis, High Wycombe, Bucks HP12 3SE, Great Britain
[2] University of Oxford, Department of Materials, Oxford OX1 3PH, Great Britain

E-mail: peter.statham@oxinst.com

Abstract. We consider a new approach for quantitative analysis in transmission electron microscopy (TEM) that offers the same convenience as single-standard quantitative analysis in scanning electron microscopy (SEM). Instead of a bulk standard, a thin film with known mass thickness is used as a reference. The procedure involves recording an X-ray spectrum from the reference film for each session of acquisitions on real specimens. There is no need to measure the beam current; the current only needs to be stable for the duration of the session. A new reference standard with a large (1 mm x 1 mm) area of uniform thickness of 100 nm silicon nitride is used to reveal regions of X-ray detector occlusion that would give misleading results for any X-ray method that measures thickness. Unlike previous methods, the new X-ray method does not require an accurate beam current monitor but delivers equivalent accuracy in mass thickness measurement. Quantitative compositional results are also automatically corrected for specimen self-absorption. The new method is tested using a wedge specimen of Inconel 600 that is used to calibrate the high angle angular dark field (HAADF) signal to provide a thickness reference and results are compared with electron energy-loss spectrometry (EELS) measurements. For the new X-ray method, element composition results are consistent with the expected composition for the alloy and the mass thickness measurement is shown to provide an accurate alternative to EELS for thickness determination in TEM without the uncertainty associated with mean free path estimates.

1. Introduction

When an electron beam strikes a bulk specimen, emitted X-ray intensity depends on element composition. A comparison of intensities obtained when the same beam is focussed on a bulk unknown specimen and a bulk standard of known composition gives a direct measure of element concentration. Furthermore, if an energy-dispersive X-ray spectrometer (EDS) is well-characterised, quantitative element composition analysis of bulk specimens in a scanning electron microscope (SEM) can be achieved with only a single measurement on a bulk pure element reference standard using the same electron beam conditions [1]. The benefit of using a reference standard measurement is that results do not have to be normalised so that the analytical total provides a useful diagnostic to detect errors such as missed or misidentified elements or inaccurate matrix corrections.

For a thin specimen in a transmission electron microscope (TEM), X-ray intensity depends on both element composition and specimen thickness so that even if the same electron beam is used on an unknown specimen and a bulk reference standard, a diagnostic analytical total cannot be obtained but if element mass fractions are assumed to sum to unity the measurement on a bulk reference allows the mass thickness of the specimen to be determined. This approach was used by Dijkstra *et al.* [2] but TEM beam voltages are much higher than in SEM and at a beam current necessary to give adequate X-ray intensity from a thin sample, the X-ray yield from a bulk standard not only caused overload of the EDS but also the high penetration depth for the incident beam introduced inaccuracy because of the very large correction for X-ray absorption in the standard [2]. Boon solved the overload problem by devising a beam current monitor linear over 3 - 4 decades so that a bulk standard could be analysed at much lower current than the thin specimen and intensities corrected for the large difference in beam current but huge X-ray absorption in the bulk standard was still an issue [3].

Watanabe and Williams [4] avoid bulk standards completely in the ζ-factor (zeta-factor) method. However, determination of zeta-factors for a particular instrument requires thin film standards of known thickness, density and composition and such standards are not readily available. Furthermore, a critical requirement is an accurate method of measuring incident current but beam current monitors are not always available, convenient, or of sufficient accuracy.

In what follows, we consider a new method for quantitative analysis in TEM that uses a well-characterised EDS and a single thin film reference standard to achieve measurement of mass thickness and elemental concentration with no requirement to measure beam current.

2. New method

For analysis of thin specimens in TEM, X-ray intensity for the emission line of element A is given by:

$$I_A = (\rho.t\,)_{sp}.\,C_A.\ [\,N_0/A_A\,.\,(Q_A.f_A)\,].\ f(\chi)_A.\ D_{sp}\,.\,(\Omega/(4\pi)).\ \varepsilon_A \qquad (1)$$

for element mass fraction C_A, Avogadro constant N_0, atomic weight A_A, ionisation cross-section Q_A, fractional emission f_A, self-absorption $f(\chi)_A$, electron dose D_{sp}, detector solid angle Ω, and conversion efficiency ε_A. As in previous methods for TEM quantitative analysis, $(Q_A.f_A)$ can be calculated from theory where f_A is usually expressed as the product of fluorescence yield and Coster-Kronig enhancement for the shell and ratio of line emission intensity to total for all emissions from that shell. $f(\chi)_A$ can be evaluated if the composition, mass thickness and geometry are known. Thus, I_A depends not only on beam current (governs D_{sp}) and composition (C_A), but also on density (ρ) and thickness (t). The mass thickness $(\rho.t\,)_{sp}$ is proportional to the aerial density of atoms in the direction of the electron beam. If we have a reference standard where $(\rho.t)_{ref}$. C_{ref} is known and an X-ray detector that has been characterised so that conversion efficiency ε is known as a function of energy [5] then we can determine $D_{sp}.(\Omega/(4\pi))$ from the measurement on the reference standard in the TEM. Provided we use the same beam current and kV when a spectrum is recorded from the specimen, eq. (1) can be then be used to determine $(\rho.t)_{sp}.C_A$ for each element and $(\rho.t\,)_{sp}$ is obtained by assuming that the mass fractions sum to unity. Self-absorption can be ignored to obtain first estimates of mass thickness and composition which are then used to calculate $f(\chi)_A$ and further improved estimates obtained by iteration.

Doing a "beam measurement" on the reference standard at the start or end of any session thus replaces explicit measurement of beam current. Besides enabling the analytical results to be corrected for X-ray absorption, the mass thickness value can also be used for lineal thickness measurement if density is known, calculation of electron beam scattering effect on spatial resolution, modelling of image contrast, electron energy-loss spectrometry (EELS) quantification and determination of defect densities for example.

3. Thin film reference standard

For the reference standard we use a self-supporting 100 nm thick silicon nitride TEM support film covering a 1 mm x 1 mm aperture within a standard 3 mm diameter disk of silicon 0.1 mm in thickness. This type of film is usually grown on top of a silicon wafer and the aperture is formed by etching to remove the silicon support. Although the film is uniform in composition and thickness, the density and stoichiometry depend on the process used to manufacture the film [6]. Therefore, we characterise the thin film standard using an EDS and SEM. To determine the mass thickness for Si, we modelled X-ray emission from bulk Si and thin Si_3N_4 films at 10 kV and 30 kV using Monte Carlo simulation to establish a relationship between mass thickness and k-ratio (intensity from film divided by intensity from bulk). Using a transmission holder for the thin film standard and constant beam current, we measure X-ray intensity from the reference thin film and from a pure bulk Si wafer to obtain *k*-ratios and thus obtain $(\rho.t)_{ref}.C_{Si}$ values at a series of points all over the support film and confirm uniformity over at least the central 0.6 mm x 0.6 mm region.

4. Results

4.1. Instrumentation and materials
For testing the new X-ray analysis method we used a JEOL JEM-2200-MCO-FEGTEM and a low background holder, tilted to 7.2°, with a cut-out section to reduce occlusion of the line of sight to the X-ray detector.

4.2. Determination of occlusion of the detector and beam current stability
Partial occlusion of a detector affects Ω and if occlusion varies with position, any X-ray method for measuring mass thickness will give a variable result even if the thickness is uniform. Therefore, it is imperative to maintain a clear line of sight from the beam spot to the X-ray sensor. Holder cut-outs and slight tilt helps but degree of occlusion may still vary with stage X/Y position and stage tilt, particularly with a large solid angle detector that is positioned very close to the specimen. The characterised reference film covers a large 1 mm x 1 mm area and a beam measurement was taken at many stage positions to determine if the Si-K intensity was constant or affected by occlusion at any position. Repeat measurements at the same position were used to confirm beam current stability before and after specimen exchange.

Figure 1 shows an occlusion map where the bubble diameter shows the reduction of signal relative to the maximum Si-K intensity at several stage positions. It is apparent that although tilting the sample and using a cut-out has reduced any variable occlusion in the direction of the detector, the edge of the cut-out is causing about 20 % occlusion as the stage is moved to Y= + 300 μm, perpendicular to the detector direction. The occlusion map provides diagnostic information that can be used to improve design of a holder and check detector alignment or can be used as a map of sensitivity for correction of thickness measurements.

4.3. Calibrated test specimen
A wedge-shaped specimen of Inconel 600 was prepared by focussed ion beam etching. The intended wedge angle of 0.5° ± 0.07° was used to calibrate the high angle annular dark field (HAADF) signal as follows. Taking the composition and density to be uniform, HAADF signal A(x) at position x for this type of specimen was assumed to be proportional to thickness and by fitting a straight line to A(x) at a series of stage positions in the direction of the wedge, the gradient m was used to obtain thickness t from HAADF intensity as shown in figure 2.

4.4. Thickness determination by the new X-ray method
The silicon nitride reference film was first used to obtain a beam measurement. Then, without altering microscope conditions, the reference was replaced by the wedge test specimen of Inconel 600 and a series of spectra were obtained along the wedge (white crosses in figure 3). For these measurements the stage was positioned at the point marked with a blue arrow in the occlusion map of figure 1. Using the new X-ray method, measurements of mass thickness were converted to thickness in nm by assuming a constant density of 8.43 g cm^{-3}.

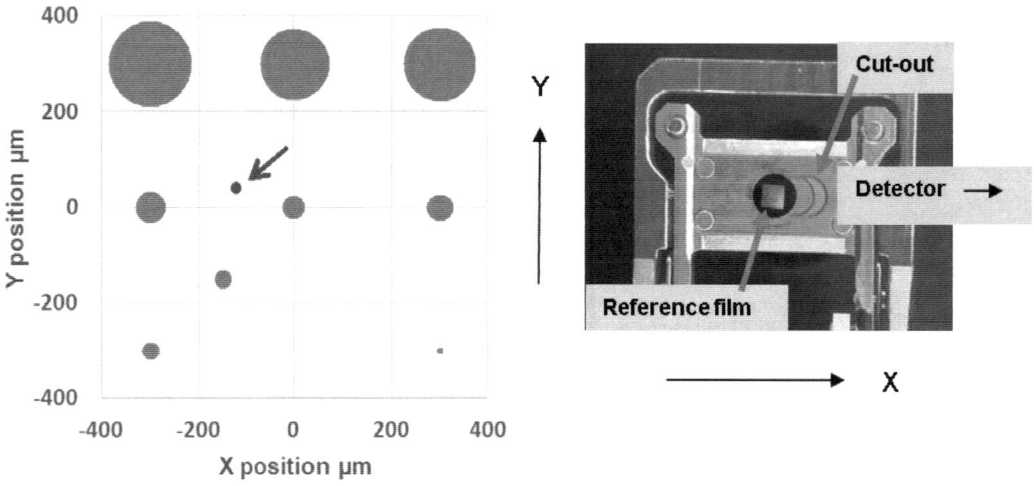

Figure 1. Occlusion bubble map where occlusion (%) = 100 x (bubble diameter µm)/800. Right hand image shows relative positions of film, cut-out and X-ray detector and axes for stage movement. Blue arrow indicates stage position that was used for X-ray measurements with the new method.

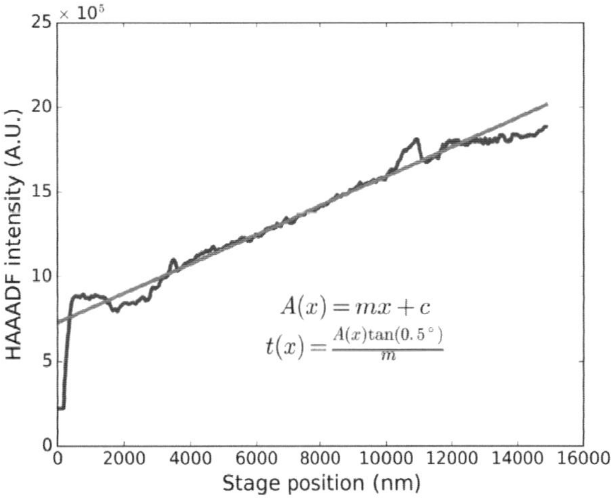

Figure 2. Calibration of HAADF signal with 0.5° wedge specimen to obtain thickness from intensity.

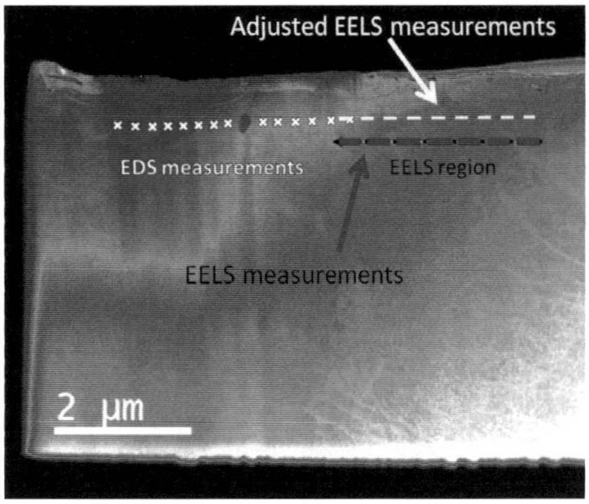

Figure 3. False-colour HAADF image (low intensity blue, high intensity red) of wedge test specimen of Inconel 600 showing regions used for acquisition of X-ray and EELS data.

4.5. Thickness determination by EELS measurements

In a separate session using the same test sample, EELS measurements of areas for total, I, and zero-loss, I_0, for the spectrum were taken along a line of points and thickness, t , estimated by the formula t = λ. ln (I/I_0). Values for the mean free path λ were calculated by published methods, 81 nm by Malis *et al.* [7], and 109 nm by Iakoubovskii *et al.* [8]. Because the lines used for EELS and X-ray measurements did not overlap, a small correction was made to the EELS results to give equivalent values along the same line as used for the X-ray data by scaling according the HAADF signal.

4.6. Thickness results

Figure 4 shows the comparison of results from the different techniques. Uncertainty in the FIB angle for the wedge dominates the error bar for HAADF measurement and uncertainty in λ dominates the EELS log ratio method. Nevertheless, it is encouraging that the results for the new X-ray method agree well with the HAADF estimates.

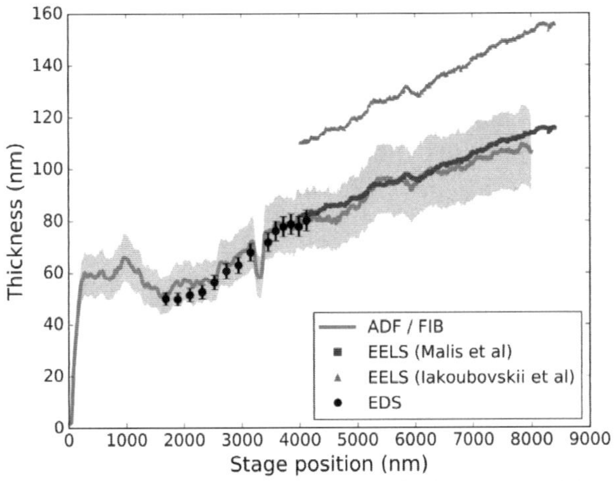

Figure 4. Thickness values at different stage positions. Red line shows the thickness from the HAADF signal calibrated using the known wedge angle with error bar showing the possible variation due to uncertainty in wedge angle. The black dots are the mass thickness results from the new X-ray method, converted to nm using a density of 8.43 g cm^{-3}. The green and blue lines show the results from EELS using estimates of λ obtained by references [7] and [8] respectively.

4.7. Composition results

The X-ray method also produces quantitative results for element weight% and these have been plotted in figure 5 for a series of stage positions. The quantitative results are within the range of compositions expected for Inconel 600 alloy, as shown by the dotted lines.

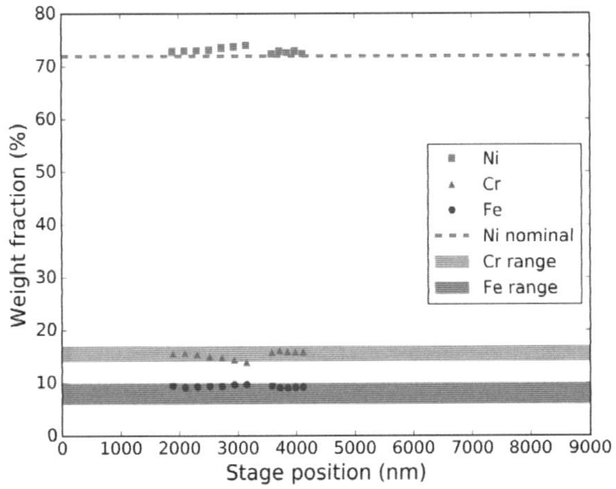

Figure 5. Element composition weight% obtained by the new X-ray method at the analysis positions corresponding to those used in figure 3.

IOP Conf. Series: Materials Science and Engineering **304** (2017) 012017 doi:10.1088/1757-899X/304/1/012017

5. Conclusion

The new reference standard with a large (1 mm x 1 mm) area of uniform thickness reveals regions of X-ray detector occlusion that would give misleading results for any X-ray method that measures thickness. With just a single measurement on the silicon nitride reference standard to calibrate the beam, many quantitative measurements can be made on a specimen in an analysis session by simply acquiring an X-ray spectrum at each point. The new X-ray method delivers equivalent accuracy in mass thickness measurement without the need for the accurate beam current monitor that is required for previous methods and quantitative compositional results are automatically corrected for specimen self-absorption. The mass thickness measurement also provides a convenient alternative to EELS for thickness determination in TEM without the uncertainty associated with mean free path estimates.

References

[1] Statham P 2009 *Microsc. Microanal.* **15** (Suppl. 2) 528-529
[2] Dijkstra J M, Bastin G F, Heijligers H J M and Klepper D 1994 *Microchim. Acta* **114/115** 277-284
[3] Boon G 2000 *Simultaneous determination of specimen composition and thickness using the transmission electron microscope.* Thesis. (ISBN 90-386-2781-5)
[4] Watanabe M and Williams D B 2006 *J. Microscopy* **221**(Pt 2) 89-109
[5] Statham P J 2010 *Microsc. Microanal.***16** (Suppl. 2) 1304-1305
[6] Zaluzec N J 2015 *Microsc. Microanal.* **21** (Suppl. 3) 959-960
[7] Malis T, Cheng S C and Egerton R F 1988 *J. Elec. Microsc. Tech.* **8**, 193-200
[8] Iakoubovskii K, Mitsuishi K, Nakayama Y and Furuya K 2008 *Phys. Rev. B* **77** 104102

IOP Conf. Series: Materials Science and Engineering **304** (2017) 012018 doi:10.1088/1757-899X/304/1/012018

Information observed in Ti-Lα,β and Ti-Lℓ,η emission lines of Ti and its oxides

M Terauchi[1], S Koshiya[2] and K Kimito[2]

[1] Tohoku University, Institute of Multidisciplinary Research for Advanced Materials, 2-1-1 Katahira, 980-8577 Sendai, Japan
[2] National Institute for Materials Science, Research Center for Advanced Measurement and Characterization, 1-1 Namiki, 305-0044 Tsukuba, Japan

E-mail: terauchi@m.tohoku.ac.jp

Abstract. Ti L-emission lines of metal-Ti and its oxides are measured by using a soft-X-ray emission spectrometer attached to a scanning electron microscope. Lα,β emissions due to electron transitions from valence 3d states to core 2p levels show a variety of intensity distributions for Ti oxides with different crystal structures reflecting different bonding states. Lℓ,η-emissions are due to transitions from 3s level to 2p levels and show simple intensity distribution. The energy shift of Lℓ,η-emission peak should reflect a relaxation energy in the final state with a hole in 3s level. The information of L-emission spectrum is discussed based on a chemical shift of photoemission experiment.

1. Introduction

Recent developments of soft-X-ray emission spectroscopy (SXES) instruments for conventional transmission electron microscope (TEM) enable us to obtain partial density of states (DOS) of valence bands (VB, bonding electron states) from identified specimen areas by transmission electron microscopy [1]. SXES method has two advantages in practical applications as follows, 1) SXES is not a surface sensitive method as photoemission spectroscopy and/or electron spectroscopy for chemical analysis (ESCA) and does not need ultra-high vacuum condition due to a larger absorption length of soft-X-ray photons than a few tens nm [2], and 2) the spectrum energy position is not affected by charging of specimen because of soft-X-ray emission originates from electronic transitions inside a specimen. The latest SXES spectrometer for TEM has an acceptable energy range from 50 eV to 4,000 eV by using four varied-line-spacing (VLS) gratings. It can detect from Li K-emission (55 eV) to Ta L-emission (3.8 keV) [3, 4]. Major problem of this method is extremely low emission efficiency lower than 1 % [5]. Thus, an application to electron probe microanalyser (EPMA) as a commercial instrument has been done due to its large probe current [6]. The commercial instrument has a detectable energy range from 50 to 210 eV for first order diffracted spectrum observation. It realised a shorter detection time than one minute except a spatial resolution is lower than that of TEM. The 50 - 3,800 eV spectrometer was also applied to a conventional scanning electron microscope (SEM) for testing capabilities of the spectrometer [7].

X-rays originating from electronic transitions from valence bands to inner-shell electron levels inform us energy states of bonding electrons, L-emissions of 3d transition metal elements. Those range from 400 to 1,000 eV, are very important to assign chemical states of those elements in compounds.

Lα- and Lβ-emissions due to $3d_{5/2,3/2} \rightarrow 2p_{3/2}$ and $3d_{3/2} \rightarrow 2p_{1/2}$ transitions, respectively, are a suitable probe of valence states (bonding states). On the other hand, Lℓ- and Lη-emissions correspond to transitions from a shallow inner-shell level of $3s_{1/2}$ to $2p_{3/2}$ and $2p_{1/2}$ levels, respectively. Thus, Lℓ,η-lines can give us different information from that of Lα,β-emissions. Furthermore, the charge state is also important when discussing physical properties because of the amount of 3d electrons closely related to magnetism and conductivity of 3d transition metal compounds.

SXES analysis across a wide variety of materials can be conducted by using a SEM-SXES instrument as a tool of chemical state analysis for bulk materials. In this paper, Ti L-emission lies obtained by a SXES-SEM instrument and the information obtained are discussed not only the DOS of VB but also the chemical shift based on the discussion in photoemission spectroscopy.

2. Experimental

Present experiments were done by using an improved SEM-SXES instrument with an energy resolution of 80 meV at aluminium L-emission [8]. Figure 1 shows a photo of the SEM-SXES instrument used. For obtaining Ti L-emission, a grating of JS2000 with a groove density of 2,400 lines/mm was applied. A multichannel plate (MCP) detector was operated in photon counting mode. Then, relative intensity of emission lines is directly related to emitted number of photons. An energy resolution for Ti L-emission is estimated to be 0.2 - 0.3 eV. The energy of the spectrum was calibrated by using a fifth order curve determined by using energies of Ti-Lα- and Lℓ-lines of metal-Ti, Cu-Lα and -Lℓ and its higher-order lines of metal-Cu. Those calibration data were obtained under the same experimental set up of present investigation. The electron probe current and acquisition time are 190 nA and 20 minutes, respectively. Figure 2 shows a comparison of X-ray emission spectra of TiO_2 (rutile) obtained by a conventional energy-dispersive X-ray spectrometer (EDS) attached to a TEM (left, 0.0 - 5.5 keV) and the SEM-SXES instrument used in this experiments (right, 350 - 580 eV). An asymmetrical intensity distribution of O K-emission in EDS data is clearly resolved into three sharp peaks of O-K, Ti-Lα,β and Ti-Lℓ,η lines with the present SXES instrument. In a simple ionic model, TiO_2 is considered as $Ti^{4+}O^{2-}{}_2$ and cannot emit Lα,β-intensity because Ti^{4+} ion does not have any 3d electron. However, Lα,β-emissions are clearly seen. Thus, Ti atom in TiO_2 is not a pure Ti^{4+} ion via covalent bonding between Ti and surrounding O atoms. A broad intensity in the lower energy side of Ti-Lℓ,η peak is a plasmon satellite (ps).

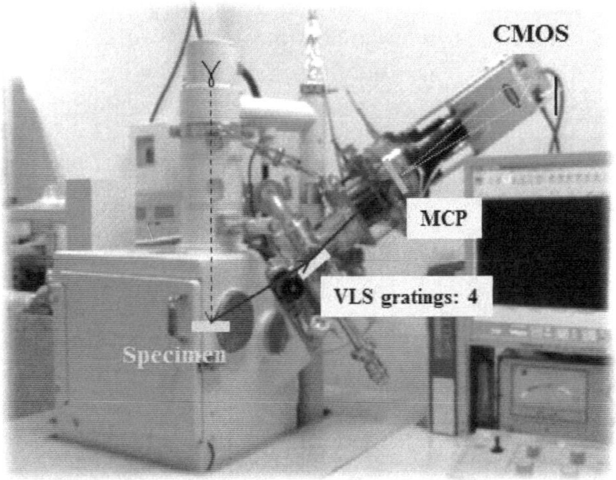

Figure 1. Photo of SEM-SXES instrument used [8].

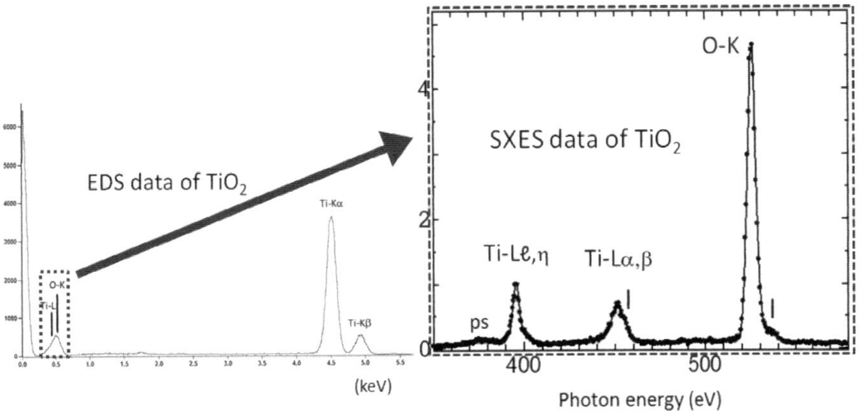

Figure 2. EDS and SXES spectra of TiO_2. Asymmetric intensity of O-K in EDS spectrum is clearly resolved into three sharp peaks in SXES measurement.

3. Results and discussion

Figure 3 shows Lα,β- and Lℓ,η-emissions of metal-Ti, ε-TiO, new crystalline phase recently reported [9], and TiO_2 (rutile). Those spectra were obtained from those bulk specimens. Spectral intensities are normalised in each figure. Lα-peak of metal-Ti shows asymmetric intensity distribution originating from DOS of 3d electrons of VB. Those intensity profiles of ε-TiO and TiO_2 (rutile) show a peak and a shoulder structures. However, those positions of peaks and shoulders are different for two crystalline materials, reflecting different bonding states. Ti atom is surrounded by six O atoms forming an octahedron in TiO_2, and surrounded by trigonal prismatic (six-fold coordination) and trigonal planar (three-fold coordination) arrangements of O atoms in ε-TiO. Those characteristic intensity profiles of ε-TiO and rutile are well reproduced by theoretical calculations of Ti-3d DOS of VB [9, 10]. When discussing the relative energy positions of those spectra, chemical shifts of Ti-2p level and band gap energies of those materials should be taken into account. For the analysis, assignment of the top of VB position is important as a reference energy position. However, it is difficult to deduce it from present Lα,β-intensity distribution, because of a presence of Lβ-intensity and Lα-absorption edge exists between Lα- and Lβ-emissions. In contrast, Lℓ-emission due to transitions between inner-shell levels should show symmetric intensity distribution about the peak positon. Each vertical line for Lℓ-emission peak is placed at the middle of each peak width. Compared to metal-Ti, TiO_2 shows an apparent shift to higher energy side by 0.7 eV. Lℓ-peak position of ε-TiO is almost the same with metal-Ti, but seems a little shift to higher energy side by 0.1 - 0.2 eV. Those differences in Lℓ-peak position should be related to local screening energy of 3s core-hole in the final state (see following discussion).

Figure 4 shows Lα,β- and Lℓ,η-emissions of three TiO_2 crystals with different crystal structures of rutile, anatase and brookite. Those spectra were obtained from those bulk specimens. Spectral intensities are normalised in each figure. Crystalline anatase and brookite of natural minerals are used in this measurement. The rutile used was an artificial gemstone. Although these crystalline materials are composed of similar TiO_6 polyhedra, distortion and arrangement of TiO_6 and resulting space group of the materials are different. Each Lα-peak shows a peak with a shoulder in higher energy side. However, energy separations of the two structures of the materials are different each other. From a theoretical study on the electronic structure of those three TiO_2 materials, the main peak and the shoulder structures of Lα-intensity can be assigned to σ-hybrid state between Ti and O atoms and lone pair π-state, respectively [10]. Lℓ-emission peak of each TiO_2 material indicated by a vertical line at the middle of the peak width is in higher energy positon relative to that of metal-Ti. Rutile and anatase show almost the same energy position of Lℓ. That of brookite is a little higher energy position than those of rutile

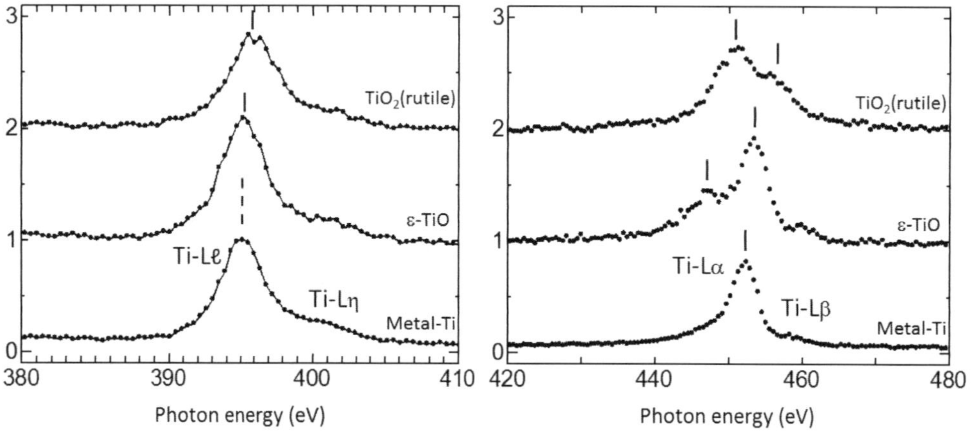

Figure 3. Lα,β (right) and Lℓ,η (left) mission intensity distribution of metal-Ti, ε-TiO and TiO_2 (rutile).

and anatase. It should be noticed in spectra of TiO_2 materials that Lα peaks are shifted to lower energy and Lℓ-peaks are shifted in higher energy side compared with those of metal-Ti. This suggests that chemical effects on Ti atoms appear in different ways for Lα- and Lℓ-emissions. Lβ and Lη should show similar information to those of Lα and Lℓ, respectively, but those are too weak to discuss in those spectra.

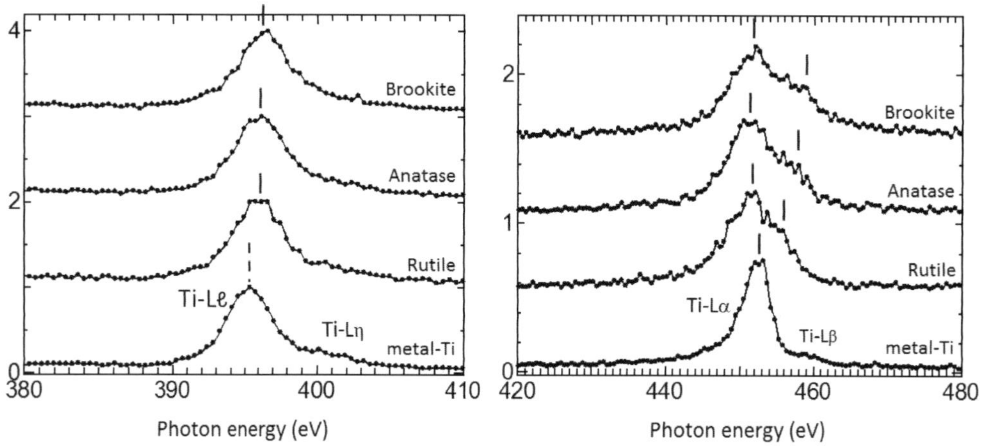

Figure 4. Lα,β (right) and Lℓ,η (left) emission intensity distribution of TiO_2 materials of rutile, anatase and brookite. Those crystalline materials are composed of similar TiO_6 polyhedra but different distortions.

For discussing chemical effects on Lα- and Lℓ-emissions, an expression on binding energy shift due to a change of chemical effects in X-ray photoelectron spectroscopy is helpful [11]. A change in binding energy referred to Fermi level ΔE^F_B is written as follows:

$$\Delta E^F_B = \Delta q/r - \Delta V - \Delta E_R - \Delta\phi$$

where $\Delta q/r$ represents a core-level shift due to a change of valency, a chemical effect in an initial state. ΔV is a change of solid state effects of crystal-field, density of states, bandgap and etc. This term usually works for valence states. ΔE_R is a change of relaxation effects, a final state effects with a hole. $\Delta\phi$ represents a change of work function. For the case of X-ray emission, two electron levels inside a material are involved and thus $\Delta\phi$ is zero. As Lα-emission is due to electronic transitions from valence states to core $2p_{3/2}$ states, there remains a hole in valence states in the final state. This final state is an excited state from the ground state with a rather low energy. Then, relaxation energy due to this valence hole should be small and/or negligible. Thus, chemical effects on a change of Lα-emission energy, $\Delta E_{L\alpha}$, is written as follows:

$$\Delta E_{L\alpha} = \Delta q/r - \Delta V.$$

As already presented in figures 3 and 4, solid state effect ΔV is dominated. The initial state effect $\Delta q/r$ is included in Lα-emission but difficult to deduce accurately from experimental data. As Lℓ-emission due to transitions from a shallow core-state $3s_{1/2}$ to deeper core-level $2p_{3/2}$, ΔV term should be ignored. There remains a core-hole in $3s_{1/2}$ state in the final state. Then, ΔE_R term should remain, core-hole effect in the final state. A theoretical study showed the valency effect $\Delta q/r$ on different core-levels, binding energy shifts, are almost uniform for all core-levels [12]. Thus, $\Delta q/r$ can be omitted and chemical effects on a change of Lℓ emission energy, $\Delta E_{L\ell}$, is written as follows:

$$\Delta E_{L\ell} = -\Delta E_R.$$

This means that Lα- and Lℓ-emissions gives different chemical information. To evaluate the discussion of $\Delta E_{L\ell}$, theoretical calculations were conducted by using WIEN2k code for metal-Ti and TiO_2 (rutile) with a core-hole in $3s_{1/2}$ level, which corresponds to the final state for Lℓ-emissions. A 3 x 3 x 3 super-cell was used in the calculation. Energy differences between $3s_{1/2}$ and $2p_{3/2}$ of rutile was larger than that of metal-Ti by 0.64 eV. This value is almost equal to the observed shift for higher energy side by 0.7 eV.

The above discussion on $\Delta E_{L\ell}$ for Lℓ-emission is also applied for other emissions caused by electronic transitions between two core-levels. For example, Kα-emissions of atoms later than sodium atom in the periodic table are due to transitions between core-levels. Chemical shifts of Al-Kα emissions of metal-Al and its compounds including an oxide were reported and discussed a relation with amounts of valence charge of the materials [13]. A more precise scheme of the discussion should be as follows that chemical shifts of Kα are directly related to relaxation energy (core-hole effect or local screening effect), which is related to a valence charge and/or local dielectric properties of materials. Similar information on relaxation energy is also obtained by taking differences between two core-level binding energies in X-ray photoelectron spectroscopy (XPS) experiments. Relation between relative chemical shifts of 1s and 2p in XPS spectra and dielectric constants are reported for Si and its compounds [14] and Al and its compounds [15].

4. Concluding remarks

Ti-Lα,β- and Ti-Lℓ,η-emission spectra of metal-Ti and its oxides obtained by using a SEM-SXES instrument were discussed. As Lα,β-emissions caused by electron transitions from VB to core 2p levels, variety of intensity distributions were observed for Ti oxides with different crystal structures reflecting different bonding states. Thus, Lα,β-emissions is useful for chemical bonding analysis. Simple intensity profiles of Lℓ,η-emissions due to transitions between core levels is suited for examining peak energy shifts, which indicate the relaxation energy (core-hole effect or local screening effect) in the final state. It should be noted that integrated intensity of Lα,β should be related to the number of 3d electrons and that of Lℓ,η-intensity is to the number of atoms included in the analysed volume. This information is important in microbeam analysis of materials including 3d transition metal elements.

Acknowledgments
This work was partly supported by the Research Programme of "Dynamic Alliance for Open Innovation Bridging Human, Environment and Materials" in "Network Joint Research Center for Materials and Devices".

References

[1] Terauchi M 2014 Valence electron spectroscopy for transmission electron microscopy. in: *Transmission electron microscopy characterization of nanomaterials.* (Kumar C S S R; Ed.) (Berlin, Heidelberg: Springer-Verlag) 287-331
[2] Attwood D 2000 *Soft X-rays and extreme ultraviolet radiation.* (New York: Cambridge University Press) 3
[3] Terauchi M, Takahashi H, Handa N, Murano T, Koike M, Kawachi T, Hasegawa N, Imazono T, Koeda M, Nagano T, Sasai H, Oue Y, Yonezawa Z and Kuramoto S 2012 *J. Electron. Microsc.* **61** 1
[4] Terauchi M, Takahashi H, Handa N, Murano T, Koike M, Kawachi T, Imazono T, Koeda M, Nagano T, Sasai H, Oue Y, Yonezawa Z and Kuramoto S 2013 *Microscopy* **62** 391
[5] Krause M O 1979 *J. Phys. Chem. Ref. Data* **8** 307
[6] Murano T, Takahashi H, Handa N, Terauchi M, Koike M, Kawachi T, Imazono T, Hasegawa N, Koeda M, Nagano T, Sasai H, Oue Y, Yonezawa Z and Kuramoto S 2012 *Microsc. Microanal.* **18** (Suppl. 2) 776
[7] Terauchi M, Koshiya S, Satoh F, Takahashi H, Handa N, Murano T, Koike M, Imazono T, Koeda M, Nagano T, Sasai H, Oue Y, Yonezawa Z and Kuramoto S 2014 *Microsc. Microanal.* **20** 692
[8] Terauchi M, Takahashi H, Takakura M, Murano T, Koike M, Imazono T, Nagano T, Sasai H and Koeda M 2016 *Microsc. Microanal.* **22** (Suppl. 3) 414
[9] Amano S, Bogdanovski D, Yamane H, Terauchi M and Dronskowski R 2016 *Angew. Chem.* **55** 1652
[10] Landmann M, Rauls E and Schmidt W G 2012 *J. Phys. Condens. Mat.* **24** 195503
[11] Kim K S and Winograd N 1975 *Chem. Phys. Lett.* **30** 91
[12] Cole R J, Machonald B F and Weightman P 2002 *J. Electron Spectrosc. Relat. Phenom.* **125** 147
[13] Koshiya S and Terauchi M 2015 *Philos. Mag.* **95** 1462
[14] Hirose K, Kihara M, Okamoto H, Shinagawa S, Nohira H, Ikenaga E, Higuchi M, Teramoto A, Sugawa S, Ohmi T and Hattori T 2006 *Appl. Phys. Lett.* **89** 154103
[15] Hirose K, Suzuki H, Nohira H, Ikenaga E, Kobayashi D and Hattori T 2008 *J. Phys.: Conf. Series* **100** 01211

IOP Conf. Series: Materials Science and Engineering **304** (2017) 012019 doi:10.1088/1757-899X/304/1/012019

Monte Carlo simulation of the effect of shape and thickness on SEM-EDS microanalysis of asbestos fibres and bundles: the case of anthophyllite, tremolite and actinolite

G Valdrè, D Moro and G Ulian

University of Bologna, Interdisciplinary Research Centre of Biomineralogy, Cristallography and Biomaterials, Department of Biological, Geological and Environmental Sciences, Piazza di Porta San Donato 1, 40126 Bologna, Italy

E-mail: giovanni.valdre@unibo.it

Abstract. Asbestos is a generic term used for six types of silicate minerals that are found in fibres or bundles of fibres, which can be easily cleaved into thinner ones. Scanning electron microscopy energy-dispersive X-ray spectrometry (SEM-EDS) quantitative microanalysis of asbestos mineral fibres still represents a complex analytical issue because of the variable fibre shape and small thickness (< 5 μm) compared with the penetration depth of the incident electron beam. Following previous work on chrysotile, crocidolite and amosite, here we present a study by means of Monte Carlo simulations of the thickness and shape effect on SEM-EDS microanalysis of anthophyllite, tremolite and actinolite asbestos. Realistic experimental conditions, such as sample geometry, SEM set-up and detector physics were taken into account. We report the results obtained on 100 μm long fibres and bundles of circular and square section and thicknesses from to 0.1 μm to 10 μm, for electron beam energies of 5, 15 and 25 keV. A strong influence of the asbestos mineral fibres and bundles shape and thickness on the detected EDS X-ray intensity was observed. In general, the X-ray intensities as a function of fibre thickness showed a considerable reduction below about 0.5 μm at 5 keV, 2 μm at 15 keV, and 5 μm at 25 keV for all the elements and minerals, with a non-linear dependence. Correction parameters, *k*-ratio, for the thickness effect were calculated and proposed.

1. Introduction

Asbestos is a generic term used for six types of silicate mineral fibres that are or have been widely exploited in many types of technology. They belong to two mineral groups: serpentines and amphiboles. Chrysotile is the only asbestiform member of the serpentine group. The amphibole group contains five asbestiform members: crocidolite (asbestos variety of riebeckite), amosite (asbestos variety of grunerite), anthophyllite, tremolite and actinolite. They are found in fibres or bundles of fibres, which can be easily separated into thinner ones. Accurate chemical analysis of asbestos mineral fibres is of paramount importance for their identification and in depth mineral characterisation. Fibre composition is related to the minerals' environmental and health impact and potential toxicological effects [1-3].

Asbestos fibres toxicity was demonstrated by *in vitro* and *in vivo* studies to be associated to the chemical composition, and there have been suggestions that the Fe content and its oxidation state may play a role [4]. In addition, the generation of reactive oxygen species and other radicals, which are catalysed by iron ions at the fibre surface, is also thought to be a mechanism of cytotoxicity and genotoxicity [5]. The mineralogical composition of fibres with bonded metal ions induces a strong biochemical reaction. Techniques able to accurately quantify the chemistry of mineral fibres in situ at the micrometre- and sub-micrometre scale will make important contributions to fundamental cell biology research on the interaction of cells with fibres containing highly reactive chemical species [6].

Among the microanalytical techniques, scanning electron microscopy equipped with energy-dispersive X-ray spectrometry (SEM-EDS) has proven to be of great importance for minerals, materials and biomaterials characterisation, and furthermore in evaluating the phenomena and possible mechanism of interaction between materials and bio-entities [7-9]. In particular, EDS is by far the most routine method for asbestos chemical analysis [1]. Nevertheless, SEM-EDS quantitative micro-analysis of asbestos fibres may be subjected to systematic errors due to the variable shape of the fibre and small thickness (< 5 μm) with respect to the penetration of the incident electron beam. The classical correction methods for the correction of matrix effects (e.g., ZAF or $\phi(\rho z)$ procedures) used in quantitative microanalysis of bulk samples assume both standard and sample to have flat polished surfaces and be infinitely thick compared to the penetration of the electron beam.

The size and shape of micrometric and sub-micrometric particles may cause large errors in the quantification because of particle effects on the generation and measurement of X-rays from the sample. These effects are associated with the elastic scattering of electrons in the finite size (mass) of the particle, which is strongly influenced by the average atomic number. The particle thickness is key, for a given mean atomic number, with a shape component affecting both the absorption and fluorescence contributions to the correction routine. X-ray absorption path length and emergence angle cannot be determined as for flat polished samples in the case of micrometric and sub-micrometric particles analysis [10, 11]. Furthermore, since EDS systems are often set to normalise the results, they are prone to hide poor analytical results like high or low totals, which in turn hide any problem in the analytical protocol. Empirical methods were developed to overcome these issues, however they are cumbersome and need characterised standards for thickness, geometry and composition [11-13].

The Monte Carlo method is an effective tool for examining the basis, the reliability and the limitations of quantification procedures. In the context of asbestos fibres analysis, Monte Carlo simulations can be of paramount importance for analysing the electron transport and X-ray generation under unconventional measurement conditions or in samples with complex geometry [14-16].

We recently proposed a study by means of Monte Carlo simulations to investigate the effect of thickness and shape on SEM-EDS microanalysis of the three asbestos with the highest industrial exploitation: chrysotile, crocidolite and amosite asbestos [17]. Here, we further extend this work to the investigation of anthophyllite, tremolite and actinolite asbestos. The Monte Carlo programme NIST DTSA-II [18, 19] was used to simulate electron transport, X-ray generation and detection in cylindrical fibres and bundles of square section.

Although about 95 % of asbestos used in commercial products was chrysotile, amphibole exposure in the natural environment is greater than chrysotile, since amphiboles are more widespread than chrysotile [20-24]. Furthermore, anthophyllite, tremolite and actinolite may be found as a contaminant in other fibres or in other industrial minerals. In addition, amphiboles generally contain higher levels of iron than chrysotile.

2. Methods and simulation models

2.1. Methods

The Monte Carlo simulations were performed using the programme NIST DTSA-II Jupiter 2017-2-28 revision [18], which simulates electron transport, X-ray generation and transmission in complex sample geometries. Accurate quantitative analysis by X-ray spectroscopy is possible through the understanding of electron scattering, X-ray generation, absorption and fluorescence. The Monte Carlo method applied to electron and X-ray transport allows one to simulate electrons trajectories and X-rays through the asbestos mineral fibre and to the X-ray detector.

NIST DTSA-II carries out a single scattering model to track the trajectories of electrons as they interact with matter. It implements three elastic scattering models: a basic screened Rutherford model [25], the Mott scattering cross-section of Czyzewski and co-workers [26], and the Mott cross-section of Jablonski and co-workers [27]. The energy loss is modelled using the Joy-Luo expression [28] that is an empirical modification of the Bethe energy loss equation [29]. The ionisation cross-section is modelled using the expression of Bote and Salvat [30]. The mass absorption coefficients are those of Chantler and co-workers [31], and the fluorescence yields are tabulated experimental values [32].

The electron source is defined as a Gaussian beam. A response function that mimics the energy resolution of an EDS detector is convolved with the emitted X-ray events. The effect of shape and thickness was evaluated by simulating EDS spectra of single cylindrical asbestos fibres and of bundles of fibres. Peak intensities were integrated and compared as a function of thickness, shape, and beam energy for all minerals, after background subtraction.

Finally, correction factors, *k*-ratios, for the asbestos fibre thickness effect are calculated and proposed.

2.2. Simulation models

Fibres and bundles of fibres of variable thickness and shape placed on a pure carbon holder were simulated taking into account realistic experimental conditions, such as sample geometry, detector physics and SEM set-up.

We investigated anthophyllite, tremolite and actinolite asbestos with chemical formulae as specified in Table 1, and mass density of 3.21, 3.00 and 3.23 g/cm^3, respectively. This is an average chemical formula since there can be a range of Mg to Fe ratio within each species of amphibole.

Table 1. Average chemical formulae of the simulated asbestos.

Mineral name	Chemical formulae
Anthophyllite	$Mg_{5.25}Fe^{2+}{}_{1.75}Si_8O_{22}(OH)_2$
Tremolite	$Ca_2Mg_{4.75}Fe^{2+}{}_{0.25}Si_8O_{22}(OH)_2$
Actinolite	$Ca_2Mg_{3.5}Fe^{2+}{}_{1.5}Si_8O_{22}(OH)_2$

Each mineral was modelled taking into account both a three-dimensional fibre of circular section and a three-dimensional bundle of square section, deposited on a flat graphite support (see figure 1 for a 2D view of the different geometries). For both models the mineral was 100 μm long and with thicknesses ranging from to 0.1 μm to 10 μm. The influence of thickness and shape on the simulated spectrum was investigated for electron beam energies of 5 keV, 15 keV and 25 keV.

An electron probe of 40 nm in diameter was considered. The probe was focussed in parallel illumination onto the surface of the bundle or fibre, in a mid position with respect to the edges. The modelled EDS detector used to generate the spectrum had an ultra-thin polymer window (Moxtek AP 3.3 film), a gold layer of 7 nm, a dead layer of 10 nm, a detector diode thickness of 3 mm, a sample-to-detector distance of 45 mm, a detector area of 10 mm^2, 4,096 channels each of 10 eV and a resolution of 130 eV (FWHM at Mn Kα). The detector elevation angle was set to 40°, with an azimuthal angle of 0°. The bundle or fibre was oriented with its long axis in the same direction as the detector.

IOP Conf. Series: Materials Science and Engineering **304** (2017) 012019 doi:10.1088/1757-899X/304/1/012019

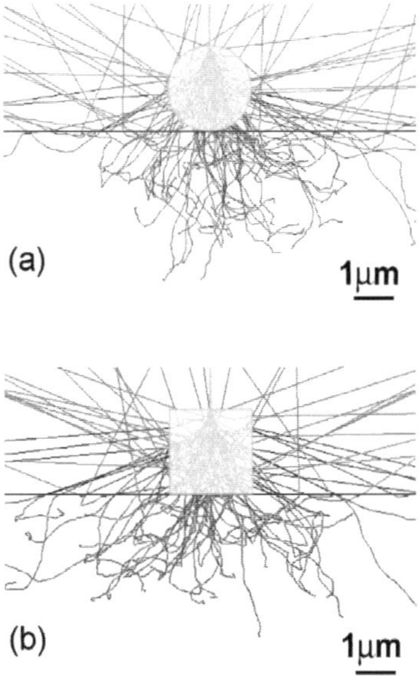

Figure 1. Simulated trajectory of electrons (E_0 = 25 keV) as they interact with the sample (tremolite) and the substrate (graphite). a) Tremolite fibre of circular section. b) Tremolite bundle of fibres of square section. 2D view of the 3D simulated trajectory of electrons (green: interacting with the sample; black and blue: towards and penetrating the substrate, respectively; red: exiting) as they interact with a tremolite fibre (marked with a green line) deposited on a bulk graphite substrate (marked with a blue horizontal line). Red trajectories are related both to backscattered electrons, as well as to electrons inelastically and subsequent elastically scattered outside the body. The view is from the direction parallel to the long axis of the fibre.

3. Results and discussion

The size of the electron beam interaction volume and X-ray generation volume in conventional SEM-EDS microanalysis (with beam energy between 10 keV and 30 keV) are of the order of several μm^3 in most materials. When the primary electron beam penetration depth is greater than the particle size, a fraction of the electron beam escapes the particle before exciting X-rays (finite size effect). As a consequence, a portion of the electron beam scatters out of the particle volume and can excite X-rays from adjacent particles or the substrate. Furthermore, the standard is typically a much larger size and so there are more X-rays generated in the standard. Thus the *k*-ratio is additionally diminished because the intensity of the element X-ray in the standard (denominator of the *k*-ratio) is larger. These phenomena can affect the analytical results.

The simulated trajectories of electrons for a 25 keV electron beam focussed onto the surface of a tremolite cylindrical fibre (figure 1a) and onto the surface of a tremolite bundle of fibres of square section (figure 1b) in a mid position with respect to the edges are shown in figure 1. The fibre and the bundle have a diameter and a thickness of 2 μm, respectively, and are assumed deposited on a graphite substrate. The primary electrons (green vertical beam) interact with the mineral as visualised by the green trajectories. At this energy level (25 keV) a fraction of the electrons (> 40 %) penetrates into the graphite substrate (blue trajectories) for several micrometres, eventually exiting (red lines). Furthermore, a fraction of the electrons (> 40 %) scatters out of the mineral volume from its sides, both toward the substrate, black trajectories, and leaving the mineral, red lines.

We report in table 2 some reference values for anthophyllite, tremolite and actinolite bulk mineral samples: the nominal range value of electron penetration depth and X-ray intensity (generated and emitted) for the three beam energies considered in this work (5 keV, 15 keV, 25 keV). The calculation was performed with the same detector physics and SEM set-up reported in section 2.2 considering 1,000 electron trajectories and a probe dose of 60 nA·s. Table 2 reports for each X-ray line the absolute magnitude of generated and emitted X-ray intensity calculated on bulk samples, in arbitrary units as reported by Ritchie [33].

IOP Conf. Series: Materials Science and Engineering **304** (2017) 012019 doi:10.1088/1757-899X/304/1/012019

Table 2. Electron penetration depth (e.p.d., in μm) and absolute magnitude (in arbitrary units, as reported by Ritchie [33]) of generated (g) and emitted (e) X-ray intensity for each X-ray line calculated on bulk samples. Fe L = Lα+Lβ+Lγ.

	Anthophyllite			Tremolite			Actinolite		
	5keV	15keV	25keV	5keV	15keV	25keV	5keV	15keV	25keV
e.p.d. [μm]	0.26	1.74	4.11	0.26	1.74	4.55	0.26	1.81	3.85
Ca K_α	—	—	—	(g) 1	(g) 87	(g) 247	(g) 1	(g) 85	(g) 238
				(e) 1	(e) 81	(e) 207	(e) 1	(e) 80	(e) 200
Ca K_β	—	—	—	(g) 0	(g) 9	(g) 26	(g) 0	(g) 9	(g) 25
				(e) 0	(e) 9	(e) 23	(e) 0	(e) 9	(e) 22
Mg K_α	(g) 61	(g) 410	(g) 848	(g) 56	(g) 367	(g) 776	(g) 61	(g) 410	(g) 848
	(e) 57	(e) 240	(e) 276	(e) 52	(e) 230	(e) 276	(e) 57	(e) 240	(e) 276
Fe K_α	(g) 0	(g) 31	(g) 129	(g) 0	(g) 4	(g) 19	(g) 0	(g) 26	(g) 107
	(e) 0	(e) 31	(e) 125	(e) 0	(e) 4	(e) 18	(e) 0	(e) 26	(e) 102
Fe K_β	(g) 0	(g) 4	(g) 16	(g) 0	(g) 1	(g) 2	(g) 0	(g) 3	(g) 13
	(e) 0	(e) 4	(e) 15	(e) 0	(e) 1	(e) 2	(e) 0	(e) 3	(e) 12
Fe L	(g) 23	(g) 107	(g) 204	(g) 3	(g) 15	(g) 29	(g) 19	(g) 88	(g) 170
	(e) 17	(e) 23	(e) 16	(e) 2	(e) 3	(e) 2	(e) 14	(e) 18	(e) 12
Si K_α	(g) 65	(g) 588	(g) 1291	(g) 65	(g) 582	(g) 1304	(g) 63	(g) 570	(g) 1253
	(e) 62	(e) 422	(e) 615	(e) 62	(e) 435	(e) 662	(e) 61	(e) 424	(e) 632
O K_α	(g) 353	(g) 1503	(g) 2802	(g) 354	(g) 496	(g) 948	(g) 345	(g) 486	(g) 910
	(e) 300	(e) 581	(e) 476	(e) 281	(e) 418	(e) 292	(e) 276	(e) 421	(e) 298

3.1. Anthophyllite asbestos

A bundle of square section and a fibre of circular one with a composition and mass density as reported in table 1 were taken into account to investigate the significance of the thickness and shape effects on anthophyllite asbestos. See section 2.2 for further details on the simulation model. Figure 2 shows the trend of the integrated intensities (counts) obtained from the simulated EDS spectra of a bundle of fibres of square section as a function of bundle thickness for each element (Mg, Fe, Si, O) at beam energies of 5 keV, 15 keV and 25 keV (see the legend in figure 2f). The simulations refer to a bundle thickness from 0.1 μm to 10 μm.

Generally speaking, for all the elements of both models (circular and square) a strong dependence of the simulated X-ray intensity on the mineral thickness was observed. The intensity values reported for thickness of 10 μm can be considered as a reference value valid for a bulk sample (infinite thickness). Indeed, by comparing the intensity values for a 10 μm thick sample with those calculated for a massive one, we verified that a thickness of 10 μm represents, at our simulated experimental conditions, a bulk sample for all energies and minerals here considered. Table 3 reports the reference intensity values for massive samples calculated at an electron beam energy of 25 keV. In general, the integrated intensity starts to have a reduction for thickness below 5 μm. Specifically, for all the considered elements, Mg, Fe, Si and O, and both fibre and bundle geometry, the integrated intensity strongly decreases starting from thickness of 0.5 μm at 5 keV, 2 μm at 15 keV and 5 μm at 25 keV.

IOP Conf. Series: Materials Science and Engineering **304** (2017) 012019 doi:10.1088/1757-899X/304/1/012019

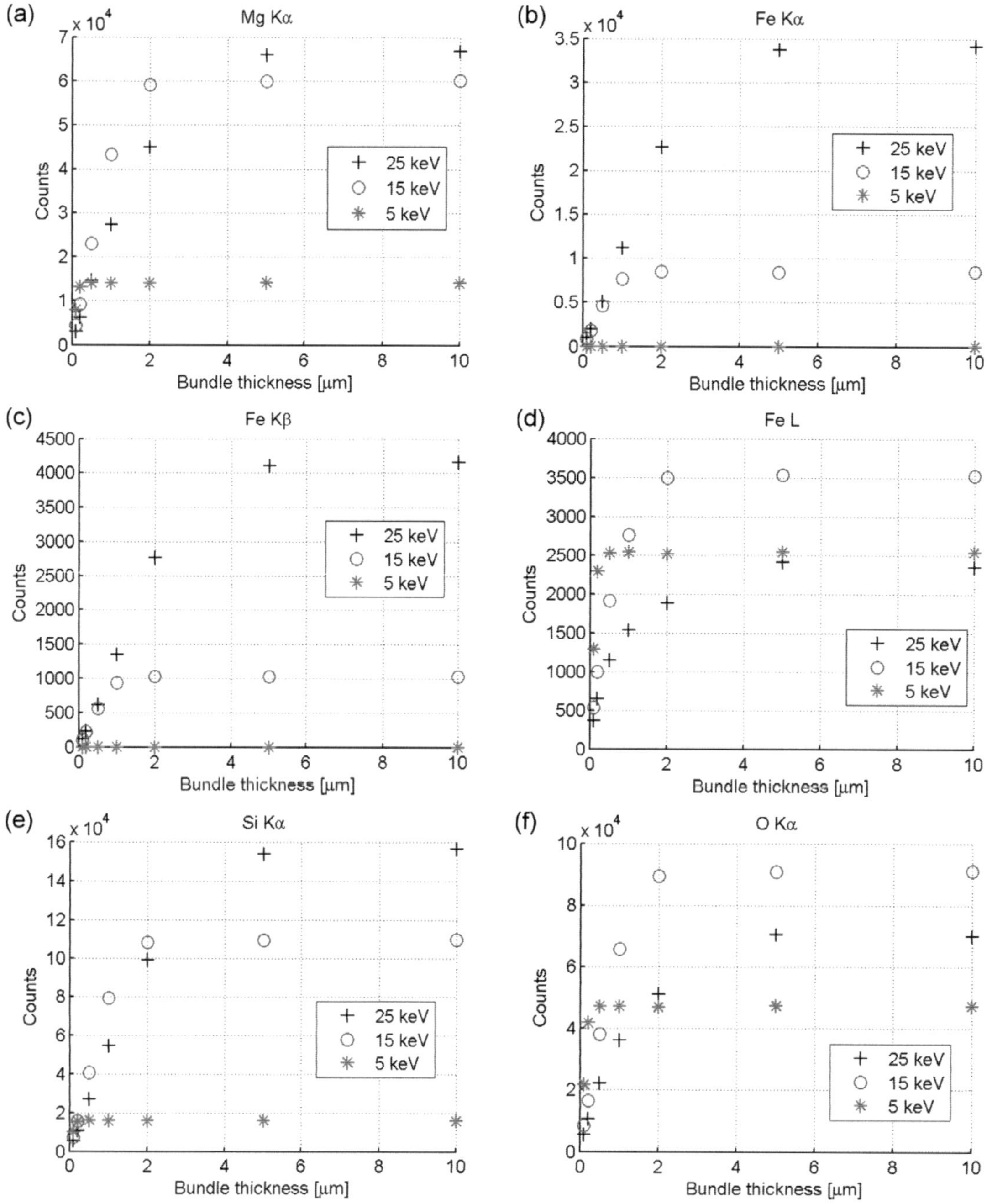

Figure 2. Simulated SEM-EDS X-ray intensity as a function of thickness for an anthophyllite bundle of fibres of square section with beam energy of 5 keV, 15 keV and 25 keV; see legend in (f).

Table 3. Reference intensity values for a bulk sample at 25 keV in our experimental set up (§ 2.2).

	$Ca_{K\alpha}$	$Ca_{K\beta}$	$Mg_{K\alpha}$	$Fe_{K\alpha}$	$Fe_{K\beta}$	Fe_L	$Si_{K\alpha}$	$O_{K\alpha}$
Anthophyllite	—	—	$6.82 \cdot 10^4$	$3.43 \cdot 10^4$	$4.18 \cdot 10^3$	$2.51 \cdot 10^3$	$1.60 \cdot 10^5$	$7.40 \cdot 10^4$
Tremolite	$5.45 \cdot 10^4$	$6.04 \cdot 10^3$	$7.08 \cdot 10^4$	$4.75 \cdot 10^3$	$5.91 \cdot 10^2$	$3.29 \cdot 10^2$	$1.73 \cdot 10^5$	$5.09 \cdot 10^4$
Actinolite	$5.35 \cdot 10^4$	$5.94 \cdot 10^3$	$4.23 \cdot 10^4$	$2.80 \cdot 10^4$	$3.43 \cdot 10^3$	$1.85 \cdot 10^3$	$1.63 \cdot 10^5$	$4.61 \cdot 10^4$

3.2. Tremolite and actinolite asbestos

A fibre of circular section and a bundle of square one were also simulated for tremolite and actinolite asbestos, with a composition as reported in table 1. See section 2.2 for further details on the simulation model. Figure 3 reports the results for a single cylindrical fibre. It shows the trend of the integrated intensity (counts) calculated from the simulated EDS spectra, as a function of fibre diameter (thickness), for each element (Ca, Mg, Fe, Si, O) at beam energies of 5 keV, 15 keV and 25 keV (see legend in figure 3h). Fibres were simulated with a diameter from 0.1 μm to 10 μm. Bundles with a thickness from 0.1 μm to 10 μm were also simulated (data not shown, but available on request).

Generally speaking, for all the elements of both the models (circular and square) a strong dependence of the simulated X-ray intensity on the mineral thickness was observed. Intensity values calculated for thicknesses of 10 μm were found to be very close to those calculated for a bulk sample. Table 3 reports, as an example, the reference intensity values for massive samples calculated at an electron beam energy of 25 keV. In general, the integrated intensity starts to have a reduction for thickness below 5 μm. In particular, for all the considered elements, Ca (Kα), Ca (Kβ), Mg (Kα), Fe (Kα), Fe (Kβ), Fe (L), Si (Kα), O (Kα), and both fibre and bundle, the integrated intensity strongly decreases starting from thickness of 0.5 μm at 5 keV, 2 μm at 15 keV and 5 μm at 25 keV.

Both the square and circular section models present a similar effect of intensity enhancement before the reduction starting point. In terms of the previously cited finite size (thickness) and reduced absorption (shape component) effects, the reduced absorption effect initially prevails on the finite size one, which in turn becomes predominant for smaller thicknesses. Generally, in the case for "soft" X-rays, when there is high absorption, the magnitude of the reduced X-ray absorption effect is largest, in agreement with the experimental results of Paoletti and co-workers, and Small [10, 11].

3.3. Correction factors, k-ratios

Our investigation was performed to develop specific thickness correction parameters for SEM-EDS quantitative analysis of asbestos minerals. We report in table 4 the *k*-ratios (intensity ratio of X-rays emitted by the asbestos mineral fibre to those produced by a massive sample) calculated for single cylindrical fibres of anthophyllite, tremolite and actinolite with diameter of 0.1 μm, 0.5 μm, 1 μm and 2 μm, in the case of an electron beam energy of 15 keV.

Figure 4 shows the trend of *k*-ratios calculated in the case of anthophyllite asbestos as a function of some specific parameters. Figure 4a shows the *k*-ratios calculated for the Kα X-ray lines of a cylindrical fibre of anthophyllite with diameter of 1 μm, as a function of the electron beam energies considered in this work (5 keV, 15 keV, 25 keV). The *k*-ratios have specific values depending on the X-ray line and electron beam energy. At a beam energy of 5 keV, the intensity of X-rays emitted by the asbestos mineral fibre is comparable to those produced by a massive sample, for the excitable lines. As a comparison, figure 4b shows the results of the same calculations in the case of an anthophyllite bundle of fibre of square section. In general, a higher value of the *k*-ratios calculated for the bundle of square section compared to the circular fibre can be noted (from percentage fractions to several % depending on the beam energy). In figure 4c the trend of the *k*-ratios for an anthophyllite cylindrical fibre at an electron beam energy of 15 keV is reported as a function the fibre diameter in

IOP Conf. Series: Materials Science and Engineering **304** (2017) 012019 doi:10.1088/1757-899X/304/1/012019

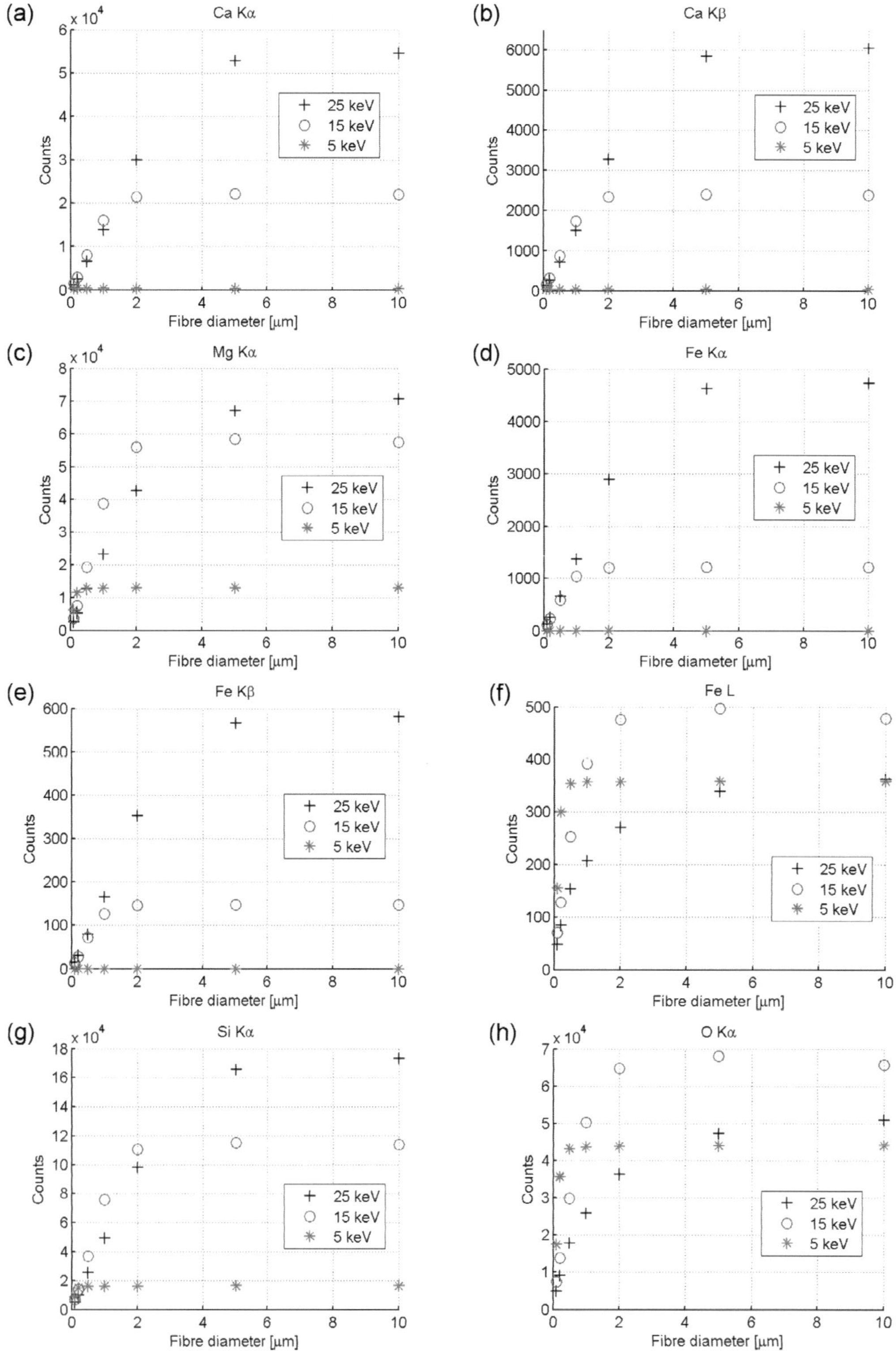

Figure 3. Simulated SEM-EDS X-ray intensity as a function of thickness for a single tremolite cylindrical fibre with beam energy of 5 keV, 15 keV and 25 keV; see legend in (h).

IOP Conf. Series: Materials Science and Engineering **304** (2017) 012019 doi:10.1088/1757-899X/304/1/012019

Table 4. *k*-ratios calculated at 15 keV for cylindrical fibres of different thickness.

Mineral	Thickness	X-ray lines							
		$Ca_{K\alpha}$	$Ca_{K\beta}$	$Fe_{K\alpha}$	$Fe_{K\beta}$	Fe_L	$Mg_{K\alpha}$	$Si_{K\alpha}$	$O_{K\alpha}$
Anthophyllite	2.0 μm	—	—	0.99	0.99	1.06	0.99	0.98	1.02
	1.0 μm	—	—	0.88	0.88	0.83	0.71	0.70	0.73
	0.5 μm	—	—	0.50	0.50	0.53	0.36	0.34	0.39
	0.1 μm	—	—	0.10	0.10	0.15	0.07	0.07	0.09
Tremolite	2.0 μm	0.98	0.98	0.99	0.99	0.98	0.97	0.97	0.98
	1.0 μm	0.73	0.73	0.86	0.86	0.81	0.67	0.67	0.76
	0.5 μm	0.36	0.36	0.49	0.49	0.52	0.33	0.32	0.45
	0.1 μm	0.07	0.07	0.09	0.09	0.15	0.06	0.06	0.11
Actinolite	2.0 μm	0.96	0.96	0.98	0.98	1.04	0.97	0.96	1.01
	1.0 μm	0.76	0.76	0.88	0.88	0.84	0.71	0.70	0.78
	0.5 μm	0.38	0.38	0.51	0.51	0.55	0.36	0.34	0.46
	0.1 μm	0.07	0.07	0.10	0.10	0.16	0.07	0.06	0.12

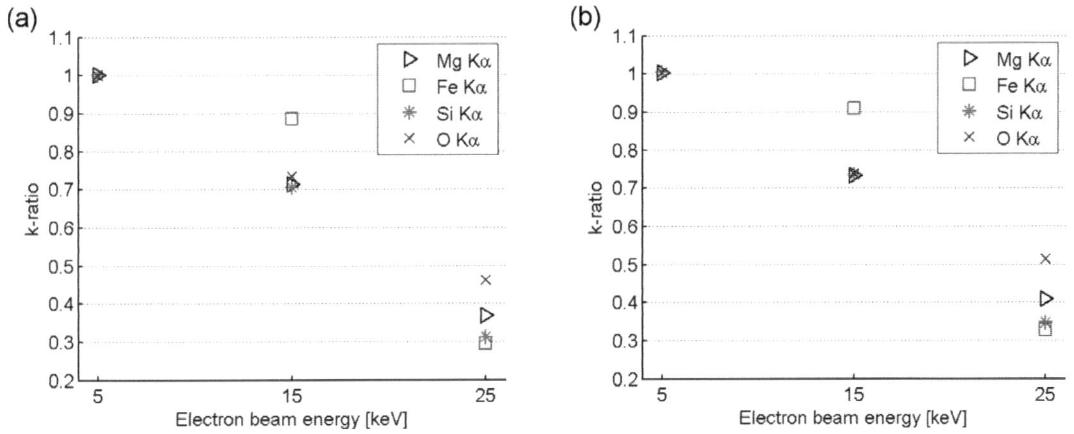

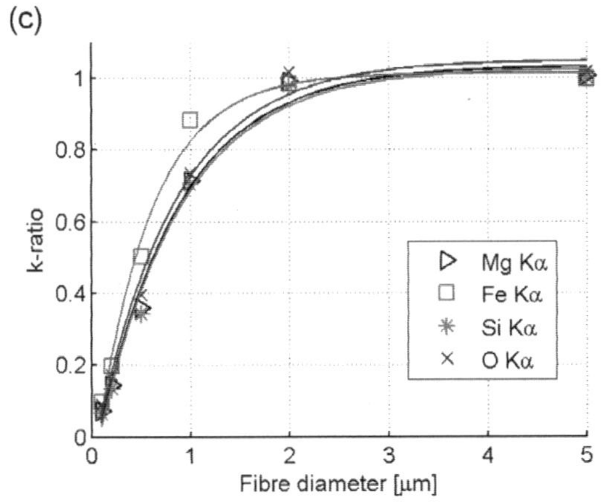

Figure 4. Calculated k-ratios. a) Anthophyllite cylindrical fibre with a diameter of 1 μm. b) Anthophyllite bundle of fibres of square section and thickness of 1 μm. c) Anthophyllite cylindrical fibres at 15 keV in function of thickness.

the thickness range (0.1 – 5 μm) which is affected by the lowering of the X-ray intensity. The calculated data are highlighted by markers, whereas an exponential fit is shown by a solid line for each element. It should be remembered that the k-ratios were calculated for data simulated in our specific experimental conditions for the mineral chemistries presented in table 1, taking into account the detector characteristics and SEM experimental set-up.

4. Conclusions

The calculation of detected EDS X-ray intensity in the case of anthophyllite, tremolite and actinolite asbestos bundles and fibres revealed a strong influence of the mineral thickness and shape. In particular, the mineral fibre thickness was found to be the main factor affecting the variation of the simulated X-ray intensity. It should be noted that an X-ray intensity enhancement (reduced absorption effect) was often observed before the X-ray intensity reduction starting point, with a magnitude that depends on the mineral type, shape, specific element and beam energy. A non-linear dependence of the X-ray intensity trends as a function of thickness was observed for all the elements and minerals. In general, the X-ray intensity showed a considerable reduction for thicknesses below about 5 μm at 25 keV, 2 μm at 15 keV and 0.5 μm at 5 keV, as the excitation volume is greatly reduced by using low accelerating voltages. The Monte Carlo simulation method has proved to be an effective tool to demonstrate the effects related to the size and shape of micrometric and sub-micrometric particles in SEM-EDS microanalysis allowing the calculation of correction *k*-ratios factors and functions. The simulation approach appears less complex and cumbersome than experimental ones.

References

[1] Gunter M E, Belluso E and Mottana A 2007 Amphiboles: environmental and health concerns. in: *Amphiboles: crystal chemistry, occurrence, and health sssues.* (*Reviews in Mineralogy & Geochemistry* vol. 67) (Hawthorne F C, Oberti R, Della Ventura G and Mottana A; Eds.) (Chantilly, VA: Mineralogical Society of America) chapter 12, 453-516
[2] Berstein D M and Hoskin J A 2006 *Regul. Toxicol. Pharmacol.* **45** 252
[3] Berstein D M 2014 Serpentine and amphibole asbestos. in: *Inhalation toxicology. Third edition.* (Salem H and Katz S A; Eds.) (Boca Raton, FL: CRC Press) chapter 14, 295-326
[4] Hardy J A and Aust A E 1995 *Chem. Rev.* **95** 97
[5] Gazzano E, Turci F, Foresti E, Putzu M G, Aldieri E, Silvagno F, Lesci I G, Tomatis M, Riganti C, Romano C, Fubini B, Roveri N and Ghigo D 2007 *Chem. Res. Toxicol.* **20** 380
[6] Yao S, Della Ventura G and Petibois C 2010 *Anal. Bioanal. Chem.* **397** 2079
[7] Bocchi G and Valdrè G 1993 *J. Inorg. Biochem.* **49** 209
[8] Borgia G C, Brown R J S, Fantazzini P, Mesini E and Valdrè G 1992 *Nuovo Cimento* **14** 745
[9] Gatti A M, Valdrè G and Tombesi A 1996 *J. Biomed. Mater. Res.* **31** 475
[10] Small J A 2002 *J. Res. Natl. Inst. Stand. Technol.* **107** 555
[11] Paoletti L, Bruni B M, Gianfagna A, Mazziotti-Tagliani S and Pacella A 2011 *Microsc. Microanal.* **17** 710
[12] Paoletti L, Bruni B M, Arrizza L, Mazziotti-Tagliani S and Pacella A 2008 *Per. Mineral.* **77** 63
[13] Armstrong J T 1991 Quantitative elemental analysis of individual microparticles with electron beam instruments. in: *Electron probe quantification.* (Heinrich K J F and Newbury D E; Eds.) (New York, NY: Plenum Press) 261-315
[14] Merlet C and Llovet X 2012 *IOP Conf. Series: Mater. Sci. Engng.* **32** 012016
[15] Merlet C and Llovet X 2011 *X-ray Spectrom.* **40** 47
[16] Salvat F, Llovet X, Fernández-Varea J M and Sempau J 2006 *Microchim. Acta* **155** 67
[17] Moro D and Valdrè G 2016 *IOP Conf. Series: Mater. Sci. Engng.* **109** 012011
[18] NIST DTSA-II by Ritchie N W M: www.cstl.nist.gov/div837/837.02/epq/dtsa2/index.html
[19] Ritchie N W M 2009 *Microsc. Microanal.* **15** 454
[20] Skinner H C W 2003 *Indoor Built Environ.* **12** 385
[21] Wiewióra A 1990 *Clay Min.* **25** 93

[22] Hawthorne F C and Oberti R 2007 Amphiboles: crystal chemistry. in: *Amphiboles: crystal chemistry, occurrence, and health issues.* (*Reviews in Mineralogy & Geochemistry* vol. 67) (Hawthorne F C, Oberti R, Della Ventura G and Mottana A; Eds.) (Chantilly, VA: Mineralogical Society of America) chapter 1, 1-54

[23] Hawthorne F C and Oberti R 2007 Classification of the amphiboles. *Amphiboles: crystal chemistry, occurrence, and health issues.* (*Reviews in Mineralogy & Geochemistry* vol. 67) (Hawthorne F C, Oberti R, Della Ventura G and Mottana A; Eds.) (Chantilly, VA: Mineralogical Society of America) chapter 2, 55-88

[24] Hawthorne F C, Oberti R, Harlow G E, Maresch W V, Martin R F, Schumacher J C and Welch M D 2012 *Am. Miner.* **97** 2031

[25] Myklebust R, Newbury D and Yakowitz H 1976 *NBS Special Publication* 460 (Heinrich K, Yakowitz H and Newbury D; Eds.) (Washington, DC: National Bureau of Standards) 105

[26] Czyzewski Z, MacCalium D O'N, Romig A and Joy D C 1990 *J. Appl. Phys.* **68** 3066

[27] Jablonski A, Salvat F and Powell C J 2010 *NIST electron elastic-scattering cross-section database*. (Gaithersburg, MD: National Institute of Standards and Technology)

[28] Joy D C and Luo S 1989 *Scanning* **11** 176

[29] Bethe H A and Ashkin J 1953 Passage of radiations through matter. in: *Experimental nuclear physics, vol. 1.* (Segrè E; Ed.) (New York, NY: Wiley)

[30] Bote D and Salvat F 2008 *Phys. Rev. A* **77** 042701

[31] Chantler C T, Olsen K, Dragoset R A, Chang J, Kishore A R, Kotochigova S A, Zucker D S 2005 *NIST Standard Reference Database* version 2.1. Available at http://physics.nist.gov/ffast

[32] Bambynek W, Crasemann B, Fink R W, Freund H U, Mark H, Swift C D, Price R E, Rao P V 1972 *Rev. Mod. Phys.* **44** 716

[33] Ritchie N W M 2010 *Microsc. Microanal.* **16** 248

IOP Conf. Series: Materials Science and Engineering **304** (2017) 012020 doi:10.1088/1757-899X/304/1/012020

Synthesis and characterisation of $PuPO_4$ - a potential analytical standard for EPMA actinide quantification

K E Wright[1,2], K Popa[1] and P Pöml[1]

[1] European Commission, Joint Research Centre, Directorate G – Nuclear Safety and Security, P.O. Box 2340, 76125 Karlsruhe, Germany
[2] Idaho National Laboratory, P.O. Box 1625 MS 6157,US-83415 Idaho Falls, ID, U.S.A.

E-mail: karen.wright@inl.gov

Abstract. Transmutation nuclear fuels contain weight percentage quantities of actinide elements, including Pu, Am and Np. Because of the complex spectra presented by actinide elements using electron probe microanalysis (EPMA), it is necessary to have relatively pure actinide element standards to facilitate overlap correction and accurate quantitation. Synthesis of actinide oxide standards is complicated by their multiple oxidation states, which can result in inhomogeneous standards or standards that are not stable at atmospheric conditions. Synthesis of $PuPO_4$ results in a specimen that exhibits stable oxidation-reduction chemistry and is sufficiently homogenous to serve as an EPMA standard. This approach shows promise as a method for producing viable actinide standards for microanalysis.

1. Introduction

Nuclear fuels are being developed for next-generation nuclear reactors (GEN-IV) that have the ability to burn Pu, Am, and Np from spent nuclear fuels. Upon irradiation, these so-called transmutation fuels convert long-lived minor actinide elements (MA) to shorter-lived fission products, thereby decreasing the radiotoxicity and heat load of spent nuclear fuel.

The thermodynamic behaviour of MA is unlike that of uranium in conventional nuclear fuels. Moreover there is much greater variation in the chemical and isotopic composition of transmutation fuels, such that they cannot be considered to be a simple extension of extant fuels [1].

Because of these features, considerable research and development is required for both fabrication and analysis of transmutation fuels. Integral to the fabrication and analysis of irradiated fuel is electron probe microanalysis (EPMA). This technique has been used since at least 1961 [2] for the analysis of nuclear fuels because wavelength dispersive spectrometers permit better separation of complex peaks, such as those exhibited by the actinide metals group, than is possible with energy dispersive detectors.

Reference materials are necessary to calibrate the instrument in order to obtain an accurate MA measurement. While some progress has been made with regard to creating models for standardless analysis [3], such efforts are still in their infancy. Therefore, nuclear fuels researchers still rely on the use of physical standards for the analysis of MA. Unfortunately, fabrication and use of such standards is fraught with difficulty. All fabrication and polishing must be done in special nuclear-grade facilities. Such facilities typically include glove boxes with prescribed atmospheres, and remote handling tools. Metal specimens are relatively easy to handle, but when exposed to oxygen, will oxidise very rapidly.

This causes a large dimensional change to the specimen, and it will begin to spall, resulting in radioactive contamination. Oxides have been commonly employed because they are more stable under atmospheric conditions; however, they are difficult to fabricate. Actinide elements can exist in many oxidation states [4], therefore the atmosphere and temperature control during fabrication is crucial for creating stable oxide compounds [5, 6]. Incorrect atmosphere or temperature control during synthesis can result in an oxide compound that is not homogeneously oxidised and/or will not remain at that oxidation state when removed from the glove box atmosphere.

Lanthanide orthophosphates can crystallise in two different forms. The lighter lanthanides, $LaPO_4$ - $GdPO_4$ precipitate as a monoclinic structure analogous to monazite (a naturally occurring light rare earth mineral) while the heavier lanthanides, $TbPO_4$ - $LuPO_4$ precipitate as a tetragonal structure, analogous to zircon [7]. Experiments with Pu, Am, Cm, Bk, Cf and Es show that these elements precipitate with a monoclinic structure [7].

Several methods exist to make monoclinic orthophosphates. Begun [8] described dissolving and reacting lanthanide (Ln) oxides with molten lead orthophosphate at high temperature and then separating the crystals from the lead phosphate matrix by dissolving the matrix in boiling nitric acid. Cherniak *et al.* [9] created Ln-phosphates by dissolving Ln-nitrates in distilled water, adding $NH_4H_2PO_4$ and heating the solution on a hot plate. The ensuing precipitate was then mixed with a flux consisting of MoO_3 and either Li_2CO_3 or Na_2CO_3 and heated to at least 1,280 °C. Bjorklund [10] created monoclinic $PuPO_4$ via the precipitation and high temperature dehydration of hexagonal $PuPO_4(0.5\ H_2O)$, and Jardin *et al.* [11] prepared $PuPO_4$ by sol-gel reaction starting with a Pu (+4) solution that was mixed with $(NH_4)_2HPO_4$ to create PuP_2O_7. This compound was then reduced with H_2 to create $PuPO_4$.

While actinides sometimes do behave like lanthanides, Pu is known to decompose above 1,000 °C [11], thus the Begun [8] and Cherniak *et al.* [9] methods were not likely to create a homogeneous $PuPO_4$ specimen. A solid state method based on Bregiroux *et al.* [12] was chosen because PuO_2 was available as a starting material and this method demonstrated that single phase monoclinic $PuPO_4$ could be fabricated relatively easily via sintering in an argon atmosphere. While this method has been employed for structural investigations [13], it has not been tested for its ability to make crystals that are homogeneous on a micrometre spatial scale.

Herein we describe a method for fabricating stable, highly homogeneous actinide phosphate specimens for use as EPMA standards.

2. Methods

2.1. *$PuPO_4$ synthesis*

PuO_2 with ~2 % AmO_2 resulting from beta decay of ^{241}Pu was ground and mixed with $(NH_4)_2HPO_4$, placed into an alumina crucible, and heated slowly to 1,000 °C in an argon glove box (99.96 % purity with < 7 ppm H_2O). Heating and cooling ramps of 100 °C/h were employed with a 12 hour dwell at the target temperature. A dark violet powder was produced at the conclusion of the heating cycle.

X-ray diffraction of the powder shows that the material is a pure phase and has a monoclinic structure (figure 1), with valence states of Pu, P, and O atoms approximating +3, +5, and -2, respectively [13]. This was confirmed with X-ray absorption near edge structure (XANES) spectra [13]. This structure is similar to that of monoclinic monazite.

2.2. *Preparation of the standard*

A 7-mm disk of $PuPO_4$ was prepared by pressing. A second thermal treatment was applied wherein the pressed disk was sintered at 1,000 °C for 72 hours in an argon atmosphere. The sintered disk was then sectioned and a small piece was selected, mounted in epoxy, and polished to 1 μm (figure 2a). The sample was coated with approximately 20 nm aluminium prior to electron probe examination. While an aluminium coating is not a traditional choice for microprobe analysis, it is useful for nuclear applications for several reasons. Carbon coatings are known to be inadequate under high beam currents

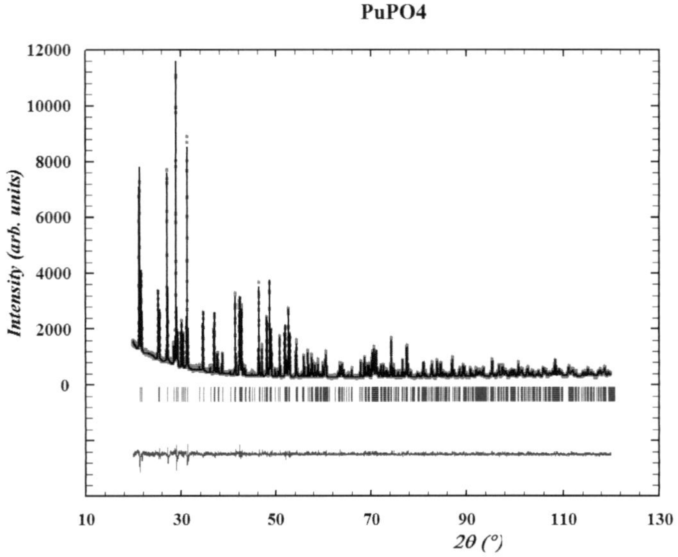

Figure 1. Comparison between with observed (red) and calculated (black) X-ray diffraction pattern of the $PuPO_4$ phase. Blue vertical ticks represent Bragg reflections, while the observed - calculated pattern is shown in blue at the bottom of the figure.

used in trace element acquisition [14], therefore, a more robust coating (one that has higher thermal and electrical conductivity) is required. While gold would be adequately robust, the porous nature of nuclear materials sometimes necessitates the use of thicker coatings (e.g., 20 nm or more) to ensure adequate conductivity across the entire surface. Such thick gold coatings would highly absorb lower energy X-rays, therefore aluminium is a more suitable choice. Since nuclear specimens are often coated and handled in an inert gas environment, the coating is not expected to oxidise during the analysis. Nuclear materials also have the benefit of not usually having aluminium as a constituent, therefore aluminium is a feasible choice for nuclear scientists, whereas it would be more problematic for geologists.

2.3. Analysis of the Standard

A CAMECA SX100R EPMA was used to analyse the sample. Five 120 μm x 50 μm regions were selected for quantitative point analysis (figure 2a), with 30 points measured in each region. An accelerating voltage of 25 kV and current of 50 nA was used on each point, and the beam was focussed. Pu-Mα and Am-Mα were measured at each point. The Pu-peak was measured for 40 seconds while the Am-peak was measured for 200 seconds. P and O were not measured, but were included in the analysis as specified constituents. Probe for EPMA v. 11.7.1 with full PAP [15] phi-rho-z correction was used. Analysis locations were not chosen randomly, but rather were selected based on the size of the crystal and proximity to obvious defects (e.g., pores) (figure 2b). Despite the fact the P and O were not measured, the average deviation (N = 150) from perfect charge balance was 0.16 %.

IOP Conf. Series: Materials Science and Engineering **304** (2017) 012020 doi:10.1088/1757-899X/304/1/012020

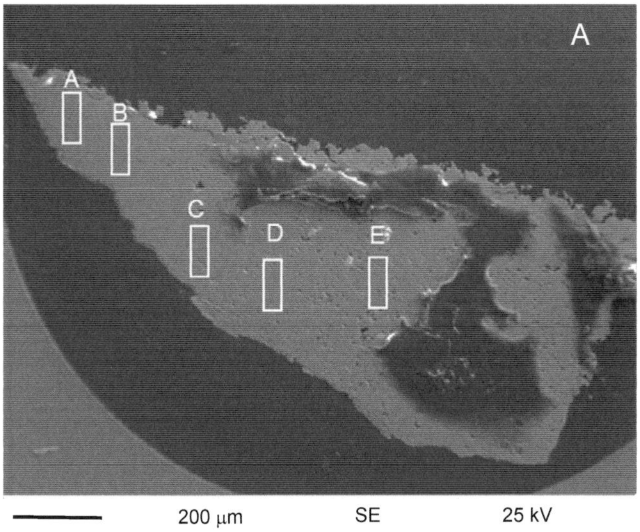

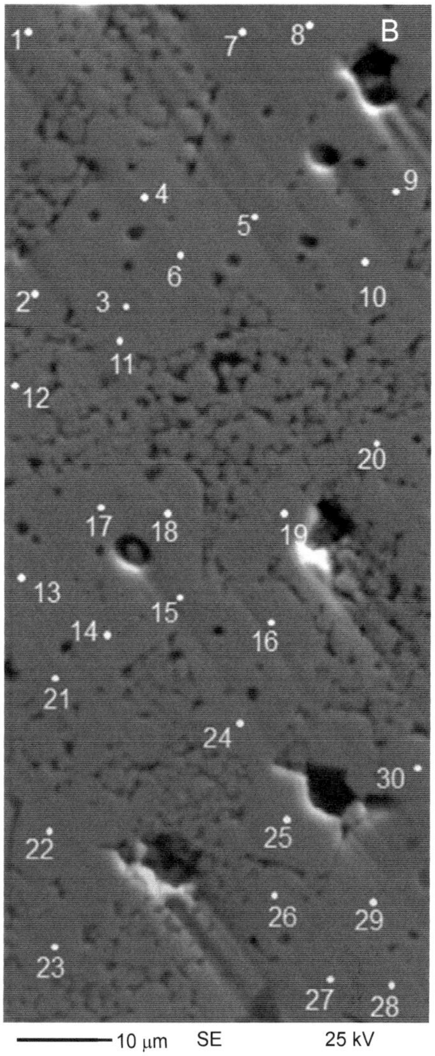

Figure 2. a) Secondary electron image of the mounted, polished specimen showing five regions selected for analysis. Each region (A-E) contained 30 analysis points. Some epoxy remains on the surface. b) Shows region D from figure 2a, with specific locations for the 30 selected points. Note the presence of relatively large grains, which provide numerous locations for reliable analysis.

3. Results

Table 1 shows a summary of statistics for the measured points. The level of homogeneity, (H) in weight percent is calculated from equation (1)

$$(H) = \pm \frac{100\left(t^{1-a}_{n-1}\right)Sc}{\sqrt{n}\,\bar{N}} (\%) \qquad (1)$$

where $\left(t\begin{matrix}1-a\\ n-1\end{matrix}\right)$ is the student's *t*-value for a 1-α confidence level and *n*-1 degrees of freedom, *n* is the number of measurements, *Sc* is the standard deviation of the number of counts, and $\bar{N}$ is the average number of counts for the set of measurements. A homogeneous sample should have a level of homogeneity <1 [16], which is true for Pu-Mα in this sample. In addition, under 50 nA of current for 200 seconds, the sample demonstrated no visible damage, unlike LnP_5O_{14} specimens, which exhibit obvious pitting with a 20 nA, 30 second, diffuse beam.

Table 1. Shows the average number of counts, standard deviation, average weight percent, a comparison of measured to predicted concentrations, and level of homogeneity for Pu-Mα and Am-Mα.

Parameter	Pu Mα	Am Mα
Mean Counts (150 points)	130565	10271
Standard Deviation	1665	738
Mean Weight Percent (150 points)	70.20	1.50
Measured Concentration/Predicted Concentration	0.9952	1.042
Standard Deviation	0.85	0.11
Level of Homogeneity	0.21	1.15

While the $PuPO_4$ is slightly less homogeneous than the PuO_2 previously prepared and analysed by the Institute for Transuranium Elements (level of homogeneity of 0.12 [17]), the Pu oxidation state in $PuPO_4$ is extremely stable at temperatures < 1,000 °C and in oxygen atmospheres. In addition, oxide standards often have small grains and are more porous than $PuPO_4$, which makes finding suitable analysis locations difficult.

4. Implications and future work

While the standard produced is sufficiently homogeneous to serve as a Pu reference material, there is opportunity to improve the quality of the crystals. Many researchers (e.g., Bregiroux *et al.* [12]; Cherniak *et al.* [9]) suggest cooling rates on the order of 3 - 4 °C per hour, rather than the 100 °C per hour that was used in this study. Unfortunately the latter rate was the slowest possible using the available furnace. We do think that a slower cooling rate could produce significantly larger crystals, thus improving the quality of the specimen.

Given the success in creating homogeneous $PuPO_4$, naturally the question arises as to whether other MA standards can be fabricated in this manner. Bregiroux *et al.* [12] showed that $^{241}AmPO_4$ could indeed be made by this route. However, ^{241}Am, which has a half-life of 432.2 years, caused the crystal to become amorphous within 300 days. If the crystal remains reasonably homogeneous and robust despite the lack of crystallinity and ingrowth of decay products, this method could be viable. Tetravalent actinides, such as Np and Th can be synthesized as a cheralite analogue $(M^{2+} M'^{4+})(PO_4)_2$ [18]. Testing is required to determine if the cheralite analogue is sufficiently homogeneous to serve as a microprobe reference material.

5. Conclusions

- The plutonium concentration is homogeneously distributed in this standard; however, care needs to be taken to select large, non-porous regions in which to make measurements.
- Americium's level of homogeneity was >1, suggesting that the element is not homogeneously distributed. However, significantly more counts are required to draw a robust conclusion.
- The grains are larger, and the porosity lower than that of actinide oxide standards that are typically used. Using actinide phosphates like the present example is a promising new method to obtain high-quality standards for difficult-to-prepare actinide materials.

Acknowledgements

The authors would like to thank Mairead Murray for sample preparation. They would also like to express their gratitude to their respective institutions, Directorate G – Nuclear Safety and Security (European Commission, Joint Research Centre) and Idaho National Laboratory for providing the means for K. Wright to serve as a visiting scientist to the former. Finally we thank Dr. Julien Allaz, University of Colorado, Boulder, and an anonymous reviewer for their excellent suggestions, which greatly improved this manuscript.

References

[1] Carmack J and Pasamehmetoglu K 2008 *Review of transmutation fuel studies.* Idaho National Laboratory Report, INL/EXT-08-13779

[2] Lévy V, Kirianenko A, Brebec G and Adda Y 1961 Contribution á l'étude de la précipitation des gaz rares dans les métaux. *C.R. Acad. Sci. Paris* **252** 876-878

[3] Moy A, Merlet C and Dugne O 2015 Standardless quantification of heavy elements by electron probe microanalysis. *Anal. Chem.* **87** 7779-7786

[4] Lieser K 2000 *Nuclear and radiochemistry. Second edition.* (Darmstadt, Germany: Wiley-VCH) 462

[5] Tanaka K, Yoshimocki H, Obayaski H and Koyama S 2015 Oxidation behavior of Am-containing MOX fuel pellets in air. Energy Procedia 71 282-292

[6] Vaucy R, Belin R, Robisson A, Lebreton F, Aufore L, Scheinost A and Martin P 2016 Actinide oxidation state and O/M ratio in hypostoichiometric uranium-plutonium-americium $U_{0.750}Pu_{0.246}Am_{0.004}O_{2-x}$ mixed oxides. *Inorg. Chem.* **55** 2123-2132

[7] Hobart D, Begun G, Haire R, and Hellwege H 1983 Raman spectra of the transplutonium orthphosphates and trimetaphosphates. *J. Raman Spectrosc.* **14** 59-62

[8] Begun G, Beall G, Boatner L, and Gregor W 1981 Raman spectra of the rare earth orthophosphates. *J. Raman Spectrosc.* **11** 273-278

[9] Cherniak D, Pyle J, and Rakovan J 2004 Synthesis of REE and Y phosphates by Pb-free flux methods and their utilization as standards for electron microprobe analysis and in design of monazite chemical U-Th-Pb dating protocol. *Am. Mineral.* **89** 1533-1539

[10] Bjorklund C 1957 The preparation of PuP_2O_7 and $PuPO_4$. *J. Am. Chem. Soc.* **79** 6347-6350

[11] Jardin R, Pavel C, Raison P, Bouëxiére D, Santa-Cruz H, Konings R and Popa K 2008 The high-temperature behaviour of $PuPO_4$ monazite and some other related compounds. *J. Nuc. Mater.* **378** 167-171

[12] Bregiroux D, Belin R, Valenza P, Audubert F and Bernache-Assollant D 2007 Plutonium and americium monazite materials: solid state synthesis and X-ray diffraction study. *J. Nuc. Mater.* **366** 52-57

[13] Popa K, Raison P, Martel L, Martin P, Prieur D, Solari P, Bouëxiére D, Konings R and Somers J 2015 Structural investigations of Pu^{III} phosphate by X-ray diffraction, MAS-NMR and XANES spectroscopy. *J. Solid State Chem.* **230** 169-174

[14] Jercinovic M and Williams M 2005 Analytical perils (and progress) in electron microprobe trace element analysis applied to geochronology: background acquisition, interferences, and beam irradiation effects. *Am. Mineral.* **90** 526-546

[15] Pouchou J and Pichoir F 1991 in: *Electron probe quantitation.* (Heinrich K F J and Newbury D E; Eds.) (New York, NY: Plenum Press) 31-76

[16] Goldstein J , Newbury D , Echlin P, Joy D, Romig Jr. A, Lyman C, Fiori C and Lifshin E 1992 *Scanning electron microscopy and X-ray microanalysis, 2nd edition.* (New York and London: Plenum Press)

[17] Ritter X 2015 Micro-analytical investigations on actinide reference materials. Master's Thesis. (Münster, Germany: Westfälische Wilhelms Universität)

[18] Bregiroux D, Popa K and Wallez G 2015 Crystal chemistry of $M^{II}M'^{IV}(PO_4)_2$ double monophosphates. *J. Solid State Chem*, **230** 26-33

IOP Conf. Series: Materials Science and Engineering **304** (2017) 012021 doi:10.1088/1757-899X/304/1/012021

A new life for the wavelength-dispersive X-ray spectrometer (WDS): incorporation of a silicon drift detector into the WDS for improved quantification and X-ray mapping

R Wuhrer[1] and K Moran[2]

[1] Western Sydney University, Advanced Materials Characterisation Facility (AMCF), P.O. Bag 1797, 2751 Penrith, NSW, Australia
[2] Moran Scientific Pty Ltd, 4850 Oallen Ford Road, 2580 Bungonia, NSW, Australia

E-mail: richard.wuhrer@westernsydney.edu.au

Abstract. The wavelength-dispersive X-ray spectrometer (WDS) has been around for a long time and the design has not changed much since its original development. The electron microprobe operator using WDS has to be meticulous in monitoring items such as gas flow, gas purity, gas pressure, noise levels of baseline and window, gas flow proportional counter (GFPC) voltage levels, count rate suppression, anode wire contamination and other detector parameters.

Recent development and improvements of silicon drift detectors (SDD's) has allowed the incorporation of a SDD as the X-ray detector in place of the proportional counter (PC) and/or gas flow proportional counter (GFPC). This allows minimal mechanical alteration and no loss of movement range.

The superiority of a WDS with a SDD, referred to as SD-WDS, is easily seen once in operation. The SD-WDS removes many artefacts including the worse of all high order diffraction, thus allowing more accurate analysis. The incorporation of the SDD has been found to improve the light and mid element range and consequently improving the detection limit for these elements. It is also possible to obtain much more reliable results at high count rates with almost no change in resolution, gain and zero-peak characteristics of the energy spectrum.

1. Introduction

The electron probe microanalyser (EPMA) instrument has been around for a long time, which utilises the X-ray microanalysis techniques of wavelength-dispersive and energy-dispersive X-ray spectrometry (WDS and EDS). The WDS design has not changed much since its original development. WDS exhibits far better spectral resolution than the conventional EDS. The combination of EDS and WDS produces a very powerful analytical technique allowing for excellent quantitative X-ray mapping (QXRM). However, the electron microprobe operator using WDS has to be meticulous in monitoring items such as gas flow, gas purity, gas pressure, noise levels of baseline and window, gas flow proportional counter (GFPC) voltage levels, count rate suppression, anode wire contamination and other detector parameters.

Over the last couple of years we have been developing and testing alternative configuration WD spectrometers, using Amptek PIN (p-intrinsic-n type diode) and silicon drift detectors (SDD's) in place of the sealed proportional counter (SPC) and gas flow proportional counters (GFPC's) respectively.

The new system configurations allow the SDD detector to be incorporated within the WD spectrometer with minimal mechanical alteration and no loss of movement range by directly replacing the gas counter with the SDD sensor and connecting to the electronics outside the spectrometer housing using a 9-pin feedthrough.

Our current research demonstrates the dramatic superiority of SDD over proportional counters (PC), whether gas flow or sealed. The very low energy resolution of the PC means that it is not really known what X-rays are being counted by the detector, for example from secondary fluorescence and other X-rays (especially high order lines). Even the application of pulse height discrimination can usually only partially remove higher order lines, and at the expense of greatly increased sensitivity of the gas counter to temperature or pressure changes. Secondary fluorescence X-rays are often similar on the peak as on the background and so may be removed by background subtraction, but the peak/background (P/B) ratio can be significantly reduced. The superior energy resolution of a SDD, typically an order of magnitude better than PC, immediately brings to light such problems, and allows for them to be more effectively removed from the measured signal.

The superiority of a WDS with a SDD, referred to as SD-WDS, is easily seen once in operation by removing many artefacts, allowing for more accurate analysis. The incorporation of a SDD, in the light and mid element range, allows improvement of the detection limit for these elements. It is also possible to obtain much more reliable results at high count rates with almost no change in resolution, gain and zero-peak characteristics of the energy spectrum.

2. Microanalysis (EDS and WDS)

Electron probe microanalysis (EPMA) utilises the X-ray microanalysis techniques of wavelength-dispersive X-ray and energy-dispersive X-ray spectrometry (WDS and EDS) to measure the characteristic X-rays generated from a sample bombarded with electrons and thereby identify the elements existing in the sample. The combination of EDS and WDS produces a very powerful analytical technique also allowing for excellent quantitative X-ray mapping (QXRM) [1-5]. Traditionally, the EDS is used for fast elemental analysis to determine what elements are present and the WDS then utilised to acquire precise major and trace elemental quantitative analysis. The WD spectrometers exhibit far better spectral resolution than the conventional EDS, as can be seen in figure 1 [1, 6].

For major element analysis, the EDS and WDS have the same electron volume for X-ray production, but the X-rays arrive at each detector through different azimuthal paths.

3. Conventional WDS design

WDS is made up of two components; a crystal which discriminates X-rays according to their wavelength, and a detector which measures the intensity of X-rays being passed by the crystal. Each part has its own characteristics, so we will look at each in turn before considering the overall performance of the spectrometer (figure 2) [7-9].

The WD spectrometer, also sometimes referred to as Bragg's spectrometer, consists of a crystal with the diffracting planes parallel to the surface. The crystal is mounted on a shaft, with an X-ray detector on an arm geared to the shaft in the ratio 2:1 to maintain equal angles of incidence and reflection (figure 2) [6, 9].

At a given angle (ignoring high order reflections), Bragg reflection is satisfied for specific wavelengths only. The detector records the intensity of the incoming X-ray signal at that point of the spectrum. The spectrometer therefore, acts as a monochromator. Different wavelengths are selected by changing theta (θ) of the diffraction crystal [6, 9, 10].

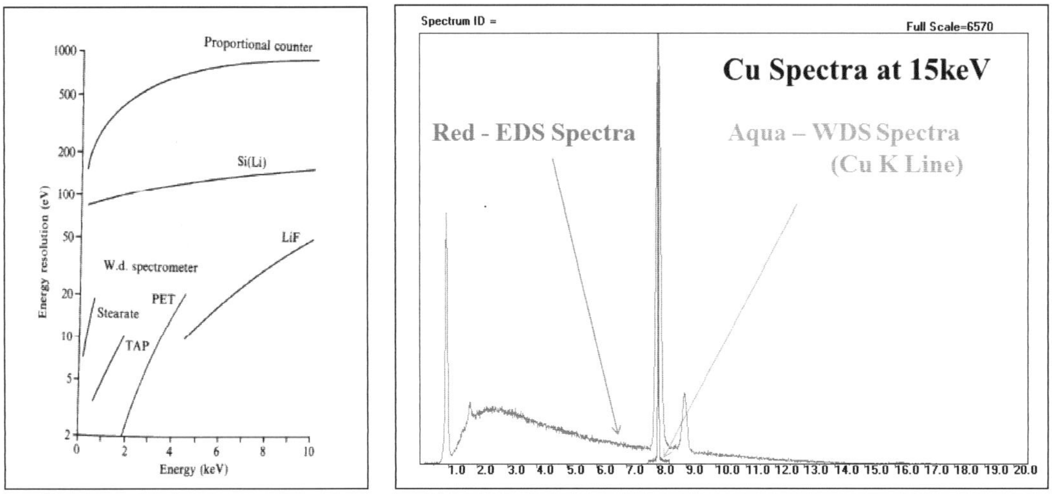

Figure 1. a) Comparison of EDS and WDS energy resolution. WD spectrometer with a proportional counter using different crystals [6], and b) EDS spectrum (red) that has a Cu-Kα line resolution of ~140 eV with a WDS spectrum overlayed (aqua) with a resolution of ~40 eV.

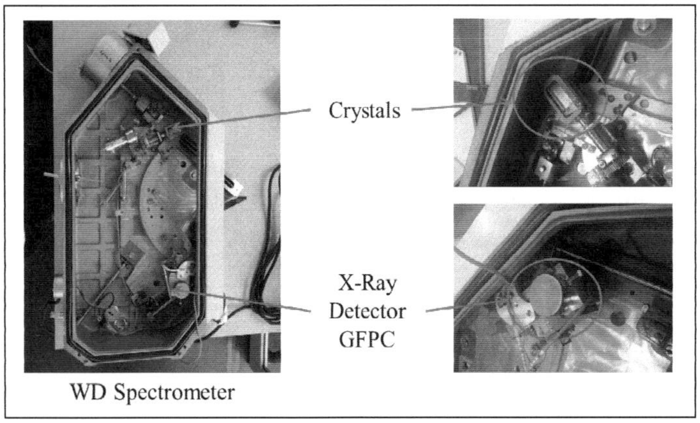

Figure 2. Images of a conventional JEOL WD spectrometer showing the internal mechanism consisting of the gas flow proportional counter (GFPC) and crystal(s).

4. Diffraction crystals

X-rays incident on a crystal are coherently scattered by the atomic electrons. At a certain angle of incidence (θ), X-rays of a given wavelength scattered by atoms in parallel planes, are in phase, causing strong "reflection" (figure 3). Thus for diffraction (electromagnetic alignment) Bragg gave us the formula shown in eq. 1, which gives us the condition at which this strong reflection occurs. It turns out that the Bragg condition is very sensitive and only a small deviation in the angle theta (θ) is enough to lose the constructive interference. This gives high spectral resolution. A number of different crystals are necessary to cover the full range of X-ray wavelengths for all elements in the range B-U [6, 9].

$$n\,\lambda = 2d\,\sin(\theta) \tag{1}$$

where n is the order of reflection (or integer multiple of wavelengths), λ is the X-ray's wavelength, d is the crystal interplanar spacing, and sin(θ) is the X-ray's angle of incidence. In this case, the angle of incidence and diffraction must be equal i.e., $\theta_i = \theta_r$. Where i is the incident and r is the reflected beam angle.

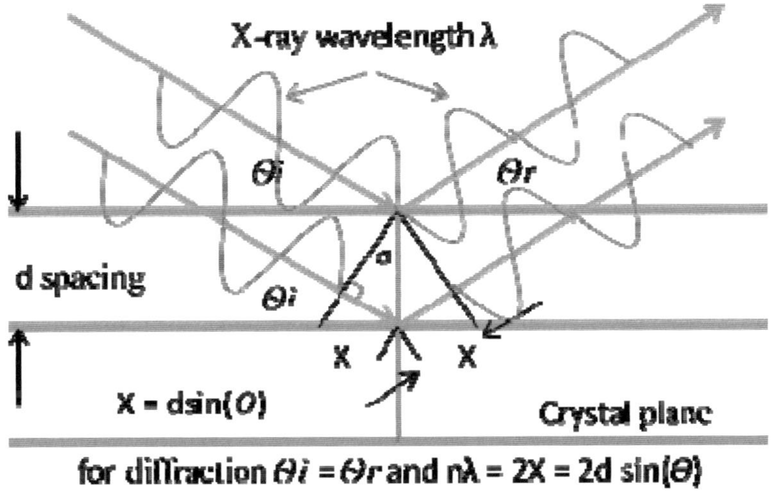

Figure 3. Diffraction according to Bragg's law. Coherent scattering of X-rays of wavelength nλ only occurs when 2 times X = the path difference = nλ = 2dsin(θ). Where incident = reflective angle theta (θ). Modified from [8, 9, 11].

Ideally the Bragg condition would only be satisfied for one wavelength, i.e., where the difference in path length between the diffracted waves is equal to one wavelength. Unfortunately, the Bragg condition is also satisfied if the path length difference is two, three, or in fact any integer value of the wavelength. This is what the 'n' term in the Bragg equation reflects. Although these higher multiple wavelength 'orders' rapidly decrease in intensity, it is not uncommon to detect 2nd, 3rd or even 4th order diffraction lines for major elements.

The spectrometer for the EMPA is arranged as a linear focussing type of spectrometer where the source to crystal distance (D) is equal to 2rsin(θ) and r is the radius of the Rowland circle. Then by substitution and assuming that n = 1, D = rλ/d. Since r and d are fixed for the spectrometer and diffracting crystal being used, this shows that λ varies proportionally with D.

Table 1 shows the four most commonly used crystals that give coverage for nearly all elements as stearate (ultra-light elements B - F), TAP, PET, and LIF. This table shows their full names, chemical formulae, energy range, elemental coverage and their nominal d-spacing's. For example, for X-ray energies higher than about 6.5 keV we can only use the LIF crystal. In other areas there is some overlap between the crystal ranges. The overlap between PET and LIF is particularly large. In these regions we can have the choice between two crystals, as can be seen in Table 1.

Another important parameter with respect to the diffraction crystal is whether it can be ground or not, as this determines the type of geometry of the spectrometer. In this selection of crystals, LIF is the only one easily ground. There are two types of analytical geometries currently in use. The first type, the Johann geometry, is constructed by bending the analytical crystal to a radius of 2R, where R is the radius of the focussing circle, the Rowland circle (figure 4) [8-11]. The second type, the Johansson geometry, is bent to a radius 2R and then ground to radius R, so that all of the points of reflection lie on the Rowland circle, which maximizes the collection efficiency of the spectrometer. Further to this, the sample, crystal and detector must lie on the Rowland circle and remain on it for all wavelengths of interest in order to focus X-rays efficiently [8-11].

Table 1. Some of the diffraction crystals used for WDS [6-17].

Crystal	Name *Formula*	2d (Å)	Wavelength range (Å)	Elemental range	Energy range keV	SDD window
LIF	Lithium Fluoride *LiF*	4.0267	0.863 – 3.595	Ca - Sr Sn - U	3.44 - 14.37	Be 0.5 mm
PET	Pentaerythritol $C(CH_2OH)_4$	8.742	1.873 – 7.805	Si - Fe Kr - Dy Hf - U	1.59 - 6.62	Be 0.3 mm
TAP	Thallium Acid Phthalate $CO_2HC_6H_4CO_2Tl$	25.78	5.525 – 23.02	F - P Cr - Nb La - Hg	0.54 - 2.24	C1 (90 nm Si_3N_4)
STE	Pb Stearate	100.7	21.57 – 89.91	B - O K - Cr	0.138 - 0.575	C2 (40 nm Si_3N_4)

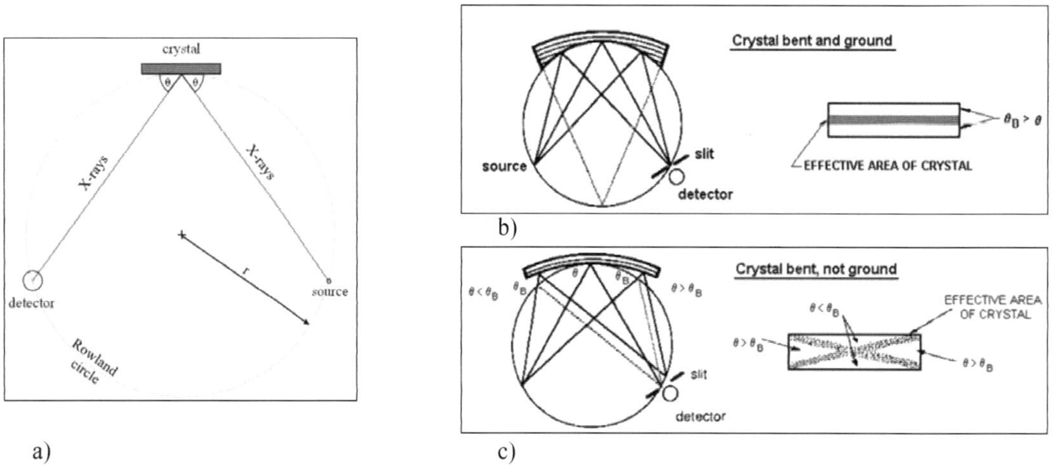

Figure 4. a) Schematic diagram illustrating the configuration of the X-ray source, analytical crystal and detector on the Rowland circle of the wavelength spectrometer [8], b) Johannson focussing geometry and effective crystal area from the source and through the slit, and c) Johann focussing geometry and effective crystal area [6, 9, 10].

A detector slit may be used at the focus of the X-rays in order to reduce stray radiation, and to block radiation of a different wavelength that may be diffracted either because of the focussing geometry or because of imperfect crystal fabrication. The resolving power of the spectrometer will increase with decreasing slit width, while the peak intensity will decrease with decreasing slit width. An adjustable slit has the advantage that the best slit width may be chosen for a given application. For maximum precision in detecting large signals, a large slit width would be utilised. For best detection limits the slit would be adjusted for maximum P^2/B. For studies of peak shifts due to chemical bonding very small slit widths would be chosen to obtain the best resolution [9].

5. X-ray detectors

A variety of X-ray detectors may in principle be used: these are the scintillation counter, Geiger-Mueller counter, gas filled proportional counter and solid state detector. Detectors used in WD spectrometers are most commonly gas proportional counter types in which incoming X-rays enter the detector through a collimator (slit) and thin window, are absorbed by atoms of the counter gas and a photoelectron is then ejected by each atom absorbing an X-ray (figure 5). The gas-filled proportional counter is preferred for use because of its efficiency, small dead time and good proportionality for the photons of interest [6, 10, 11].

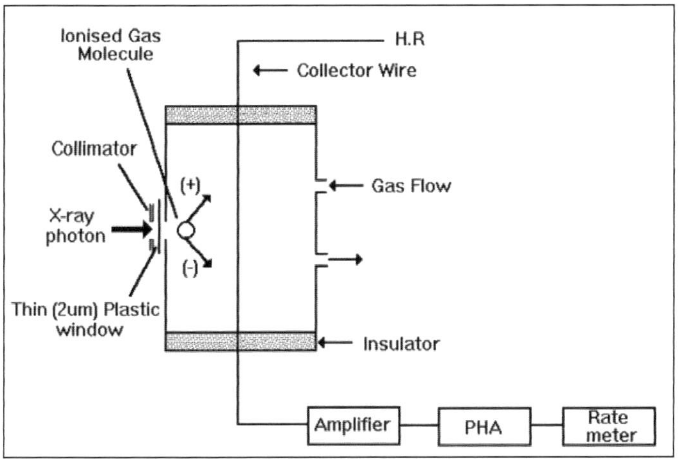

Figure 5. Basic schematic of the gas flow proportional counter (GFPC) X-ray detector for the WDS. Modified from [9]. The detector is a gas filled metal tube with an X-ray transparent entrance window. A wire runs down the central axis of the tube insulated from the walls and held at a high potential relative to the walls.

In most gas filled proportional counter designs, the cylinder body is earthed and a fine wire is biased at a positive potential of ~1,000 – 2,000 volts. An X-ray photon passing into the counter may be absorbed by an atom or molecule of the gas producing a photoelectron and an excited ion. The photoelectron or Auger electron, produced by decay of the excited ion, produce additional electron - ion pairs [9, 10-13]. The photoelectrons are accelerated to the central wire producing an electrical pulse which has amplitude proportional to the energy of the original X-ray photon.

The two general types of proportional counters are either sealed or gas-flow. The sealed proportional counters are typically used for high-energy X-ray lines and have a relatively thick Be window that prevents leakage of Xe gas in the detector. The gas-flow proportional counters are used for low-energy lines, have ultrathin windows (mylar or polypropylene), and use a counter gas (generally P-10; which is argon gas with 10 % methane) that flows through the detector at a constant rate [6, 10-13]. There are often problems with contaminants depositing on the wire, consequently affecting the output results and accuracy.

As shown in figure 6a, there are mainly three distinct gas gain behaviours. With no applied potential, there is no impetus for the ion pairs to move to the electrodes and the ion pairs simply recombine so no pulse is measured. Applying a small potential causes the ions to begin to drift toward the electrodes; some will reach before recombining. The proportion that reaches the electrodes increases with the potential until it is sufficient for all of them to be collected [6, 8, 9, 11]. This gives a gas amplification factor of 1.

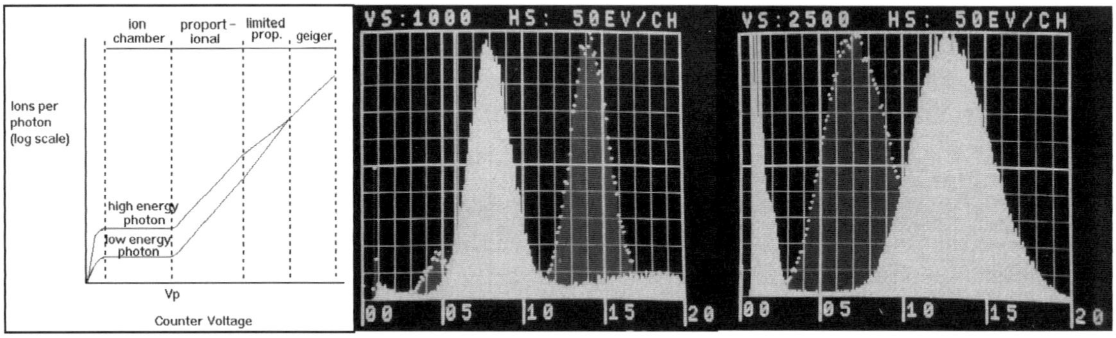

a) b) c)

Figure 6. a) Schematic showing the effect of applied counter-tube bias on the gas amplification factor (ionisation in the counter as a function of counter voltage). Modified from [8-10]. b) A comparison of drift and count rate for titanium; dots 1,000 cps, white 30,000 cps, and c) drift versus count rate for aluminium: white 1,000 cps, dots 20,000 cps (reversed dots/white due to data collected in 1975). PC X-ray output processed by an EDAX multi-channel analyser [9].

At first, increasing the potential further will not increase the gas gain because the ions will not have sufficient potential energy to cause any ionisation. When the potential has been increased enough, the electron/ion pairs will gain enough kinetic energy from the applied potential to start producing secondary ionisations and the gas amplification will start to increase. Over the range of gas amplification factors of about $10e^2$ to $10e^5$ the gas gain increases proportionally with the applied potential. More importantly, the energy pulse in this region is proportional to the energy of the absorbed X-ray. This is the proportional counter (PC) region and is the region that the detector needs to be maintained at (and this is why it is called a gas proportional counter) [6, 8].

When the energy of the ion pairs becomes large enough the quench gas can no longer prevent an avalanche occurring and the proportionality breaks down. This final region, is called the Geiger counter region, the ion pairs from a single incoming X-ray gain enough kinetic energy to cause a complete cascade ionisation of all the gas. It is difficult to determine anything about the energy of the X-ray absorbed, it simply registers a high-energy pulse.

When many pulses arrive together there is a temporary reduction of the gas amplification factor, because of the screening of the wire by the positive ion sheath formed in the chamber. This screening effect is the worst aspect of the proportional counter and can severely limit count rate performance. In figures 6b and 6c, the very bad aspects of the PC are on show. For the Ti spectrum at 10 kcps, the dotted lines show a substantial escape peak above the baseline, while the peak for 30 kcps has shifted down to nearly 50 % of the energy of the original peak. Its escape peak has gone below the detecting window and a substantial sum peak is now visible. The energy shifts on the aluminium peak are even worse, with a shift down to a 50 % energy peak from 1 kcps to 20 kcps. Some instruments have auto gain shifting to compensate for this effect, which does alleviate it to some degree.

X-ray pulses arriving close together may not be detected as separate pulses. Instead they are recorded as one pulse with the energy of both pulses added together. Where they are formed from a major peak they are called sum peaks and form a peak at twice the energy of the original peak. There is also a small possibility of getting 3 and 4 times the original peak height but this is rarely seen [9].

The job of the counter window is to isolate the counter gas from the vacuum in the spectrometer. Absorption of X-rays by the window is an undesirable effect, so the window needs to be as thin and transparent to X-rays as possible [8]. For these the window is now commonly supported by a grid. Whilst the grid is effectively opaque and therefore reduces the effective area of the window this is more than compensated by the low absorption of the very thin window. Due to the thinness of the windows, they are not fully gas impermeable and so the gas in the counter needs to be continuously replenished with a slow bleed of gas (hence why they are called gas flow counters) [8, 9].

6. Problems with a conventional WDS

As can be seen from the above discussion points of X-ray detectors, there are many variables that can affect the output and performance of the X-ray detector and the quality of the data produced from the WD spectrometer. Some of these issues include: count rate suppression, gas flow, gas purity, gas pressure, noise levels of baseline and window, GFPC voltage levels, anode wire contamination, electronic stability, high escape peak contribution when present, wavelength position error (easily solved by integrating the peak).

A WDS is an imperfect monochromator, with resolution characteristics depending on crystal, slit combination and diffraction angle. The major complications for the WDS are:

1. The gaseous detector resolution and stability, as seen in the section above;
2. High order diffraction;
3. Signal intensity as a function of beam displacement from the x, y or z alignment condition;
4. Changes in solid angle versus diffraction angle; and
5. Spectral holes from diffracting crystal absorption.

The worst of these are detector resolution and stability. This can be substantially solved by replacing the gas detector with a silicon drift detector SDD [4, 5]. The poor resolution directly affects the discrimination of the second worst problem, i.e., removal of high order diffraction.

Pulse height suppression is exacerbated by the age and cleanliness of the detector and central wire. The central wire contaminates easily and thus causes electric field shielding, which greatly enhances pulse height suppression (PHS). The absolute voltage of the wire also affects PHS. The greater the voltage, the greater the PHS. There are also affects due to localised temperature and pressure changes. These shifts can be substantial at moderate to high count rates and thus will affect the ratios of intensities from unknown to standard.

The effect of PHS can be mitigated by having the unknown and standard with similar intensities and more accurately plotting count rate versus beam current (BC) to determine a 'combined' dead-time, which includes losses due to actual dead-time, escape peaks, sum peaks and PHS. This combined dead-time is usually not linear. Xe gas counters suffer less from these effects when new, but as the Xe slowly escapes from the detector, it gets progressively worse (this may take many years to occur).

These changes are always ongoing with time. The poor resolution combined with these poor characteristics produce a detecting system that is substandard and has to be constantly monitored, to such an extent, that only very experienced operators are able to understand these effects. We are forced to introduce other means of correction, e.g., modified dead-times and substantial overlap tables. There are also escape peaks interfering with elements of energy greater than 3 keV for Ar and 4.1 keV for Xe. With the SDD replacement, the escape peak for energies above 1.74 keV (due to the Si from which the SDD is made) is substantially smaller and easily removed.

7. Incorporation of a silicon drift detector into a conventional WDS

For quantitative analysis on a WDS we must keep track of all the interferences that are possible that affect our result. For best sensitivity it is imperative that backgrounds are as low as possible. For both these conditions this is best done by careful control of excitation conditions and by proper control of pulse height discrimination (energy discrimination). Since the pulse distribution is proportional to the energy distribution of the X-rays being detected at any position on the spectrometer, then the position of the energy peak within the discriminators must be continuously adjusted for each position on the spectrometer. For a PC this is easiest to maintain with fixed baseline window settings and varying the detector voltage to keep the peak in the correct position within the discriminators. Setting a very wide window, or even to operate in integral mode, is not ideal because background and interference rejection are ignored. However, setting a very tight window (relative to the broad peaks measured in the PC) greatly increases the sensitivity of the detector to any changes in the gas (for example changes in pressure or temperature). For operation with the SDD we know exactly where the peak needs to be measured and the optimisation is so stable that it is now possible to change the integration width at each position on the spectrometer. Interference effects, such as high electron noise entering the light element SD detectors, can be easily subtracted from the X-ray count.

Over the last couple of years we have been testing the WD spectrometers using Amptek SDD's [4, 5] in place of the sealed proportional counter (PC) and GFPCs (figures 7a and 7b) [4]. Initial work used an Amptek SDD and more recently the Amptek Fast 123 system. These system configurations include the detector, preamplifier, digital processor and power supplies all in one unit. The original, proof-of-concept, work [4, 5] was not optimised and was performed under difficult circumstances, by having to place the complete Amptek 123 system inside the spectrometer (figure 7b). This severely limited the movement range of the spectrometer, but did show that the concept was worth developing.

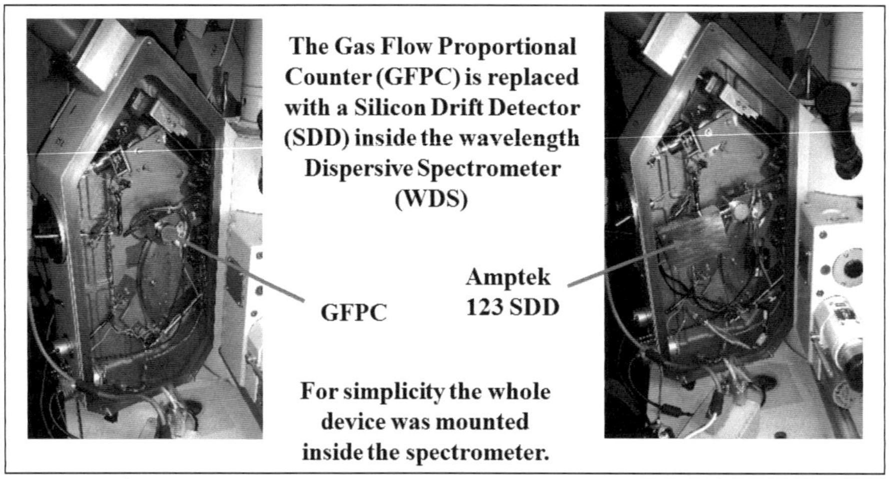

a) b)

Figure 7. Photographs of a JEOL WD spectrometer, with b) a gas flow proportional counter fitting (GFPC), and c) an Amptek 123 EDS mounted (25 mm^2) in a JEOL spectrometer through an old proportional counter fitting, which has been bored out to allow alignment of the EDS [4, 5].

7.1. Silicon drift detectors (SDD)

The silicon drift detector (SDD) is a reverse biased photodiode with an electrode structure that leads to much lower electronic noise and enables much higher count rates than traditional planar diodes. Any free charge created within the diode leads to a temporary increase in its conductivity. Near perfect semiconductor crystals show some residual conductivity due to random thermal excitation of electrons across the energy gap. To minimise this leakage current, detectors are operated at around minus 10 °C to -50 °C. The Si detector has no internal gain and 1 - 10 keV X-rays produce output pulses containing only 260 - 2,600 electrons, or about 10^{-16} coulombs. Figure 8a is a schematic sketch of a SDD, showing the electrode structure as a series of drift rings, which produce a radial field, guiding the electrons to a very small, low capacitance anode. The small value of the anode capacitance, which is practically independent of the total area of the device, translates into the signal having much lower electronic noise at short peaking times. This is the key to the excellent energy resolution and high count rates of the SDD.

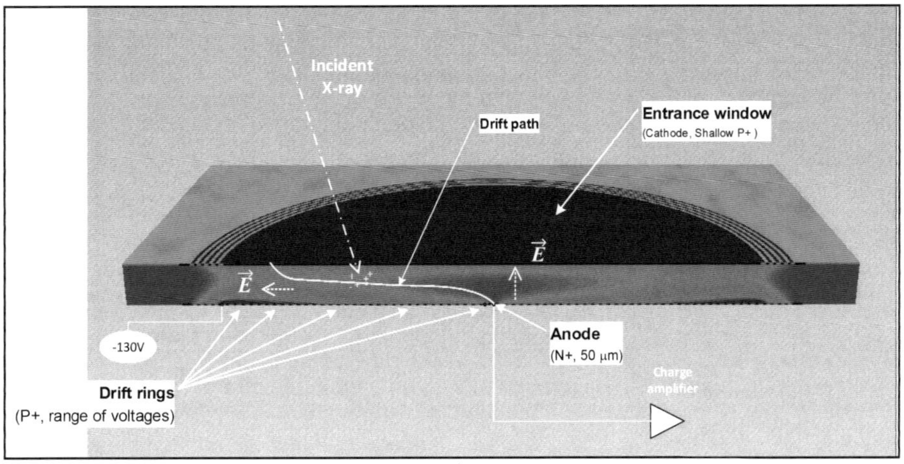

a)

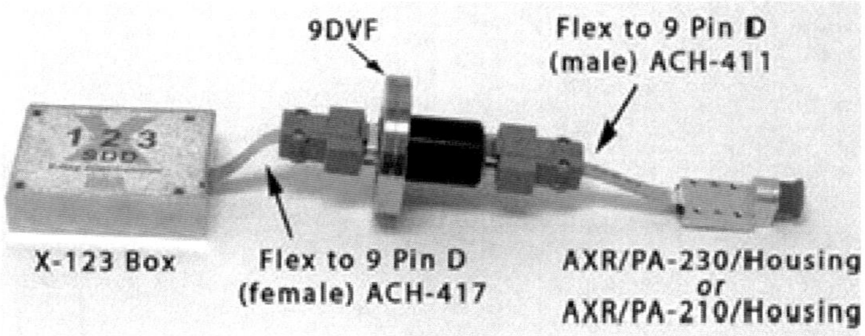

b)

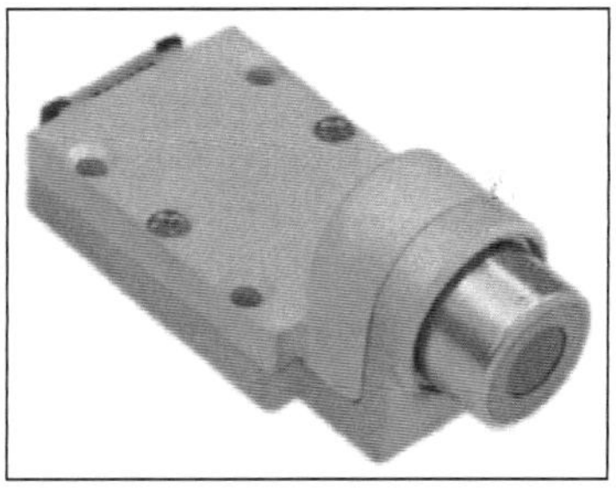

c)

Figure 8. a) Schematic sketch of Amptek silicon drift detector design [19], b) New Amptek SDD being incorporated with the control electronics moved to the outside of the WDS, and c) Amptek detector [18].

In Amptek's FAST SDD®, a low noise CMOS preamplifier is mounted inside the package alongside the photodiode, further reducing noise. The signal is further amplified by a main amplifier before being converted to digital format. The whole signal is extracted digitally from which the pulse is extracted and digitally filtered. The spectrum is then analysed using a multichannel pulse height analyser which sorts the detector pulses according to their height, and hence X-ray energy, and records the number received in each channel (a channel being a small energy range, usually 10 eV).

Resolution values of 125 eV for Mn are easily achievable. The only disadvantage with this device is that it loses efficiency rapidly over 10 keV due to its detector thinness. However, they are excellent for high speed quantitative analysis, as count rates can approach 100 kcps with almost no loss in resolution, and very small gain and zero shift of the spectrum.

More recently, Amptek have produced a different option that allows the detector/preamplifier to be incorporated within the WD spectrometer, separated from the electronics unit and connected using a 9-pin vacuum feedthrough (figures 8 and 9). This allows the detector to be operated remotely from its electronics box. Thus, a very lightweight detector is able to be mounted in the detecting position behind the secondary slit. The SDD can be operated at varying time constants and optimised for resolution or count rate. The SDD detector area is 25 mm^2 (5.6 mm diameter).

This gives a high-resolution spectrum of all the energies being diffracted, enabling observation of all the high order diffraction and other artefacts of the system. All of this is very important adding to the information about the EM probe, enabling better results.

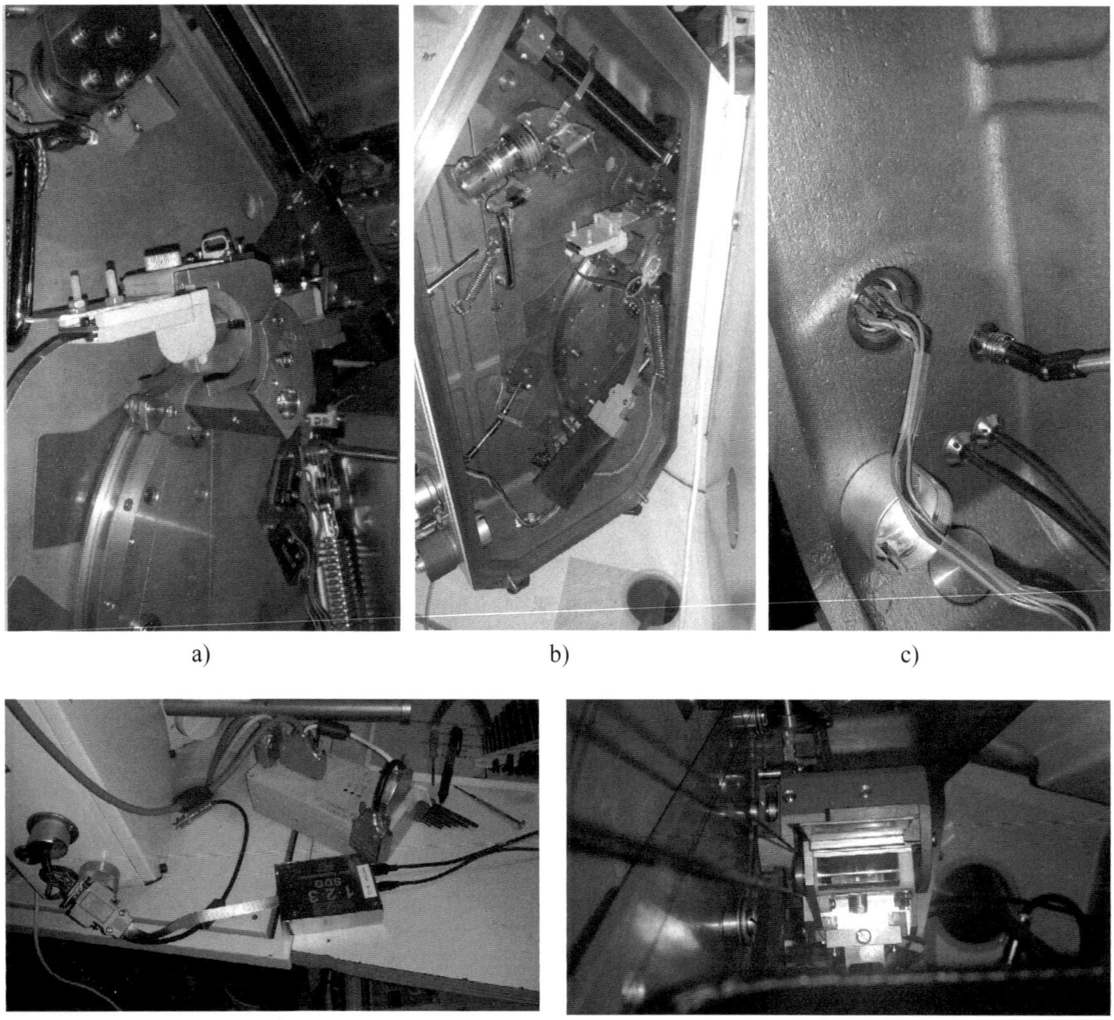

a) b) c) d) e)

Figure 9. a) New Amptek configuration SDD (SD-WDS), b) internal electronic wiring, c) feedthrough design, d) external configuration where the detector has been separated from the electronics unit, and e) Amptek SDD viewed through slit design of WDS.

8. Amptek silicon drift detectors and results from incorporation

As discussed above, gas detectors are entirely unsatisfactory for the collection of X-rays under conditions where there are large variations in count rates. The new SD detectors on the contrary are extremely stable. With the incorporation of SD detectors (SD-WDS), the ability to accurately and finely discriminate the energy spectrum means that high order diffraction becomes a problem of the past. The longer pulse conversion time for the SD detectors means more accurate dead-time correction. Because of their stability this correction is easy to determine.

SD detectors were incorporated in four spectrometers on a JEOL 733. The spectrometers have dual crystals on each spectrometer. For best performance, the best crystals of each type were chosen so that there would be only one crystal per spectrometer, maximising the efficiency of that spectrometer and allowing the best possible detector to be allocated for each spectrometer (table 1). For X-ray detectors to have satisfactory efficiency for long wavelengths, the detectors are fitted with ultra-thin windows:

1. For the LIF crystal. A SDD with a slightly thicker Be window means that it can be used at much higher beam voltages with few stray high energy electrons affecting its performance. All X-rays on this spectrometer can easily penetrate the 0.5 mm Be window.
2. For the PET crystal. A SDD with a thinner Be window is a compromise to prevent as many high-energy electrons as possible without attenuating the incoming X-rays.
3. For the TAP crystal. A SDD with a C1 window is purely to prevent attenuating the incoming X-rays. This window has better light exclusion performance than the thinner C2 window.
4. For the STE crystal. A SDD with a C2 window is to maximise the collection of the soft X-rays.

Previous preliminary work at 20 keV, 45 nA, 50 s, using the TAP crystal and SDD with Be window as shown figure 7b revealed the following:

- GFPC Si count rate was 125 kcps with P/B equalling 211 (giving a detection limit (DL) of 90 ppm).
- SDD Si count rate was 36 kcps with a P/B equalling 2,000 (giving a detection limit (DL) of 56 ppm).

In this case the SDD detection limit was lower due to the exceptional P/B ratio. If the SDD had the same area as the GFPC, then the DL would have been even lower.

After the new configuration SDD were fitted to the WDSs, the Si count rate was 45 kcps and the P/B was 1,750. Having a Be-window was better for removing electrons than having the current ultra-thin window (C1-Si_3N_4 90 nm window). The new system could be improved by installing a better electron trap.

To compare the old (PC) with the new (SDD), two elements on each spectrometer, which are reasonably well separated, were used to test the performance of the WD spectrometers. From the EDS and WDS resolution data, shown in figure 10, the resolution data has been plotted as a function of energy and is shown in figure 11. This data matches results shown in figure 1a. Figure 12 shows a direct comparison of specific elements on each spectrometer, comparing manufacturer's best specifications, with the last known data using the instrument's proportional counters and with data from the new Amptek SDDs.

Mounting the detector behind the secondary slit means that only a narrow portion, equal to the width of the detector times the width of the slit, is being used to collect the X-rays. In future, a detector will be made to make better use of the original design of the 20 mm detecting area.

Investigation of the effect of count rate has revealed almost no change in resolution or gain with increasing count rate from 12 kcps to 120 kcps (figure 13). Figure 14a shows the count rate versus beam current and figure 14b the zero shift (eV) versus beam current (nA). From Figure 14a, the graph of actual counts shows a residual dead-time. This dead-time time constant (TC) was calculated as 1.2 μs and includes losses from sum peaks.

The new SDD detector technology is now good enough to handle the high-count rates experienced by WDS and they are extremely stable and provide a simple way of checking their own integrity by being able to calibrate against a known spectrum. As an example of poor count rate stability in the PC, figure 6c shows a pulse height analysis plot of Al-Kα at 20 kcps. A second order Ag-Lα (2.98 keV) would appear at the same place as the first order Al at 1,000 cps. A very good gas counter can resolve the peaks but there will always be a major contribution of the second order Ag-Lα if the count rate changes. When this is compared to a SDD, there is no problem in separating the Al from the Ag (figure 15). With the SDD there is < 5 eV drift and < 2 eV change in resolution up to 100,000 cps (at 1.2 μs pulse processing TC). For the SDD, the dead-time can be measured carefully and is much easier to keep control of as it is no longer subject to pulse height suppression. Most dead-time is automatically corrected within the SDD processor. From the graph of "Actual" counts in figure 14a, it can be seen that a residual dead-time correction is required for this detector. This residual dead-time was measured at 1.2 μs (not to be confused with the pulse processing TC).

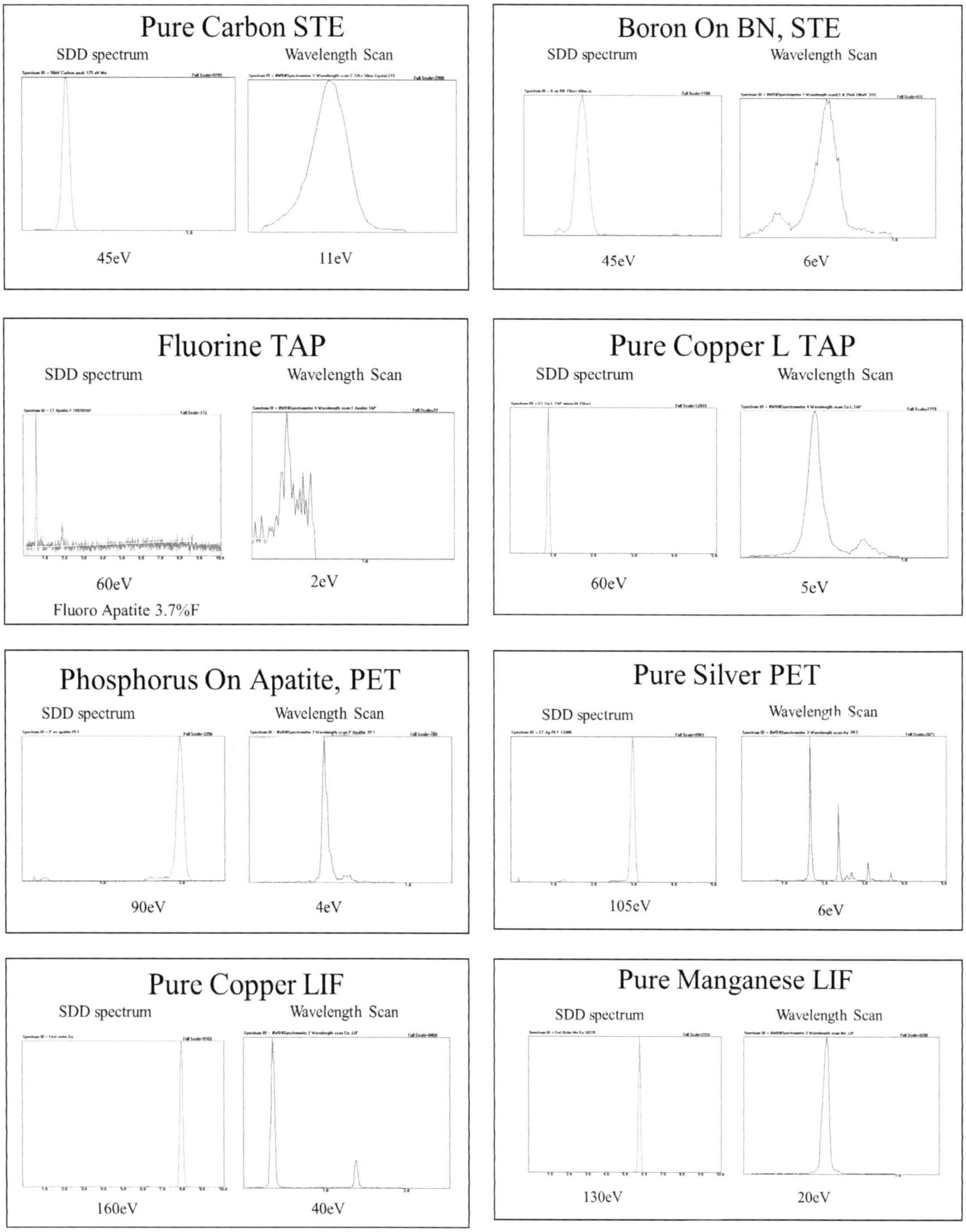

Figure 10. Figure caption at top of next page.

IOP Conf. Series: Materials Science and Engineering **304** (2017) 012021 doi:10.1088/1757-899X/304/1/012021

Figure 10. Represents the results from two elements from each spectrometer, where the elements are reasonably well separated in their respective spectrometer. Results are from SD incorporated in a WDS (SD-WDS). For each element, there is an energy distribution representing the full spectrum collected on the wavelength peak corresponding to the element and there is a wavelength scan around the wavelength peak. The energy resolution of each is indicated. For each crystal:

1) STE. Carbon K, EDS energy resolution for the C-K peak is 45 eV. The wavelength scan equivalent energy resolution is 11 eV. Boron K, EDS energy resolution for the B-K peak is 45 eV. The wavelength scan equivalent energy resolution is 6 eV.
2) TAP. Copper L, EDS energy resolution for the Cu-L peak is 60 eV. The wavelength scan equivalent energy resolution is 5 eV. Flourine K, EDS energy resolution for the F-K peak is 60 eV. The wavelength scan equivalent energy resolution is 2 eV.
3) PET. Silver L, EDS energy resolution for the Ag-L peak is 105 eV. The wavelength scan equivalent energy resolution is 6 eV. Phosphorus K, EDS energy resolution for the P-K peak is 90 eV. The wavelength scan equivalent energy resolution is 4 eV.
4) LIF. Copper K, EDS energy resolution for the Cu-K peak is 160 eV. The wavelength scan, equivalent energy resolution is 40 eV. Manganese K, EDS energy resolution for the Mn-K peak is 130 eV. The wavelength scan equivalent energy resolution is 20 eV.

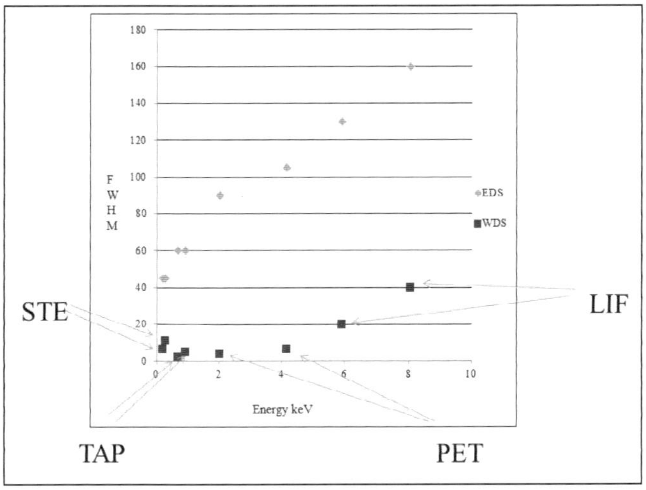

Figure 11. Experimental data obtained from EDS versus WDS full width half maximum (FWHM) resolution testing as a function of energy.

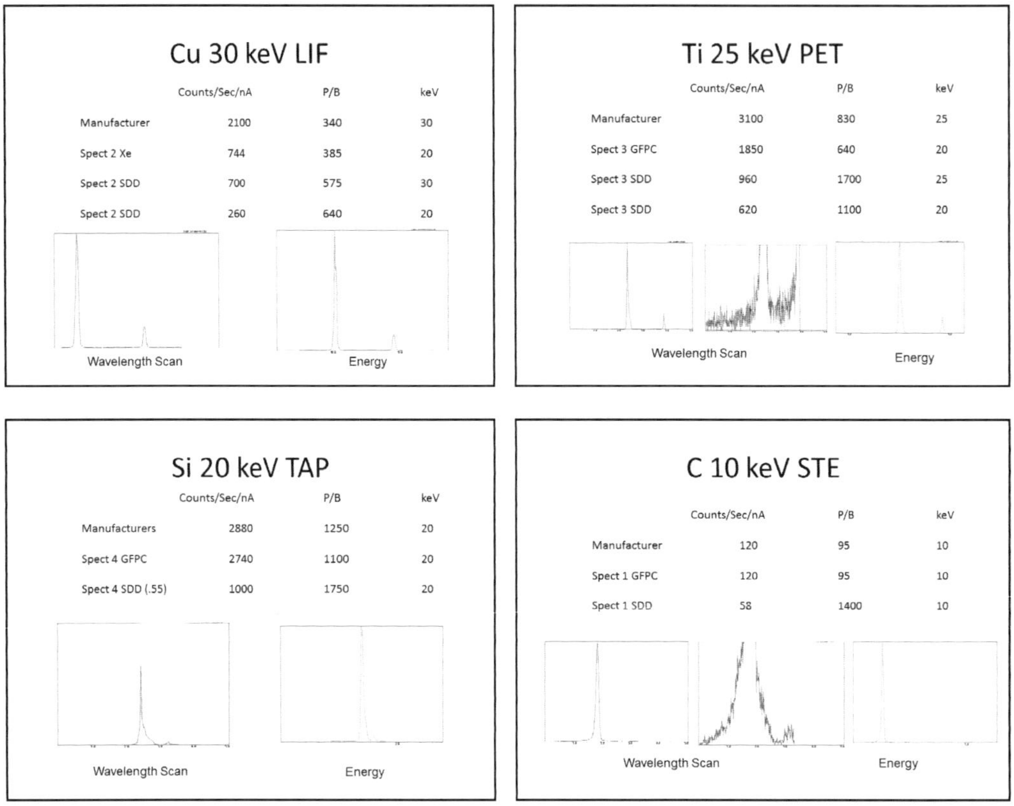

Figure 12. Spectrometer performance data for crystals LIF, PET, TAP and STE. Figures a) to d) above compare the manufacturer's best performance, with the previous proportional detector performance, and the new SDD performance (SD-WDS). In all cases the counts for the SDD are lower because the detector active collection area is 6 mm^2 compared to 20 mm^2 for the PC. For the SDD the peak to background (P/B) ratios are much better for all the crystals.

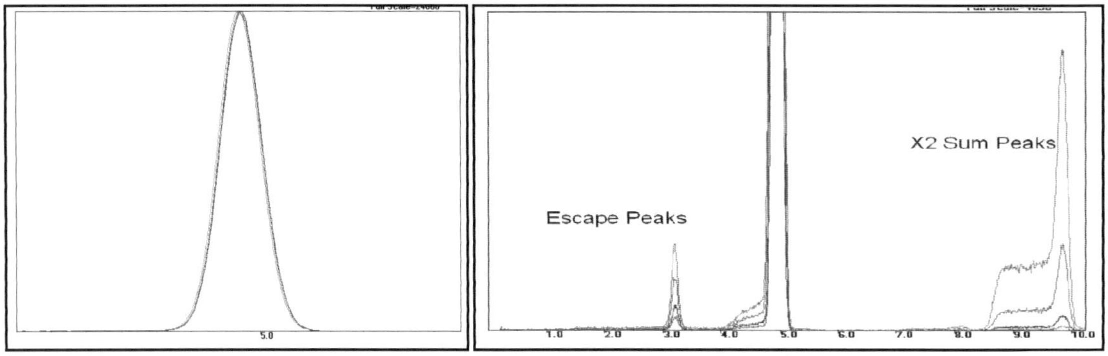

a) b)

Figure 13. a) Vanadium spectra of count rate change from 12 kcps to 120 kcps (four superimposed vanadium spectra 12 to 120 kcps. 10 nA - 12,300 cps, 20 nA - 24,600 cps, 50 nA - 61,500 cps and 100 nA - 123,000 cps vanadium), and b) Expanded range of vanadium spectra showing almost no resolution and gain shift, just a small zero shift of < 7 eV.

IOP Conf. Series: Materials Science and Engineering **304** (2017) 012021 doi:10.1088/1757-899X/304/1/012021

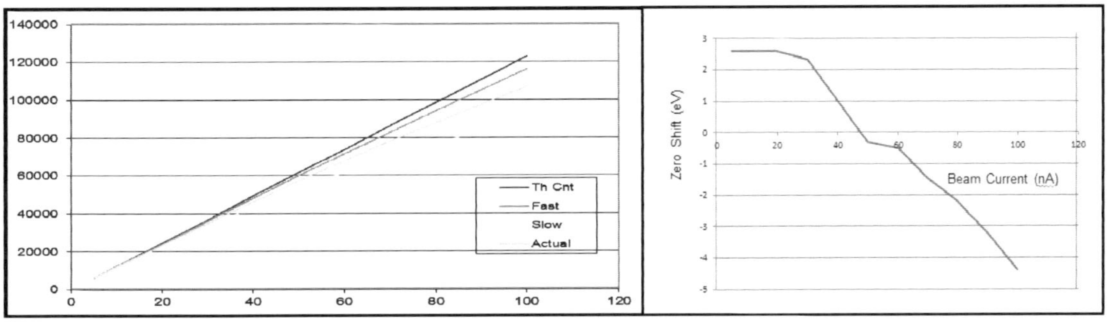

a) b)

Figure 14. a) Amptek X123 SDD count rate, using time constant of 1.2 µsec. Count rate versus beam current. The Th Cnt (dark blue) is the theoretical count rate, fast and slow are the fast and slow channel discriminators for the total spectrum. Actual (aqua) is the count integral corrected for dead-time by the system clock. b) Zero shift (eV) versus beam current (nA).

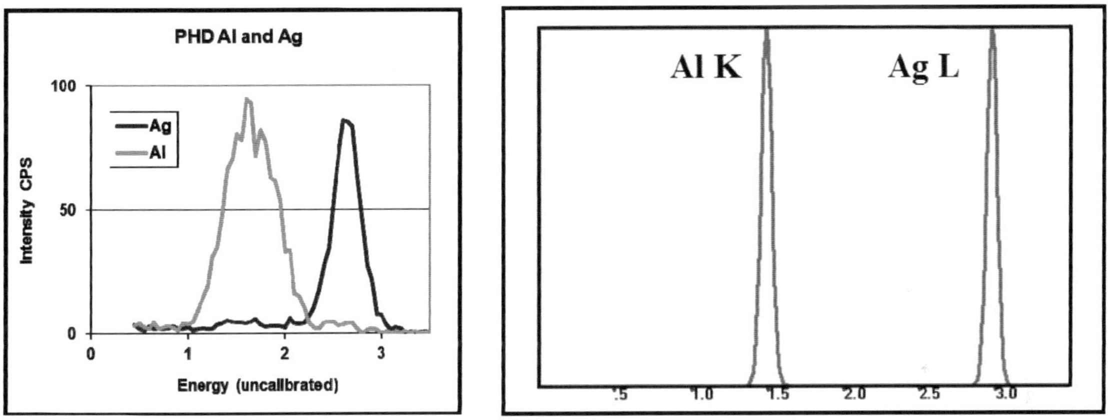

a) b)

Figure 15. Silver 2nd order overlap on Al from the SD-WDS system. a) GFPC pulse height distribution, and b) SDD PHD (theoretical 130 eV). Al-K$\alpha_{1,2}$ = 8.34 Å (1.486 keV) and 2nd order Ag-Lα*1,2* = 8.3 Å (2.981 keV). Conditions: 12 keV, 20 nA, TAP crystal.

8.1. Light element results

Figures 16a and 16b show the SDD energy spectra from the WDS peak (blue) and background (red) positions measured using the STE crystal for 8.5 % and 0.7 % carbon respectively. In both cases the red spectra represent first and second order diffracted bremsstrahlung and high energy electrons entering the detector. Figure 16c shows the comparison between 8.5 % (red) and 0.7 % (blue) carbon after the background spectra have been subtracted from both peak spectra. For analysis, only the integral on the energy spectra at the carbon peak are used in quantitation. All other artefacts are removed, especially the high order lines as shown.

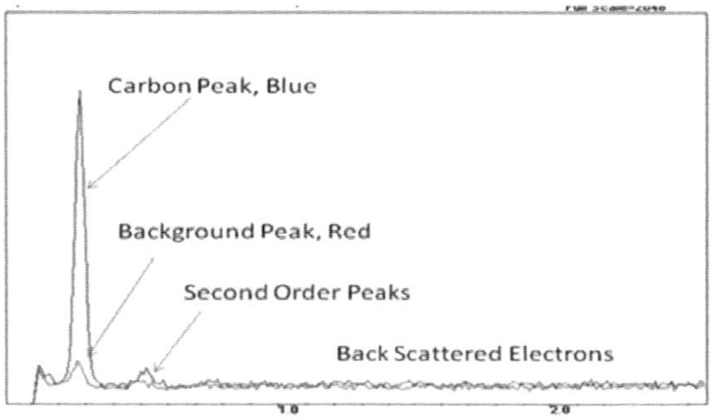

a)

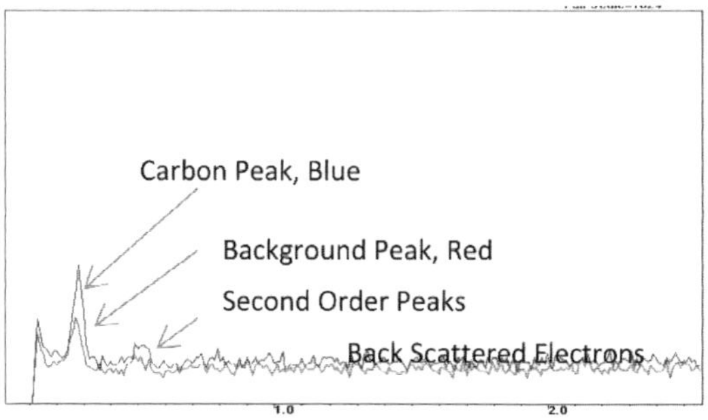

b)

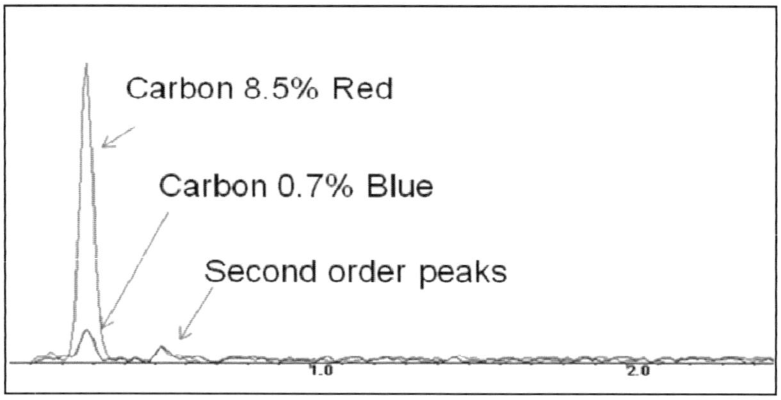

c)

Figure 16. Results from SD-WDS. a) Carbon peak (blue) and background (red) in 8.5 % carbon, b) carbon peak (blue) and background (red) in 0.7 % carbon, and c) after background removal of both spectra 8.5 % carbon is shown as red and 0.7 % carbon shown in blue. Carbon as measured using a STE crystal at 15 keV and 50 nA. DL = 0.03 % carbon in 50 sec. Carbon can now be easily calculated at the sub 0.1 % level with a lot more confidence. All spectra include high order diffraction and secondary fluorescence [4, 5].

8.2. High order interferences

Typical X-ray interference problems, including high order interferences on all crystals, can now be easily removed through use of a SDD system. These were not easily removed with the use of a proportional counter. Figures 17a and 17b show high order interferences. Figure 17a shows the fourth order magnesium peak, which dominates the spectrum and the actual bremsstrahlung is very small. Figure 17b shows high order interference on a LIF crystal by a second order copper peak, measured in an area at the point where scandium (Sc) should be. In this case the spectrum is dominated by the copper second order peak and the bremsstrahlung at this energy is about a third of the second order copper peak. Figure17b also shows fluorescence from elements like Al, Cr, Fe, Ni, Zn and Pb. These fluorescence peaks are due to the mechanical components in the spectrometers.

Figures 18-21 have been included to show the spectra from a high-density material (PbS) to maximise the bremsstrahlung and the number of high order lines, mainly those of Pb. Now that we can easily remove high order interferences on every crystal, we finally have an ***"X-ray display of first order lines only"***. No more higher order lines, unless you actually want to display them.

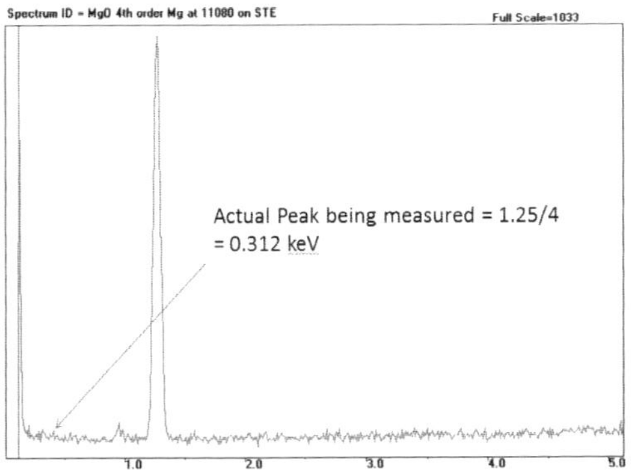

a)

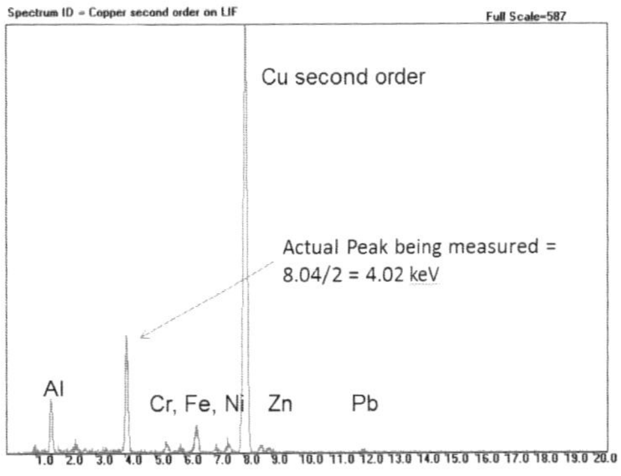

b)

Figure 17. SDD pulse height distributions (energy distributions) of bremsstrahlung measured at a) 0.312 keV on a STE crystal, and b) 4.02 keV on a LIF crystal. On spectrum a), sample MgO, the magnesium peak is a fourth order diffraction. On spectrum b), sample copper, the copper K is a second order diffraction. These energies appear at their actual energy position and thus are easily removed through proper processing of the full-spectrum at each wavelength position.

Because of the high resolution of the SDD, it is now easy to discriminate between the actual energy of the wavelength peak and multiples of that energy, which are high order diffraction peaks. Integrating only the first order energy, at the wavelength required, removes all other energies. Figures 18 to 21 show a simultaneous spectrometer scan on a PbS standard using four spectrometers with the crystals LIF, PET, TAP and STE utilising the SDD in the WD spectrometers. From the use of the SDDs, the subsequent removal of all the high order lines (which would normally be seen from a PC detector), means that there are now no high order lines displayed from the TAP crystal. This is illustrated in figures 18a and 18b. These are the same spectra but with black lines indicating where the Pb (Figure 18a) and S (Figure 18b) high order lines would have been.

Both the Pb and S first order lines should be seen from the PET crystal and if you remove the high order lines, there will be no other lines displayed from the PET crystal. Figure 19 shows this, with the black lines indicating where all the Pb high order lines would have been. Figure 20a sulphur labels show where the first order S-lines are and figure 20b lead labels, show where all the first order Pb-lines are.

Only the Pb-lines should be seen on the LIF crystal, and if you remove the high order lines, then there will be no other lines displayed on this crystal. Figure 21a shows this and the black lines indicate where all the Pb high order lines would have been.

If you remove the high order lines there will be no lines to be displayed on the STE crystal. This is illustrated in Figure 21b. The black lines indicate where the Pb and S high order lines would be. The indicators have only been displayed to the 9th order. The spectrum does display carbon from the carbon-coated sample.

IOP Conf. Series: Materials Science and Engineering **304** (2017) 012021 doi:10.1088/1757-899X/304/1/012021

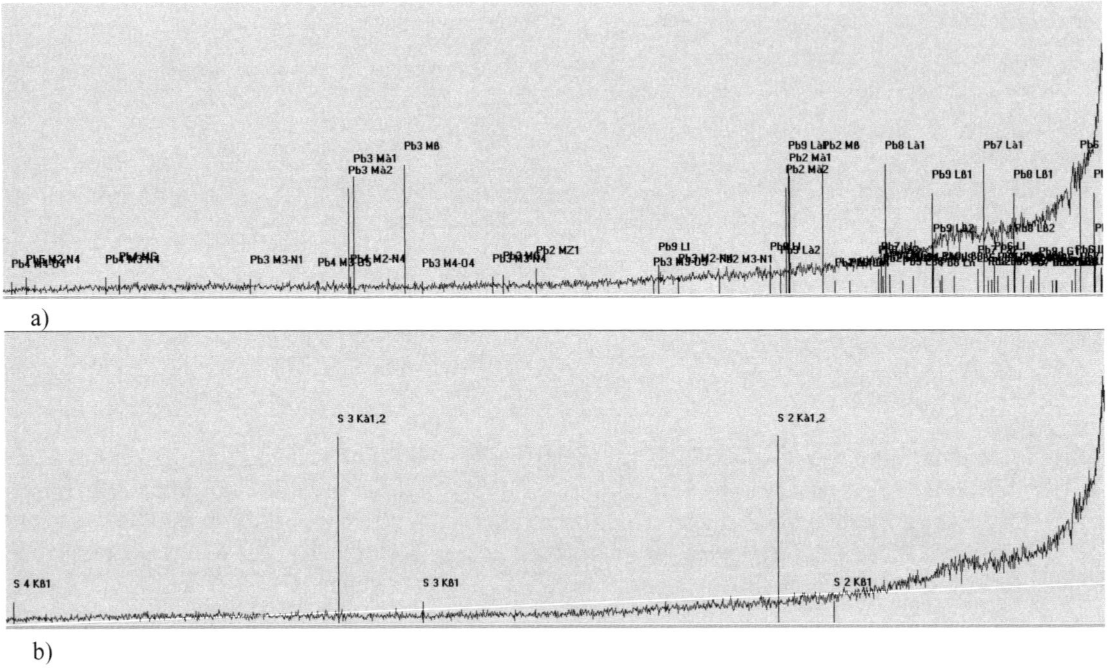

Figure 18. a) Full spectrum wavelength scan on PbS using the TAP crystal, showing where all the high order lines for lead (Pb) would normally appear, and b) the same as figure 18a, except showing high order lines for sulphur (S).

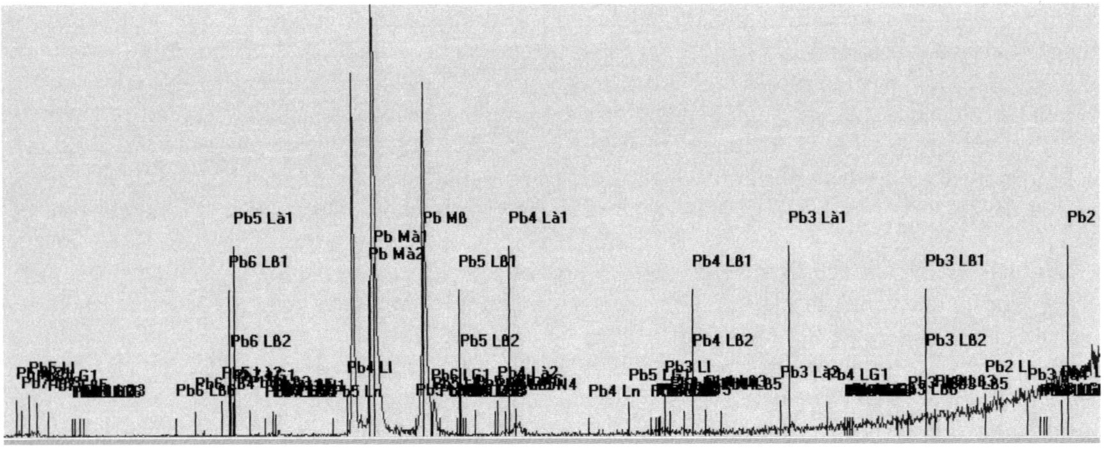

Figure 19. Full spectrum wavelength scan on PbS using the PET crystal, showing where all the lead (Pb) first order lines are shown and where the high order lines have been successfully suppressed by the SD-WDS.

IOP Conf. Series: Materials Science and Engineering **304** (2017) 012021 doi:10.1088/1757-899X/304/1/012021

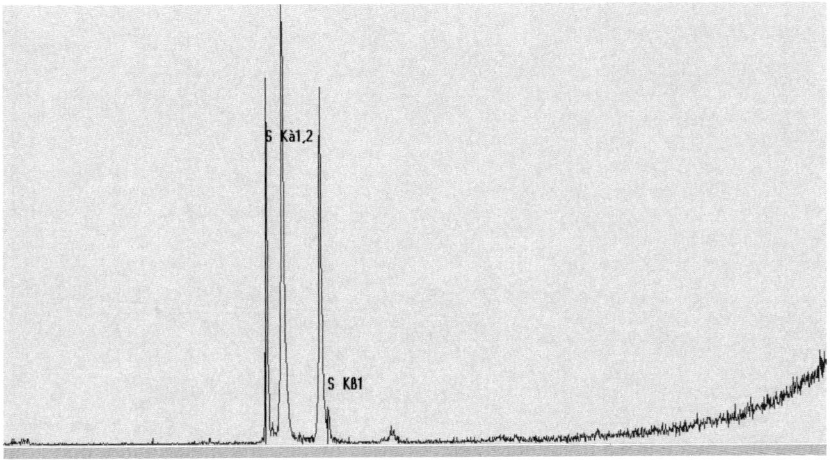

a)

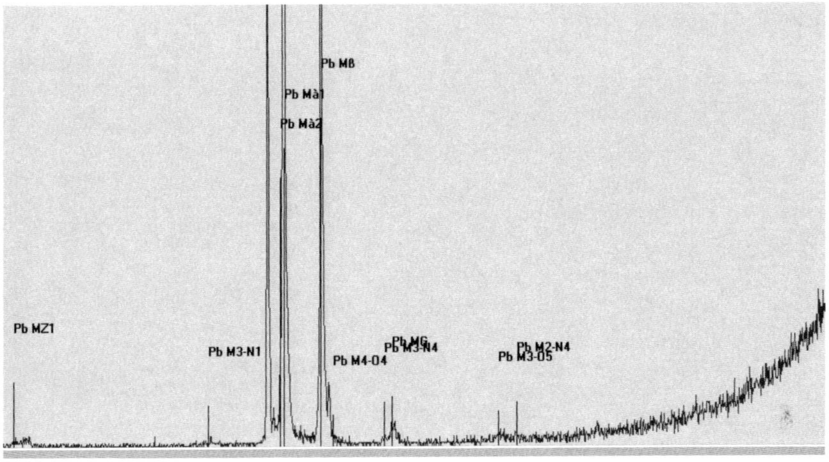

b)

Figure 20. a) Full spectrum wavelength scan on PbS using the PET crystal, labelling the first order lines for sulphur on a normalised vertical scale, and b) the same spectrum wavelength scan expanded on the vertical scale labelling all the first order lines for lead.

In the past it has been difficult to interpret the wavelength spectra when converted to an energy spectrum, due to the complication of having to explain all the high order peaks. High order peaks confuse the tools normally used for finding elements displayed in the energy spectrum. Now it is very easy to interpret the results. There are no sum peaks and no escape peaks or high order overlaps in the ED space spectrum. In the spectra overlayed in figure 22, the vertical scales of each spectrometer are normalised to the maximum height of the most significant peak in the respective spectrum. Hence the Ca-Kβ on the LIF crystal looks as high as the Ca-Kα on the PET spectrum. Note the background rising as the energy increases, which is the opposite of what happens in a “real” EDS spectrum. Figure 22b shows the log scale spectrum in figure 22a. The reason for the increase in bremsstrahlung with increasing energy is due to the rapidly increasing collection efficiency of the WDS at very low angles.

IOP Conf. Series: Materials Science and Engineering **304** (2017) 012021 doi:10.1088/1757-899X/304/1/012021

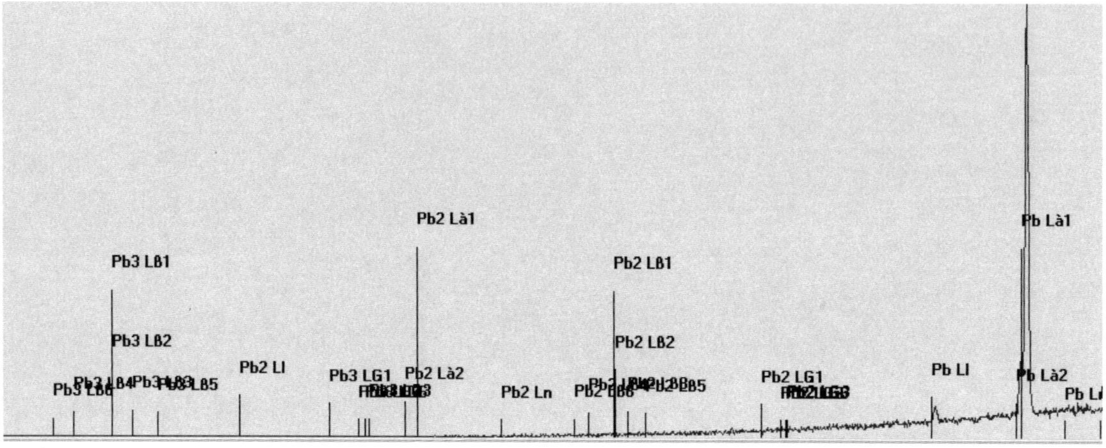

a)

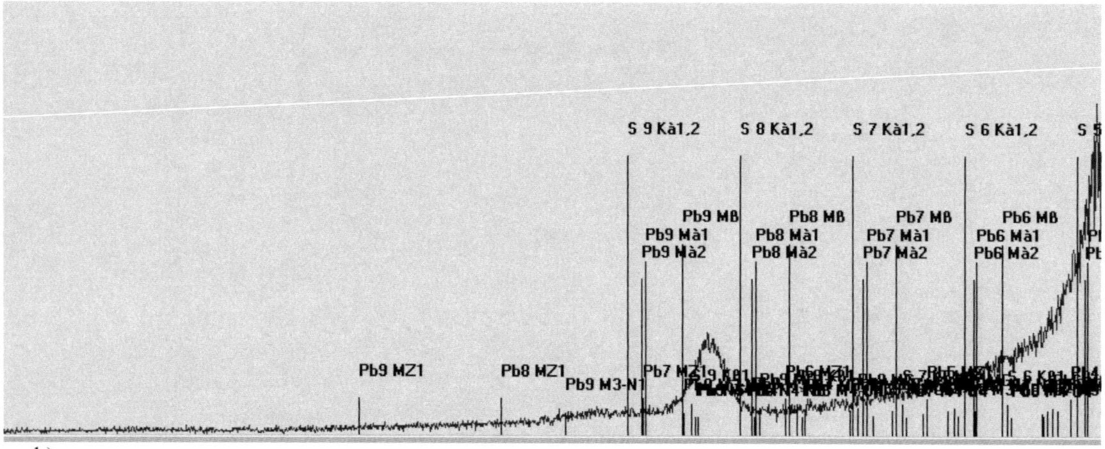

b)

Figure 21. a) Full spectrum wavelength scan on PbS using the LIF crystal, showing where all the first order lines and high order lines for lead would normally appear, and b) full spectrum wavelength scan on PbS using the STE crystal, showing where all the high order lines (up to ninth order) for lead would normally appear. Carbon is the peak shown from the carbon coating on the sample.

Every spectrometer crystal has a different energy range (table 1). When converting each wavelength scan, this range needs to be taken into account. This is done when converting to an EDS spectra of fixed eV per channel. This is why the different colour portions of the graph cover varying energy ranges, even though they cover the same spectrometer distance, as shown in figure 22b.

The biggest advantage of a SDD is stability, followed by much improved resolution (160 eV down to 70 eV for Al). There is also now the ability to actually analyse all the interferences that the detector sees, giving the opportunity to resolve these interferences or, at the very least, a method of digitally removing them (i.e., it is easy to see X-rays arising from unshielded brass crystal holders, etc.).

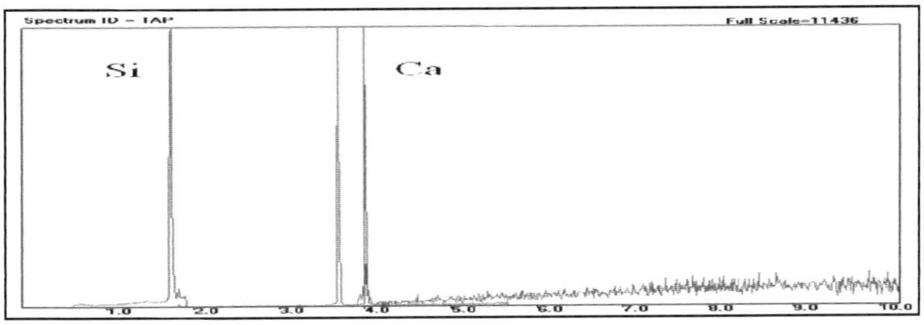

a)

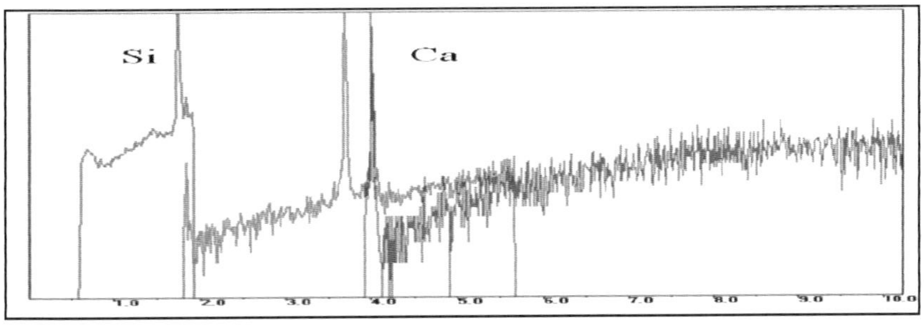

b)

Figure 22. Wollastonite sample results. a) A simultaneous wavelength scan from three spectrometers converted to an energy scan TAP = red, PET = pink, LIF = green, and b) Log scale version of figure 22a.

8.3. Trace element results

Trace elemental analysis measuring low concentrations of Si and Sr in apatite at 15 keV is shown in Figure 23. This sample has a silicon (Kα at 1.739 keV) concentration of 0.06 wt% and strontium (Lα1 1.806 keV and Lβ1 1.871 keV) concentration of 0.39 wt%. The wavelength scan for this material in the region of Si and Sr is shown in figure 23a. While the energy distribution at the peak and the background, between the two peak positions, is shown in figure 23b. It can be seen that the peak to background ratio is substantial enough to detect much smaller concentrations than those shown. From the Si spectra shown, the peak minus background is 1,800 counts for 30 seconds, giving a 2 sigma confidence level of 0.003 %. This result, showing the complete spectra at the peak and background, ensures the result is only obtained from the energy at the specific wavelength positions, giving an higher degree of confidence in the result.

8.4. Advantages and disadvantages of SDD incorporation

The major advantages and disadvantages, determined at this stage, of incorporating the SDD as a replacement for the proportional counter, can be seen in table 2.

8.5. Processing the full-spectrum data at each wavelength position

The Moran Scientific EDS software used in our investigations automatically accommodates 6 ED spectrometers, with the first spectrometer having the highest priority for EDS analysis work. When incorporating the SDD's into WDS, these detectors are mapped into the system as minor EDS systems and are available for all other software (WDS and mapping software). Currently for this early research work, we are using simple integral processing of the full energy width of the true energy position minus a background region to remove high energy electrons. Where the full energy width

IOP Conf. Series: Materials Science and Engineering **304** (2017) 012021 doi:10.1088/1757-899X/304/1/012021

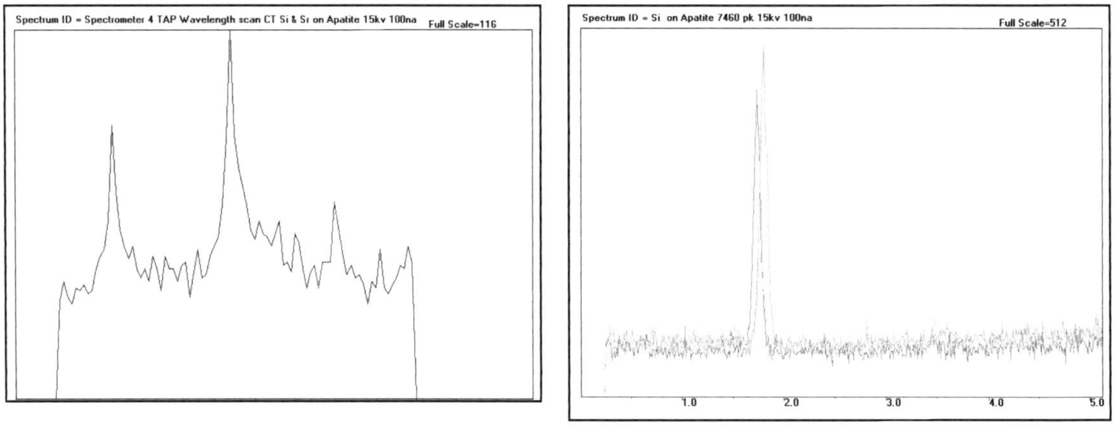

a) b)

Figure 23. a) Wavelength scan for silicon and strontium showing peaks from left to right, silicon Kα 1.739 keV (0.06 wt%), strontium $L\alpha_1$ 1.806 keV and $L\beta_1$ 1.871 keV (0.39 wt%), and b) Energy spectra for silicon and strontium. Red is silicon, grey is strontium, and green is the common background between the two peaks (30 second analysis on peak).

Table 2. Advantages and disadvantages of the SDD over the PC and/or GFPC detectors.

Advantages	Disadvantages
Stable	Heat dissipation*
High resolution	9 pin feed through*
Full spectrum	Effected by light (UTW only)**
Better dead time calculation	Magnets required (UTW only)*
Cost	Cost
Ease of mounting	
Easier to control	
No P10 gas required	
Possible theoretical BG may be obtained	
No count rate drift error	
No baseline window error	

* Requires an engineering solution, and ** Phosphorescence of the sample does not seem to affect the SDD.

(ΔE is approximately 2x FWHM), varies with the resolution of the peak energy, as shown in the EDS peak resolutions of figure 11, the integration of the energy window (ΔE) varies proportionally with the energy. To accurately position the integrals it is imperative to have minimal resolution, gain and zero changes in the acquired spectra.

Most maps today are not fully quantitative and they require the accurate measurement of the live-time at each pixel. Standardless analysis can only be considered as an approximation. Trace WDS mapping requires background stripping. For example, the P/B ratio for Si on EDS is ~ 100, WDS ~ 1,000 and for SD-WDS ~ 2,000. With background removal for both EDS and WDS this ratio can be improved. With SD-WDS, the option to estimate the background at the peak position, while at the peak position, enables an effective improvement to the mapping detection limit.

9. Conclusion

Analysis of major elements using the WDS was never much of a problem; the same applied using the EDS, with the proviso that only a restricted set of conditions is used for the analysis. Even minor elements can be accomplished with both EDS and WDS accurate overlap files. It is always important to analyse as many pure elements as possible for all the other elements in the analysis file, to obtain an idea of which overlaps occur. This process is always more difficult for trace and light elements. It is especially difficult when there were many rare earth and transition metal L-line elements present.

Our current research demonstrates the dramatic superiority of SDD over PC (whether gas flow or sealed). With a proportional counter it can be difficult to know what is being counted by the detector. As shown in previous figures, an SDD immediately brings to light problems such as secondary fluorescence, electron scattering and high order diffraction artefacts. Often with a PC the extra noise is similar on the peak as on the background, enabling it to be removed by background subtraction, but the P/B ratio can be significantly reduced. Most PCs have an order of magnitude poorer energy resolution than the SDD; this means that if interferences are present you can get a corresponding improvement in background by removing them, either by energy discrimination, or physically removing their source. It is important to know whether they are there, so that you can remove them to improve your performance. The simplicity of a WDS without a proportional counter is easily seen, once in operation, and removes many artefacts allowing for higher confidence and accurate analyses.

Only a system that had SD Detectors incorporated in the WDS (SD-WDS) would be considered for purchase today. The stability and removal of high order interferences, using SDDs, makes the operation of the WDS much more accurate and easier for the operator to use and understand.

Acknowledgements

The authors wish to acknowledge the support from Amptek, in particular John Pantazis. We would also like to thank Mike Matthews and EMAS for support to attend the workshop in 2017.

References

[1] Moran K and Wuhrer R 2006 Quantitative bulk and trace element X-ray mapping using multiple detectors. *Mikrochim. Acta* **155** 59-66

[2] Wuhrer R and Moran K 2015 Dedicated X-ray mapping system with single and multiple SDD detectors for quantitative X-ray mapping and data processing and chemical phase mapping with an Amptek SDD. *Microsc. Microanal.* **21** (Suppl. 3) 2191-2192

[3] Moran K and Wuhrer R 2006 X-ray mapping and interpretation of scatter diagrams. *Mikrochim. Acta* **155** 209-217

[4] Moran K and Wuhrer R 2016 Current state of combined EDS-WDS quantitative X-ray mapping. *Microsc. Microanal.* **22** (Suppl. 3) 92-93

[5] Moran K and Wuhrer R 2017 Replacing the gas flow proportional counter (GFPC) in a wavelength dispersive spectrometer (WDS) with an Amptek silicon drift detector (SDD). in: *14th Biennial Australian Microbeam Analysis Symposium (AMAS).* (Brisbane, Australia) 84-85

[6] Reed S J B 1993 *Electron microprobe analysis, Second Edition.* (Cambridge: Cambridge University Press)
[7] Henry D and Goodge J Wavelength-dispersive X-ray spectroscopy (WDS). http://serc.carleton.edu/research_education/geochemsheets/wds.html
[8] Matthews M and Kearns S 2015 Introduction to EDS/WDS X-ray microanalysis workshop (WDS Detectors). (Dresden, Germany: European Microbeam Analysis Society (EMAS)). (Portorož, Slovenia)
[9] Moran K and Wuhrer R 2010 Wavelength dispersive analysis (WDS and EM Probe Workshop). Course Notes. (Bungonia, NSW, Australie: Moran Scientific Pty Ltd). http://www.moranscientific.com.au/ms_courses_frame.html
[10] Williams K L 1987 *Introduction to X-ray spectrometry.* (Allen & Unwin)
[11] Goldstein J I, *et al.* 2003 Scanning electron microscopy and X-ray microanalysis. (New York, NY: Springer)
[12] Birks L S 1969 *X-ray spectrochemical analysis, Second edition.* (Interscience Publishers)
[13] Woldseth R 1973 *X-ray energy spectrometry.* (Kevex Corporation)
[14] Cameca- Ametek technical briefing 2015 *Electron probe microanalysis, First edition.* in: Microscopy and Analysis. (Chichester: John Wiley and Sons)
[15] http://www.cameca.com/literature/scientific-publications/epma.aspx
[16] Thermo Scientific technical briefing 2016 *Wavelength dispersive (X-ray) spectroscopy, First edition.* in: Microscopy and Analysis. (Chichester: John Wiley and Sons). http://www.laborexport.hu/_data/VFS_be564080af6d6a2942db1be5b9743e39.PDF
[17] Oxford Instruments Analytical-technical briefing 2002 *Wavelength dispersive X-ray microanalysis - Explained.* (High Wyciombe: Oxford Instruments). https://www.oxford-instruments.com/products/microanalysis/wavelength-dispersive-systems-wds
[18] Wuhrer R and Moran K 2016 Low voltage imaging and X-ray microanalysis in the SEM: challenges and opportunities. *IOP Conf. Ser.: Mater. Sci. Engng.* **109** 012019
[19] http://amptek.com/products/fast-sdd-silicon-drift-detector/